**Tetrahedron:** Punch out the equilateral triangles and the holes at the centers. Fold on the dashed lines and tape at the red marks. The tetrahedra are then joined by pushing a pencil through the hole of one, out the opposite vertex, through the open vertex of the other and out its opposite hole. Hence, two tetrahedra are joined tip to tip. Rotation of the two tetrahedra demonstrates rotation in butane. See text page 59. **(1,2)**

To make a model of ethene, place two tetrahedra edge to edge. The positions of the atoms in ethene (with the carbons at the center of the tetrahedra) are given in the diagram. **(3)**

A rough model of ethyne can be made by placing the two tetrahedra face to face as shown. Again remember that the carbon atoms in ethyne are located at the center of each tetrahedron. **(4)**

**"Wings":** Punch out the wings and slit the vertex of each as indicated by the red line. Join two wings together slit to slit to form a tetrahedron-shaped molecule with one, two, three, or four substituents, depending on the selection of wings.

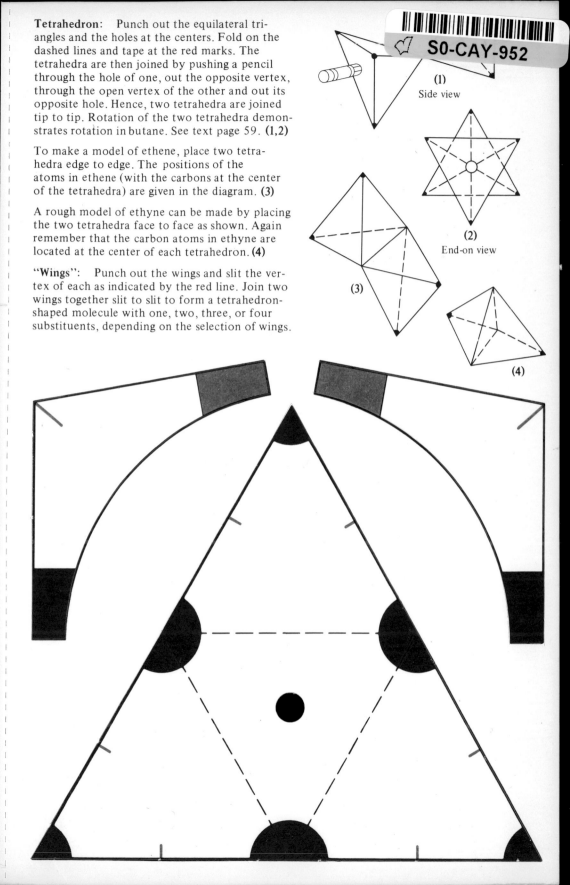

(1)
Side view

(2)
End-on view

(3)

(4)

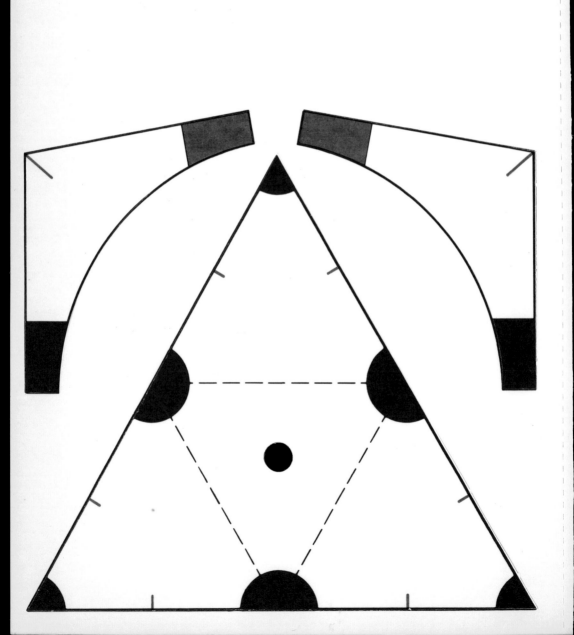

# CHEMISTRY:
# Its Role in Society

# CHEMISTRY: Its Role in Society

James S. Chickos
David L. Garin
Robert A. Rouse
*University of Missouri—St. Louis*

D. C. HEATH AND COMPANY
Lexington, Massachusetts   Toronto   London

*To Jan, and Joy, and the many people
who know the problems and are continually
trying to do something about them.*

# Preface

The general education requirements of many colleges and universities include several hours of science courses. In our technologically oriented society the educated citizen must have some knowledge of the nature and role of science and how scientific expertise can be applied to a wide range of problems that affect all of us.

Most chemistry departments, ours included, offer a limited number of courses for this purpose ranging from a lengthy treatment of a small number of scientific principles to a dilute coverage of many. For a large number of students these options are insufficient. The very reasons for requiring the course, an introduction to scientific methodology and its application to scientific problems, are lost in a maze of facts and formulas that appear irrelevant to their goals for the future.

In 1969, with departmental support, we volunteered to examine the possibility of developing for the non–science major a new course having emphasis on topical issues. Our initial approach was to outline those areas in chemistry that are fundamental to a basic knowledge of the science and those areas of social concern that are most urgent in this decade. After producing a list of topical issues, we decided to include only those chemical principles that bore directly on the problems. Indeed, the chief task was eliminating, not formulating. Once agreed on priorities, we envisioned putting together a collection of essays, short articles, and monologues on the different subjects. We shortly rejected this on the basis that we would generate a potpourri of reading material that could not be intertwined: if sophisticated, it would contain terms that required explanation; if free of new concepts, it could probably be better replaced by more demanding, specifically organized and oriented material. These readings could not be the core of the course.

The conclusion that we should have to write the core material ourselves led to the first edition of our textbook, which our students immediately began to revise. Since this was a new course with an experimental textbook, we distributed and evaluated many questionnaires requesting comments on each individual chapter and also the overall design. We were amazed! We learned that it is impossible to predict which ideas students will find interesting, challenging, and satisfying. But we asked and they told us. In turn, we were asked by the students to rewrite the book so that the mole concept would not have to be included. Anyone who has suffered the frustration of teaching the apparently simple mole concept and seen students wilt in those early weeks of the semester can appreciate our joy in learning that its omission will not sacrifice aims of the course.

The reader who glances through the text will be surprised to find many sophisticated concepts included, for example, optical activity and phase diagrams. The former was included because the students were excited about theories concerning the relationship of structure and reactivity; the latter has remained through two revisions because more than 95 percent of our students found it challenging and enjoyable. It is the only remaining subject that we feel is not essential, but its success at leading students through predictive analysis cannot be questioned.

We also learned that our students did not like hypothetical examples so we used real examples wherever possible. The simple punchout molecular models contained in the book are extremely effective in allowing the student to discover for himself what is meant by conformation and mirror-image isomerism. Putting together an asymmetric molecule in random fashion conveys to the student, in a way that no written description can parallel that two mirror-image forms are possible. In brief then, this book was originally written for our students and revised to suit their needs and desires. While we ourselves argue that this may not be the best way to write a course, in this case it has worked beyond our expectations. Our students are slightly above the national average (according to intelligence tests) and range from brilliant to marginal. A typical class taking this course included students from the schools of business, education, and arts and sciences and from all levels from freshman to senior. The course has been taught by six different instructors in our department using this text as the core material and often adding a smaller paperback text as they deemed desirable (such as *The Double Helix* and *The Social Responsibility of the Scientist*). They in turn suggested changes, which have been accommodated in this edition. While it is always possible to include additional chapters in a text "to be used at the discretion of the instructor," we have learned that our students much prefer a book they can be expected to read from cover to cover. Instructors wishing to treat additional material can use their own prepared handouts or lecture material.

The topical issues are extremely controversial and often produce high levels of emotional impact. It was fortunate that the three of us did not often agree on the severity of issues or on suggested remedial measures; indeed, the general public and well-informed specialists do not always agree. We have tried to present an unbiased, apolitical evaluation, stressing the need for a complete understanding of the facts, possible alternatives, and reasons they may or may not be implemented, so that the students themselves can reach a balance for decision. We emphasize that there are no simple universally applicable answers but instead a tradeoff of problems for which the essential ingredients are a working knowledge of all of the factors involved and an unselfish attitude, for we are truly "all in this together." This text can be read cover to cover in one semester. It contains more than enough chemistry to satisfy the original requirement of a science course in the college curriculum and more than enough topical material to render the citizen capable of evaluating, understanding, and acting on social issues. And that is why the book was written.

We wish to thank our students for their enthusiastic response, aid, and guidance through revised editions; members of our department for their assistance and support, especially Charles Armbruster, Eric Block, and Rudi Winter; Chuck Henrickson for a line-by-line review, and many other reviewers of individual chapters; the Committee for Environmental Information (St. Louis) for use of their excellent library; Bob West for encouragement; Vicky Bayliss for typing, drawing, and smiling through it all; and an anonymous salesman who said to a surprised trio of research chemists, "Have you ever thought of having it published?"

# Contents

## 1. Introduction   1

## 2. Matter and Energy   5

A Note on Chemical Symbols   10
Nuclear Structure and Radioactivity   11
Nuclear Energy Production   14
The Atomic Energy Commission   20
A Postscript. What Price Coal?   21
Atomic Structure and the Periodic Table   24
Molecular Structure   35
Problems   49
Bibliography   51

## 3. Organic Chemistry   53

Molecular Geometry (Shapes of Molecules)   57
  Chemistry of Saturated and Unsaturated Hydrocarbons   61
  Sites of Chemical Reactivity within the Molecule   63
  Chemistry of Compounds and Class   67
Drugs   72
  Molecular Composition Versus Physiology   72
  Everyday Drugs   76
  A Compendium of Drugs   80
Mirror Image Isomerism   87
Natural and Unnatural Polymers   90
  Addition Polymers   90
  Peptides and Condensation Polymers   94
  Rubber   99
  Practical Considerations of Polymer Technology   102
  Coal and Petroleum: Raw Materials for Polymers   102
What Price Dung? The Problem of Organic Wastes   103
Problems   104
Bibliography   107

# 4. The Atmosphere    109

Atmospheric Profile    111
Properties of Gases    114
Introduction to Air Pollution    116
Brief History of Air Pollution    117
Air Pollutants    119
Sources of Air Pollution    123
Effects of Air Pollution    130
An Air Pollution Day    136
Problems and Solutions    139
The Environmental Protection Agency    142
Postscript: Beware of the Ecological Con Man    143
Problems    143
Bibliography    144

# 5. Liquids, Solutions, and Problems    147

The Liquid State    149
  Humidity    152
Equilibrium    153
Changes of State    154
  Phase Diagrams    156
Desalination: Heat It or Cool It?    158
Solutions    159
  Ground Water    161
  Properties of Solutions    161
  Osmotic Pressure    162
Acids and Bases    163
The Chemistry of Water    166
  Chemical Constituents of Water    167
  Water Softening: Chemical Methods    168
  Cation Exchangers    169
  Chelating Agents    170
Water Purification    171
  What Else is in the Water?    171
  Filters    171
  Chemistry in Your Swimming Pool    172
  Sewage    173
  Treatment of Sewage    174
Water Pollution    176
  Industrial and Municipal Pollution    176
  Detergents    180

*Mercury Pollution*    *183*
*Cadmium: A forewarning*    *187*
*Oil Pollution*    *188*
Postscript    **191**
Problems    **192**
Bibliography    **195**

# 6. Biochemistry    197

Proteins    **202**
Lipids    **210**
Carbohydrates    **215**
Enzymes    **216**
Hormones    **222**
RNA and DNA    **225**
Postscript    **241**
Problems    **241**
Bibliography    **243**

# 7. Population    245

History of Population    **248**
Birth Control    **251**
Insecticides and Pesticides    **255**
Food Additives    **263**
Problems    **272**
Bibliography    **273**

**Appendix A**   *Stoichiometry*    **275**
**Appendix B**   *Nomenclature for Organic Compounds*    **285**
**Glossary**    **289**
**Directions for Constructing Models**    **293**
**Index**    **295**

# CHEMISTRY:
# Its Role in Society

We travel together, passengers on a little spaceship, dependent on its vulnerable resources of air and soil; all committed for our safety to its security and peace; preserved from annihilation only by the care, the work and, I will say, the love we give our fragile craft.

Adlai Stevenson
July 9, 1965

# Introduction

## 1

What has chemistry done for you lately? Well, when was the last time you ironed a shirt, had a low-cal beverage, walked to work, or used some cosmetic product? Chemistry touches almost every facet of the world man has created and the world that has created man. We have learned to make new and tougher fibers, longer-lasting plastics, better-tasting food, and faster cars, as well as to encourage the growth and extent of life, all through a better understanding of chemistry.

What has chemistry done to you lately? Consider the air and water around you. It is becoming safer to smoke cigarettes than to breathe the air of some of our big cities. And what about our "fresh water"? Any swimmer in Lake Erie knows how the ducks in the Santa Barbara oil spill must have felt. These too involve chemistry.

In the past few decades our standard of living has reached unprecedented levels. Partly responsible for this is our ability to grow more and more food (by fewer and fewer farmers) and thus keep pace with a growing population. A host of chemical agents, newly discovered or applied, do the work for us. Pesticides, like the chlorinated hydrocarbons, by controlling insects, have enabled us to control disease more effectively and to increase food yields. The success of these chemical agents has encouraged the use of other chemical agents. In a typical hardware store today, you can purchase a chemical to control just about anything, except the price.

These advancements are increasingly found to be not without drawbacks and hazards. Pesticides, particularly the chlorinated kind, are proving to be very persistent in the environment. Many biological processes concentrate rather than destroy them, transferring them back to us along the food chain.

Even our tougher fibers and plastics are giving us trouble. Most of the modern materials now in use have been developed as a result of our understanding of chemistry. Plastics, where they have replaced glass, offer improved safety and packaging. The use of plastics as building materials reduces the cost and time of constructing living units, making new, modern units available to all segments of society. The development of synthetic fibers has yielded no-press fabrics, easy-to-wash clothes, long-lasting and beautiful carpets, sun-resistant drapes, and many other home products having desirable qualities. The use of these plastics and fibers has helped reduce or hold down the cost of consumer products such as clothes and home appliances. However, many of these materials are not easily returned to a "natural state" by biodegradation. Other materials can be considered completely nonbiodegradable and it appears unlikely that some microorganism will adapt to chewing these materials up after they are of no further use. As a result we have also created tougher, longer-lasting garbage. Incineration of these materials often gives rise to noxious gases that ultimately affect the air we breathe.

---

(*Reprinted, by permission, from* St. Louis Post-Dispatch. *Photograph by Nicholas Sapieha.*)

A look at drugs will further our suspicion that chemistry can be viewed as a double-edged sword. In the past few decades we have developed drugs to do virtually everything except make house calls. These drugs used properly can greatly aid the sick and convalescing. Abused, they are not much different from the hard ones on the street.

As society continues to grow and consumption increases, the problems of pollution, noise, and garbage, keep pace. In the production of goods we consume, virtually every large waterway in the country has been polluted. We have contaminated the air so badly in places we can nearly "mine the sky"; and our conversion of vast acreages to asphalt and cement has endangered innumerable species of wildlife. The drawbacks associated with new chemical and other scientific discoveries and applications cause a growing wave of people to question the inherent value of these innovations. The views about the role of technology in society are widely divergent.

A disturbing aspect of this predicament is that our commitment to technology virtually forces us through social and economic pressure to seek and accept technological solutions for problems that are not technological in nature. Technology buys us time, but the problem, often in exaggerated form, returns to plague us. With each cycle furthering our commitment to technology, we may find that technology eventually creates more problems than it alleviates. What can be done? To many members of the scientific community, technological assessment appears to be the best approach. Such an assessment would involve the investigation of short- and long-range effects of the application of new and existing technology on society. Will a new factory pollute? Will a new food additive really improve shelf life enough to justify the possibility that certain individuals may be allergic to it? What are the effects of existing food preservatives? Questions like these, which evaluate the risk-benefit formula for a process, product, or factory, are posed in search of suitable and acceptable answers. Will this effort be enough?

# Matter and Energy

# 2

*Cosmic energy is love, the affinity of being with being. It
is a universal property of all life, and embraces all forms of
organized matter. Thus, the tendency to unite; the attraction of
atom to atom, molecule to molecule, or cell to cell. The forces of
love drive the fragments of the universe to seek each other so that
the world may come into being.*
                        P. Teilhard DeChardin, The Phenomenon of Man

*Energy is most usable where it is most concentrated—for
example, in highly structured chemical bonds (gasoline, sugar) or
at high temperatures (steam, incoming sunlight). Since the second
law of thermodynamics says that the* overall *tendency in all
processes is away from high temperature, it is saying that overall,
more and more energy is becoming less and less available.*
                Paul Ehrlich and Anne Ehrlich, Population, Resources,
                Environment: Issues in Human Ecology (1970)

Matter and energy are two concepts central to all science. By measuring these two quantities and how they change with the passing of time, the
laws of physics predict quite well the progress of the world around us. The
terms *matter* and *energy* are part of everyday vocabulary, but they must be
clearly understood for scientific discussion.

**Matter** is usually considered to be something that has mass and occupies
space. Mass is just a measure of how hard it is to get an object moving or
how hard it is to stop an object in motion; for example, a chair is harder
to lift than a pencil because it has more mass. Mass and weight are related
but they are not synonymous. An object has the same mass wherever it is
in the universe, but its weight may vary. Weight is a measure of how strongly
two objects attract each other through gravitational forces. A man weighs
less on the moon than he does on earth because of the stronger gravitational
pull of the earth.

The concept of force also needs discussion. **Force** is the push or pull
that causes objects to move faster, slow down, or change direction. When
you throw a ball or push a car you are applying a force to an object. There
are basically three kinds of forces: gravitational, electrostatic, and nuclear.
Gravitational forces are seen in the attraction of two objects for each other
due to their masses. The magnitude of the gravitational force depends on the
sizes of the masses and the distance between them. The greater the masses

---

*The Alchemist* by **Thomas Wyck.** This seventeenth century painting portrays a
master alchemist surrounded by the paraphernalia of his craft—two candles, a cross,
a skull, a bible, and the blood of a newly killed animal. The skeleton blowing a
trumpet is guiding the lab technician to say the proper incantations that will turn
the base metal suspended from the ceiling into gold. (*Courtesy of Alfred Bader, Aldrich
Chemical Company, Inc.*)

and the shorter their distance apart, the stronger the force due to gravity. Electrostatic forces can be either attractive or repulsive. Matter may have a positive or negative charge; charge is an attribute that describes the way a particle moves between the poles of a magnet or the electrodes of a battery. Objects with like charges repel one another; objects with opposite charges attract one another. The magnitude of electrostatic forces depends on the size of the charges of the objects involved and the distance between them. The larger the charges and the smaller their separation, the larger the push or pull due to the electrostatic force. In general, electrostatic forces are $10^{35}$ (1 followed by 35 zeros) times more powerful than gravitational forces. Nuclear forces are not well characterized to date; however, it appears that they are very large, and effective over very short distances only.

**Energy** is defined as the ability to do work. Everyday conversations contain many references to energy: food energy, the energy of a two-year-old, atomic energy, the energy of sunlight, and so on. We can't feel, see, hear, taste, or smell energy, but we can surely experience its effects. The effect of sunlight is felt; light is seen, that is, the nervous system reacts to the effects caused by light's striking the retina of the eye.

Whenever an object has the ability to do work, it has energy; water behind a dam possesses energy, and a speeding car has energy. From these two examples, it can be seen that there are different categories of energy: energy of position and energy of motion. Matter in motion is said to possess kinetic energy, or energy that it has merely by the fact that it is moving. Energy of position is exemplified by a brick held above a table and can be compared with its energy on the table. If it were released, the brick would fall, and as it did, it would stir up the air; when it hit the table it would cause a dent and send out sound waves. So while it is held above the table, it has the potential to do work (stir the air, dent the table, cause sound, and so on) and hence it is said to possess potential energy.

Another way to divide forms of energy is based on the origin of the energy. Gravitational energy results from the attraction of two masses; it is the origin of the potential energy of the brick discussed above. Electrical energy is potential energy that arises from the attraction of particles of opposite charge or the repulsion of particles of the same charge. Electrostatic forces push or pull particles through a distance and hence charged particles have the ability to perform work. Heat is a form of energy. Chemical energy arises from chemical reactions. These different types of energy can be converted from one form to another. For instance, electricity causes the heater filaments in a toaster to glow, giving off visible light, and infrared light to toast the bread. The electricity was generated by a steam turbine; the steam was generated by burning coal; the coal formed from plant life; and the plant grew by absorbing sunlight.

Light is one kind of energy with which we are familiar. Light is composed of waves somewhat analogous to those on a lake. They are a succession of peaks and valleys. See Figure 2–1. Light waves are characterized by the distance between two peaks, called the wavelength, $\lambda$ (*lambda*, Greek),

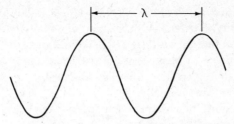

**Figure 2-1** Schematic diagram of radiation showing wavelength, $\lambda$.

and the number of peaks that pass a point in one second, or the frequency, $\nu$ (*nu*, Greek). The longer the wavelength, the smaller the frequency. The energy associated with a light wave is proportional to $1/\lambda$ or $\nu$.°

$$E \propto \frac{1}{\lambda} \propto \nu$$

The less the value of $\lambda$ (or the greater the value of $\nu$), the more energy possessed by light. Visible light consists of light of different wavelengths (hence, different energies), which are perceived as different colors. Visible light is just a part of the electromagnetic spectrum, which contains waves of longer and shorter wavelengths than visible light. Electromagnetic waves, or radiation, exhibit both electric and magnetic properties. See Figure 2–2. Other forms of electromagnetic radiation include X rays; ultraviolet (suntanning rays), infrared (heat lamp), and microwave radiation (ultrafast cooking devices); and radio and television waves. Radiation of the highest energy and smallest wavelength is from cosmic rays and is of galactic origin. High energy gamma radiation ($\gamma$ ray) is often emitted when radioactive materials decay.

How is energy measured? The unit of energy usually found in chemical discussions is the kilocalorie. One **kilocalorie** is approximately the amount of energy needed to raise the temperature of 1,000 grams (approximately 1 quart or 2.2 pounds) of water one degree Celsius.† This is about the amount of

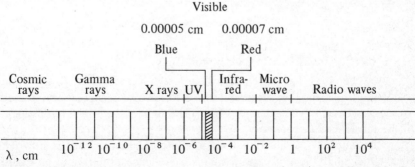

**Figure 2-2** Electromagnetic spectrum: cm $= 0.394$ inch; $10^4 = 1$ followed by four zeros; $10^{-4} = 1/10^4 = 0.0001$; UV $=$ ultraviolet.

---

° $\propto$ means "is proportional to."
† The Celsius scale is often referred to as the Centigrade scale, both are written °C, 1 °C $= 1.8$ °F.

energy a bullet from an elephant gun has. It is a million times the energy it takes to press a typewriter key. It is about one-thousandth of a good day's work by a woodcutter. One calorie is $\frac{1}{1000}$ kilocalorie.°

Energy and energy changes are of particular interest to chemists. Just as food is heated during cooking, chemists add heat (energy) to chemicals to cause reactions to occur. Actually, during cooking many chemical reactions occur. The fermenting of wine, the making of bread, the use of pectin in making preserves, and the marinating of meats are but a few examples of chemical reactions that depend on energy addition or removal during preparation. The presence of sunlight (energy) initiates chemical reactions in the atmosphere causing the formation of additional air pollutants.

The fact that some chemical reactions emit energy is very important. The burning of fuels is the main energy source for the production of electricity, heat, and motive power. Chemical reactions provide the glow exhibited by fireflies, and some mushrooms, fishes, and other life forms. They provide the energy for keeping human bodies alive; for instance, the metabolism of one gram of sucrose (sugar) in a living body emits four kilocalories of energy. Rocket fuels are chemicals that give off large amounts of energy per pound of fuel; in this way, the rocket can have as light weight as possible. This list is really endless, but the above examples are illustrative of some energy changes.

There is a close relationship between mass and energy. This famous relationship, $E = mc^2$, was developed by Albert Einstein as part of his theory of relativity; the equation relates energy and mass through the constant $c^2$, the speed of light squared.†

An idea of the building blocks of matter will help us in our further discussion of matter. Take a cube of sugar, for instance. The cube is usually composed of sugar grains compacted together; the grains composed of sugar molecules; the molecules composed of carbon, oxygen, and hydrogen atoms; the atoms composed of protons, neutrons, and electrons; and perhaps these particles, now known as the most elementary, may be composed of other smaller units of matter.

## A NOTE ON CHEMICAL SYMBOLS

As chemistry has developed historically it has acquired a certain shorthand formalism. This shorthand formalism involves symbols and is used in discussing chemistry in a compact form. Throughout the text, this shorthand will be introduced as necessary.

In a nutshell, chemistry is the study of how atoms interact with each other, either singly or in groups. Symbols have been developed to represent

---

° The kilocalorie is the "calorie" referred to in diets.

† Further discussion of the relation is beyond the scope of this text, but M. Gardner, *Relativity for the Million* will provide interesting additional reading.

the different atoms found in nature. The names and symbols of these atoms are listed for convenience at the end of this chapter in Table 2–9 and in the periodic table of the elements, Figure 2–17. For instance, the element oxygen has the atomic symbol O, and by this short form we may refer to this element or its individual atoms. Nitrogen has the symbol N; phosphorus, P; tin, Sn; and so on. When one wishes therefore to refer to oxygen, the symbol O is used. From these examples and an examination of Tables 2–9 and 2–10, it is clear that in many instances the atomic symbol is contained in the name of the element. A few, like tin and lead, have symbols of Latin or Greek or other origin. With a little practice, these symbols can be easily recognized and used.

## NUCLEAR STRUCTURE AND RADIOACTIVITY

The basic building block of matter that is of interest in chemistry is the atom. Current evidence based on sixty years of research tells us that the atom is composed of three basic particles: a **proton** with a positive charge, an **electron** with a negative charge, and a **neutron** with no charge at all.° The protons and neutrons in an atom are concentrated in a small bundle called the nucleus. The electrons are found whirling about the nucleus. Each atom is characterized by the number of its electrons, its protons, and its neutrons. If the atom is neutral, that is, if it has no net, or overall, charge, the number of electrons equals the number of protons. One element may be distinguished from another by the number of protons its individual atoms contain; this number of protons is called the **atomic number.** The atomic number is listed for each atom in Table 2–9 (p. 46) and in the periodic table of the elements, Figure 2–17. Each element, then, is characterized by an atomic number equal to its number of protons. For instance, hydrogen (H) has atomic number 1, and 1 proton in its nucleus; Radon, Rn, has atomic number 86, and 86 protons in its nucleus.

Thus each type of atom has a different number of protons in its nucleus. Atoms of a given element contain the same number of protons but may contain different numbers of neutrons in the nucleus. However, since such atoms all contain the same number of protons, they belong to the same element. Atoms that contain the same number of protons but different numbers of neutrons are called **isotopes.**

For example, hydrogen (H) has three isotopes: protium, deuterium, and tritium. (See Table 2–1.) All have one proton, but 0, 1, or 2 neutrons, respectively. Since they are neutral atoms, the number of electrons equals the number of protons. Each isotope has its own symbol, which is just the atomic symbol plus a left subscript denoting the atomic number, and a left superscript denoting the **mass number.** The mass number equals the sum of the number of protons and the number of neutrons. Most elements have several

---

° A good short history of chemistry can be found in Isaac Asimov, *A Short History of Chemistry.*

**Table 2-1**
**Hydrogen Isotopes**

| Name | Number of Protons | Number of Electrons | Number of Neutrons | Symbol |
|---|---|---|---|---|
| Protium | 1 | 1 | 0 | $^1_1H$ |
| Deuterium | 1 | 1 | 1 | $^2_1H$ |
| Tritium | 1 | 1 | 2 | $^3_1H$ |

isotopes, which occur naturally on earth; usually the isotopes of an element differ in their relative abundance, that is, they are not found in equal amounts. For instance, of all hydrogen atoms in the world, 99.985% are $^1_1H$, 0.015% are $^2_1H$ and only trace amounts of $^3_1H$ exist.

For another example, look at Table 2-2. It shows the isotopes for lithium (Li), atomic number 3. Again note the relationship between mass number and particles in the nucleus, and the equality of the number of electrons and protons in the neutral atom.

What is the structure of the nucleus? How are the protons and neutrons arranged and how do they stick together? These questions are not fully answered today and even the incomplete explanations are difficult to understand and will not be discussed here. But discovery of the answers could be the basis for the source of large amounts of energy. For our purposes, just knowing what isotopes are and how they are characterized is sufficient.

We are all familiar with the term *radioactivity*. **Radioactivity** refers to the instability of certain nuclei that leads them to disintegrate spontaneously with accompanying release of energy in the form of electromagnetic radiation and particles with high speeds. Many of these unstable nuclei occur naturally and give rise to background radiation, which accounts for 88 percent of the radiation a person receives during a year. Other unstable nuclei can be synthesized by man, using nuclear reactors and other devices. Examples of spontaneous decay are shown in Figure 2-3. Notice that no net charges are produced or destroyed in these reactions, though in reaction 2 a neutron apparently splits into a proton and an electron. In reaction 1 there are 92

**Table 2-2**
**Lithium Isotopes**

| Symbol | Number of Protons | Number of Neutrons | Number of Electrons | Half-life |
|---|---|---|---|---|
| $^5_3Li$ | 3 | 2 | 3 | less than $\frac{1}{10,000}$ sec |
| $^6_3Li$ | 3 | 3 | 3 | stable |
| $^7_3Li$ | 3 | 4 | 3 | stable |
| $^8_3Li$ | 3 | 5 | 3 | 0.85 sec |
| $^9_3Li$ | 3 | 6 | 3 | 0.17 sec |

$$^{234}_{92}\text{U} \longrightarrow {}^{230}_{90}\text{Th} + {}^{4}_{2}\text{He}. \tag{1}$$

$$^{27}_{13}\text{Al} + {}^{4}_{2}\text{He} \longrightarrow {}^{30}_{15}\text{P} + {}^{1}_{0}\text{n}. \tag{2}$$

**Figure 2–3**  Nuclear reactions. $^{1}_{0}\text{n}$ is a neutron.

protons and 142 neutrons in $^{234}_{92}\text{U}$; when $^{234}_{92}\text{U}$ decays it yields $^{230}_{90}\text{Th}$ and $^{4}_{2}\text{He}$, which together contain 92 (90 + 2) protons and 142 (140 + 2) neutrons. The second reaction in Figure 2–3 is one of the reactions used to synthesize the nucleus, $^{30}_{15}\text{P}$; a helium nucleus ($^{4}_{2}\text{He}$) is caused to smash into an aluminum nucleus ($^{27}_{13}\text{Al}$) to yield the new nucleus.

Different nuclei disintegrate, or decay, at different rates; the rate of decay is given by the **half-life** of the nuclei. The half-life is the amount of time it takes a sample of the radioactive nuclei to decay to the point where only half of the original amount remains. Take, for example, 128 grams of radioactive $^{9}_{3}\text{Li}$. After 0.17 seconds, 64 grams are left; after 0.34 seconds, 32 grams are left, after 0.68 seconds, only 8 grams are left. The longer the half-life, the longer it takes for a sample to decay.

The half-lives of many isotopes are very short and hence these isotopes are not found naturally on the earth. $^{3}_{1}\text{H}$ and $^{5}_{3}\text{Li}$ are examples of unstable isotopes. Other radioactive nuclei have very long half-lives. Carbon-14, $^{14}_{6}\text{C}$, is one of the long-lived **radioisotopes** (radioactive isotopes) of carbon. It has six protons and eight neutrons in its nucleus. The half-life of $^{14}_{6}\text{C}$ is 5,770 years. Radioactive carbon-14, $^{14}_{6}\text{C}$, is used to date objects from antiquity. How is this done? While $^{14}_{6}\text{C}$ is constantly decaying, it is also constantly produced on earth by the collision of cosmic rays with atmospheric nitrogen; the process is shown in Figure 2–4. The $^{14}_{6}\text{C}$ thus produced reacts with oxygen and forms carbon dioxide. The net result is that living plants take up $CO_2$ (carbon dioxide) that includes some radioactive carbon-14 nuclei. Through the normal channels by which plants take in carbon dioxide and convert it to building material in the plant, the radioactive carbon is deposited in the plant. This process is going on today as it has for millions of years. As long as the plant lives, it will continue to deposit radioactive $^{14}_{6}\text{C}$ within itself, the ratio of radioactive carbon-14 to stable carbon-12 being the same as in the ambient atmosphere. Once the plant dies, it stops depositing carbon, but the radioactive carbon-14 already incorporated continues to decay. By using specially designed instruments, it is possible to count the number of carbon-14 atoms disintegrating per second. So if one compares the disintegrations per second per

$$^{14}_{7}\text{N} \text{ (in atmosphere)} + {}^{1}_{0}\text{n} \longrightarrow {}^{14}_{6}\text{C} + {}^{1}_{1}\text{H}. \tag{1}$$

$$^{14}_{6}\text{C} \longrightarrow {}^{14}_{7}\text{N} + {}^{0}_{-1}\text{e}. \tag{2}$$

**Figure 2–4**  (1) Production of radioactive carbon-14 from cosmic rays. (2) Decay of radioactive carbon, in tree or animal.

**Table 2–3**
**Time vs. Decay Rate for Carbon-14**

| Elapsed Years | Decay Rate (Disintegrations per sec per gram) |
|---|---|
| None | 15.3 |
| 5,770 | 7.65 |
| 11,540 | 3.82 |
| 17,310 | 1.91 |
| 23,080 | 0.96 |
| 28,850 | 0.48 |
| 34,620 | 0.24 |

gram of carbon for an ancient wood artifact to the disintegrations per second per gram of a living wood sample today, it can be estimated how long it would take the carbon-14 in the living wood to have decayed to the amount present in the wood artifact and have a rough idea of how old it is. For example, a living tree would yield a decay rate of 15.3 disintegrations per sec per gram of carbon. After 5,770 years, a tree that died today would have a decay rate of 7.65 disintegrations per sec per gram of carbon. After 11,540 years, a rate of 3.825 disintegrations per sec per gram is observed. See Table 2–3.

To date an object, just reverse the process. If the decay rate of a sample from antiquity is 0.956 disintegrations per sec per gram, it is 28,850 years old. An implied condition for the validity of carbon-14 dating is that the ratio of carbon-14 to carbon-12 in the atmosphere today is the same as it was 30,000 years ago.

## NUCLEAR ENERGY PRODUCTION

Large amounts of energy are released when certain nuclei decay or split apart. The familiar example is uranium, which is the basic ingredient in atomic bombs. The **fission** of $^{235}_{92}U$ yields tremendous energy and produces nuclei of lower mass number, many of which are radioactive and the source of radiation hazards from fallout. Figure 2–5 shows examples of reactions involved in uranium fission. Note that each reaction produces more neutrons, which can cause two or three more reactions to occur. This snowballing, or chain reaction, releases large amounts of energy rapidly.

$$^{235}_{92}U + ^{1}_{0}n \longrightarrow ^{90}_{38}Sr + ^{144}_{54}Xe + 2\,^{1}_{0}n + \text{energy.} \tag{1}$$

$$^{235}_{92}U + ^{1}_{0}n \longrightarrow ^{93}_{36}Kr + ^{140}_{56}Ba + 3\,^{1}_{0}n + \text{energy.} \tag{2}$$

**Figure 2–5**  Uranium fission.

In addition to the uncontrolled release of nuclear energy, as in a bomb, controlled release in nuclear reactors is possible. This is a potential source of the energy that modern civilization will require if it is going to maintain its standard of living. A schematic diagram of a nuclear reactor is shown in Figure 2–6. The entire reactor is shielded to prevent the escape of radiation. The temperature is controlled by the control rods and the moderators (usually graphite), which control the rate of flow of neutrons. A coolant is circulated in a closed system through the reactor and a tank of water. The coolant, heated by the energy released in controlled nuclear fission reactions, vaporizes the water, and the steam is used to drive turbines that produce electricity. Except for the source of heat, the operation is the same as conventional fossil-fuel (oil and coal) electricity production. The coolant is most often water under extreme pressure, but gases and sodium metal (a liquid at the temperatures involved) are also used.

Nuclear energy is potentially an important source of the world's energy needs, particularly in view of the limited amounts of conventional fuels still available. Estimation of the time at which these conventional fuels will be exhausted varies from expert to expert. However, since we are using energy at a rate of about ten times that of a century ago (equivalent to 3,800 million tons of coal per year) the supply is exhaustible. This doesn't take into account the vast amount of petroleum used as a source for the chemicals that provide most synthetic products. Unless there is a technological breakthrough in the use of solar or geothermal energy sources, or in the development of shale

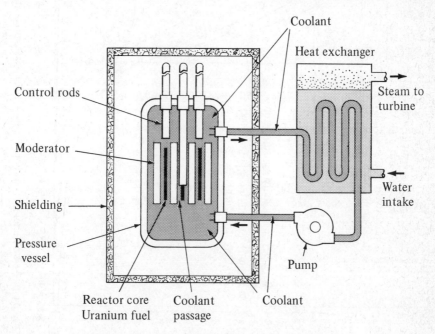

**Figure 2-6**  Schematic diagram of nuclear reactor.

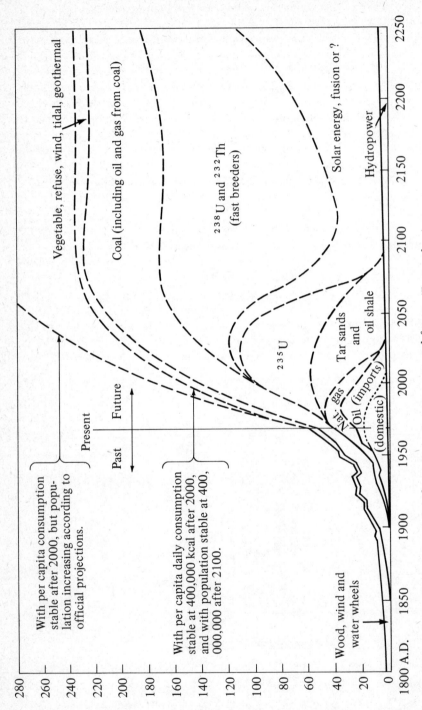

**Figure 2-7** U.S. energy consumption by source: past, present, and future. (*From a drawing by Earl Cook, Texas A. & M. University, by permission of the author.*)

oil and gas supplies, it appears that nuclear energy will be the next source tapped to provide for the demands of an ever growing world population. The likelihood of reduced or static energy demands seems to be the most remote of possibilities. See Figure 2–7 for a projection of future energy needs and sources.

Since nuclear reactions used in the production of energy for a nuclear power plant yield approximately one million times the energy derived from chemical reactions producing heat in a conventional fossil-fuel (coal, gas, oil) power plant, it is likely that nuclear power will be the next source of energy to be utilized.

Nuclear reactor power plants have several advantages over fossil-fuel plants. They produce no noxious fumes and consume only small amounts of material. There is also an essentially limitless supply of nuclear fuel since not only is fuel consumed in a nuclear reactor but it is also produced. Nonetheless, there are serious shortcomings to current nuclear technology, some of which can be overcome.

Like fossil-fuel plants, nuclear power plants could lead to destruction of scenery by construction and mining operations; but since much less nuclear fuel is required per unit of energy produced, the mining would be less serious for nuclear power. Since only about 35 percent of the energy produced by

**Figure 2-8** The Enrico Fermi atomic power plant at Logoona Beach, Michigan, operated by Power Reactor Development Company. (*Reprinted courtesy of Power Reactor Development Company.*)

fossil-fuel or nuclear power plants can be converted to useful ends, a large amount of heat is dumped into the environment around an installation. This excess heat could be put to use in heating (homes, roads, and so on) but generally it is dumped into a nearby river or lake or the atmosphere; this "thermal pollution" can raise the temperature of a stream or lake, either causing plants and fish to die or drastically changing the ecology of the body of water.

Fossil-fuel plants spew a large variety and quantity of air pollutants into the atmosphere; these combustion products, resulting from burning fuel, are responsible for about 12.5 percent of all air pollution in the United States, about 40 billion pounds per year. Nuclear plants also produce pollutants. During the nuclear reactions, which yield heat, materials are produced that are quite radioactive and hence pose a pollution threat of their own. Some of the radioactive products are used as fuel for other reactors, but most must be contained. Currently, these radioactive materials are divided into two categories: those that escape directly into the environment and those that must be stored until enough half-lives have passed and the material has cooled down (in a radioactive and temperature sense). Considerable controversy has arisen over the location and methods of containing the nuclear ashes until they are safer. The use of salt mines and concrete bunkers (with continuous cooling by water) appears to provide the most likely depositories for these materials.

The radioactive materials that escape into the environment are basically $^{85}Kr$ and tritium, $^{3}H$. The isotope $^{85}Kr$ is produced as a product of nuclear fission and is contained within the nuclear fuel, but escapes as a gas once the fuel is removed from the reactor for storage. Virtually all $^{85}Kr$ found in the environment comes from nuclear power production. Tritium is produced from a secondary reaction between neutrons and hydrogen atoms in the reactor and diffuses right through the walls of the reactor and into the atmosphere. Assuming continued growth of nuclear power production (about 50 percent of all power production by 2000) and no controls on the release of these pollutants, the amount of tritium in the environment will triple and that of $^{85}Kr$ will increase ten-thousandfold by the year 2000. Fortunately, only a small amount (0.4 percent) of the decay products from $^{85}Kr$ is damaging.

Each year living organisms receive a certain dose of radioactivity from cosmic and terrestrial sources. Terrestrial sources include carbon-14, strontium-90, and cesium-37 from nuclear weapon testing and uranium and thorium ores. In addition, human exposure in the form of medical X rays and radioactive materials in the diagnosis and treatment of disease is increasing. Table 2–4 presents the percentage exposure to radiation from various sources now and in the year 2000 (projected); the total exposure increase is 0.02 percent. Deviations from these amounts will depend on occupation and geographical location.

There is actually no safe level of radiation. As in the case of other insults on the body like those from air pollution and water pollution, drugs, and diseases, only one exposure under certain circumstances may be enough to cause an immediate fatality or a condition that produces death. But in

**Table 2-4**
**Present and Projected Total Body Dose of Radiation**
**(Percentage by Source)**

|                  | Year |      |
| ---------------- | ---- | ---- |
| Source           | 1973 | 2000 |
| Terrestrial      | 65%  | 65%  |
| Cosmic           | 23   | 23   |
| Medical          | 12   | 12   |
| Nuclear industry | 0.05 | 0.07 |

contrast with air pollution, water pollution, and drugs, disease and radiation have affected man throughout his evolution and he has developed corrective procedures for the damage they can cause to his body. Death rate statistics, calculated for the general population and based on death owing to radiation produced by the nuclear industry, vary widely since the criteria used for measuring radiation damage probabilities are subjective. But it is clear that deaths attributable to the nuclear industry are far fewer than those attributable to air pollution.

The above considerations have not included a possible catastrophe at a nuclear installation. While this could not reach the degree of a blast, large amounts of radiation would be released. Such a release would have considerable effects on the local population. Anemia, leukemia, lowered thyroid activity, embryo malformations, and fibrosis and cancer of the lungs are a few examples of damage from exposure to a catastrophic release of radiation. The magnitude and scope of these effects depend on the type of radiation emitted, the half-lives of the radioisotopes released, the amount reaching specific points in the body, and the capability of the injured individual to repair himself.

The types of radiation emitted and their associated energies play an important role in the damage incurred. Two types of subatomic particles, the alpha and beta particles, are commonly emitted as the result of the spontaneous decay of radioisotopes. An alpha ($\alpha$) particle is a helium nucleus (two protons and two neutrons) and a beta ($\beta$) particle is an electron. The faster these particles move, the more damage occurs. X rays and gamma ($\gamma$) rays are also typical decay products; again, the more energy they have, the more damage occurs.

The safety record of the nuclear industry is commendable. Since 1946, seven fatalities from nuclear accidents have occurred, only one of these in the standard operation of a power plant. No injuries owing to cumulative exposure have been reported for the large number of workers in the industry. It is impossible to evaluate long-term genetic effects at this time. Miners of uranium ore are chronically exposed to radiation, and they suffer radiation damage in addition to the typical mining maladies. At present, the mining is the only phase of the atomic industry not covered by government supervision.

$$_1^1\text{H} + {}_1^1\text{H} \longrightarrow {}_1^2\text{H} + {}_1^0\text{e}. \tag{1}$$

$$_1^2\text{H} + {}_1^1\text{H} \longrightarrow {}_2^3\text{He} + \gamma. \tag{2}$$

**Figure 2-9** Fusion reactions: ${}_1^0\text{e}$ is a particle the size of an electron with a positive charge; $\gamma$ indicates gamma ray.

Nuclear energy may in the future be harnessed by fusion reactors, which are safer than fission reactors. It is possible to cause two light nuclei to unite to form a heavier nucleus; this process is called **fusion** and is the source of energy in the hydrogen bomb. It could be a successful substitute for fission reactors since not as much radioactive waste would be produced. Figure 2–9 shows examples of fusion reactions; Figure 2–10 portrays an atomic power plant.

## THE ATOMIC ENERGY COMMISSION

The Atomic Energy Commission (AEC) was established by the Atomic Energy Act of 1946 to develop, use, and control atomic energy so that it would make a maximum contribution to defense and security, general welfare, and world peace. It directs research, development, manufacturing, and the promotional efforts with regard to the use of atomic energy. Its five civilian

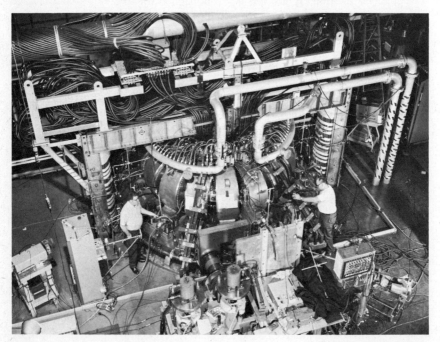

**Figure 2-10** Controlled nuclear fusion research device, located at Princeton University Plasma Physics Laboratory and sponsored by the U.S. Atomic Energy Commission. (*Reprinted by permission of Princeton University.*)

members are chosen by the president with the advice and consent of the Senate. Typical tasks directed by the AEC would include uranium mining, development and testing of nuclear weapons, development of nuclear reactors, collaboration with NASA in designing nuclear rocket systems, basic research in medical, physical, and biological sciences, dissemination of technical and scientific information, and arrangement of international cooperation in the peaceful use of atomic energy.

In addition to promoting nuclear energy applications, the AEC has the job of licensing and regulating the civilian use of nuclear materials and the construction and operation of nuclear reactors. This includes setting up safeguards for the use of atomic materials. Inspection of licenses is carried out by five regional and three district safeguards offices.

The double role of promotion and regulation has often brought the AEC under attack by wary citizens who feel the roles are incompatible. They want to know how an agency, whose reputation is on the line, can effectively regulate a project it has supported financially and administratively for years. With few exceptions, however, the AEC's performance and safety record for twenty-six years is outstanding. The expertise that the agency has developed and is developing will undoubtedly lead to cleaner use of nuclear reactors, to improved medicinal applications, and to national security.

## A POSTSCRIPT. WHAT PRICE COAL?

All forms of energy production have drawbacks associated with them, either inherent in their design or as a result of other factors, which are perhaps more controllable but are also subject to social, political, and economic pressures. Coal is one source of energy. Almost 50 percent of the coal used in the U.S. is obtained by strip-mining. This means that the topsoil surface is torn away to reveal coal deposits, instead of tunnels being made in the earth. With massive equipment and dynamite, the coal is gouged from the earth, and gaping holes are left. When the coal is depleted, the land is abandoned. It is unusable and unsightly. More than 1.8 million acres of land have already been adversely affected not to mention contamination of streams by sediments and by acid draining from the scarred surface.

In 1967, the U.S. Department of the Interior reported to Congress on the abuses of surface mining, of which strip-mining is one method. They recommended immediate measures to control the rapidly growing process that had already affected an area of land approximately equal in size to the state of Delaware. Since that report, the amount of land affected by surface mining has more than doubled, but we still lack meaningful controls.

Proponents argue that strip-mining is more productive than deep mining and safer, since cave-ins and "black lung" disease are not accompanying hazards. The obvious solution is to reclaim the land after stripping the coal—restore it to the condition in which it was found with rocks, subsoil, and topsoil. The mining industry is reluctant to make this expenditure. Some laws have been passed forcing the industries to reclaim the land *when the mining is finished.* Because of the possibility of further stripping of the same land, mining industries hesitate to begin reclamation immediately. They may leave a small parcel of unstripped land in the middle of the wasted area to emphasize that mining is not finished, forestalling reclamation and bypassing the intent of the law. See Figure 2–11(a).

The amount of money now being spent for reclamation is 5¢ per ton of coal. It is estimated that $1–$2 per ton is necessary for the job to be done correctly. This cost would be borne by the consumer but it does not represent the huge increase that is often erroneously attributed to such a move. If $1–$2 per ton were charged for reclamation, it would increase the cost of coal approximately 10%, which would result in an increase in the cost of electricity of only 2%–3%, since the cost of coal is just one factor in the cost of producing electricity. In some areas, reclamation is not possible, or it is too costly to make the process competitive. Thus, there is a growing movement for the complete abolishment of strip-mining. (See Figures 2–11(b) and 2–11(c).)

**Figure 2-11(a)**  Earth scarred by strip-mining. (*Reprinted, by permission of Time/Life Syndication Service. Photograph by Bob Gomel. Copyright Bob Gomel,* Time *Magazine,* © *1971,* Time *Inc.*)

**Figure 2-11(b)** Mined before reclamation was required in surface mining operations, this East Tennessee strip mine remained almost barren. (*Reprinted courtesy of Tennessee Valley Authority.*)

**Figure 2-11(c)** The same site as in Figure 2–11(b) eight years after tree planting and other reclamation measures were carried out in a demonstration project. While mining today is subject to reclamation, large areas of these abandoned "orphan" mines remain unreclaimed in the Appalachian coal fields. (*Reprinted courtesy of Tennessee Valley Authority.*)

## ATOMIC STRUCTURE AND THE PERIODIC TABLE

The composition of the nucleus of the atom has just been discussed. The nucleus is surrounded by electrons, which course about the nucleus according to the laws of physics. The atom's neutrality, or lack of net charge, indicates that the number of electrons equals the number of protons. The mass of the electron is about $\frac{1}{2,000}$ of the mass of the proton or neutron. Scientists cannot describe the exact trajectory of the electrons, but they are able to determine the arrangement of the electrons in an atom.

Objects tend to glow when they are heated; the terms *white hot* and *red hot* refer to visible light emitted from heated material. If hydrogen atoms, which exist as a gas, are heated to high temperatures, the gas glows with a bluish-white color. The light that hydrogen emits when heated provides a hint to the way the electron is arranged in the hydrogen atom.

Ordinary visible light—the kind emitted by the sun—has wavelengths that correspond to the visible colors, violet through red. Each different wavelength produces a sensation in humans that is symbolized by the words *blue* or *green* or *orange*. If white light is allowed to hit a prism, a triangular block of glass, the white light is separated and produces a band of separate colors, red to violet. See Figure 2–12. This is because white light consists of all the wavelengths in the visible region of the electromagnetic spectrum. The prism separates the light into its component parts.

Now if the light from the hot hydrogen gas hits the prism, only four specific colors, or lines, representing light of only four different wavelengths,

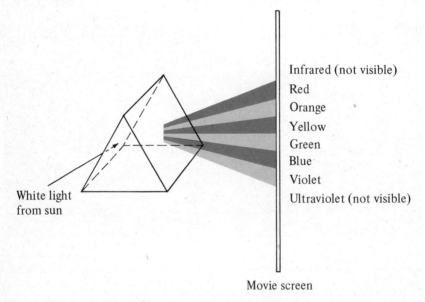

Figure 2-12  Components of white light.

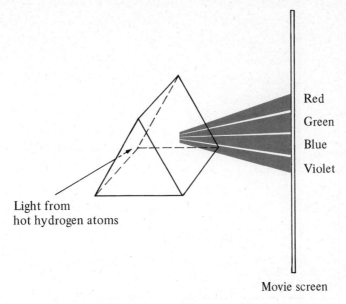

**Figure 2-13**   Discrete components of hydrogen.

are seen on the screen, not a continuous rainbow of color representing all wavelengths, as in white light. See Figure 2–13.

What model of the hydrogen atom can explain the color pattern given off? First consider an analogy—a cliff with notches hacked into it in which a ball may be held; see Figure 2–14. The ball has the least energy (least ability to do work) when it is at sea level. If the ball is raised to level 1 by doing a little work, then the ball could perform some work if it were released, that is, stir the air, make a dent in the ground, and so on. The ball could be raised to level 2, 3, or 4; a ball dropped from level 4 would stir more air and make a bigger dent and hence have the ability to do more work or have greater energy than a ball dropped from a lower level. Sea level would be the lowest energy level, or *ground* level, of the ball, and levels 1, 2, 3, 4 would be higher, or *excited*, energy levels. Also note that the ball could fall from level 4 to level 2 instead of to ground level. But whenever the ball goes from a higher level to a lower level, it does some work, that is, it gives up some energy.

In the model of the hydrogen atom, there are various energy levels; see Figure 2–15. Ordinarily the atom is in the ground level, but if some work is done to the atom by heating it up, the atom can be raised to a higher, excited energy level. Now if the atom starts to return to its ground level, it gives up the energy it has by giving off light. The atom can return to the ground state by giving off all its energy at once or it can do so in jumps, stopping at lower energy levels. For instance, an atom in the fourth excited energy level can drop to the first excited energy level and give off light corresponding to the blue color observed on the screen. Since energy is related

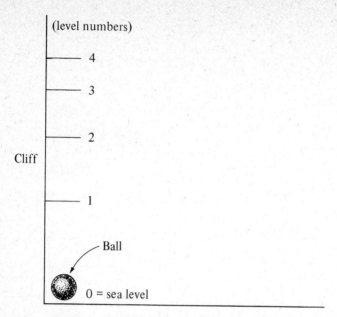

**Figure 2-14**  Cliff analogy for energy levels in the hydrogen atom.

to $1/\lambda$, blue light (having a smaller $\lambda$ value) represents greater energy than red light does.

It is common for chemists to look at the hydrogen atom as if the electron is a "ball" that is jumping from level to level. So in the lowest energy level of the atom, the electron is said to be in the ground state or level. The same sort of picture holds for atoms with more than one electron occupying energy levels. Therefore, the current model for atoms presupposes that there exists within atoms various distinct energy levels that are occupied by electrons. The permissible energy levels are called shells. Therefore electrons occupy shells within an atom. Shells differ not only in their energy but also in the number of electrons they may hold and the average distance the

4th excited level    _____

3rd excited level    _____

2nd excited level    _____

1st excited level    _____

Ground level    _____

**Figure 2-15**  Hydrogen atom energy levels.

**Table 2–5**
**Shells in Atoms**

| Shell Number | Number of Electrons |
|---|---|
| 1 | 2 |
| 2 | 8 |
| 3 | 8 (10) |
| 4 | 8 (24) |
| 5 | 8 (24) |
| 6 | 8 (10) |
| 7 | 8 (10)[a] |

[a]See page 34 for discussion of the electrons in parentheses.

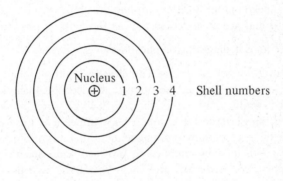

electrons in a given shell are from the nucleus. Table 2–5 shows these shells, the number of electrons each holds, and a schematic diagram of the atom. (The numbers in parentheses represent electrons added in each shell for the transition metals, rare earths, and actinides.* In any atom, the electrons occupy the levels with the lowest shell numbers first, up to their full complement of electrons.

Consider the following examples of the atomic model, with reference to Figure 2–16. Hydrogen (H) atoms contain one proton and one electron; the electron goes into shell number one. Helium (He) contains two protons and two neutrons; both electrons go into shell number one. Lithium (Li) contains three protons, three neutrons, and three electrons; two electrons go into shell number one, and one electron goes into shell number two, now the lowest available shell. The schematic diagrams in Figure 2–16 show this information pictorially. The schematics for carbon, oxygen, fluorine, neon, sodium, and phosphorus are also shown in Figure 2–16. Normally it is sufficient to know only what the charge on the nucleus is; and it should be remembered that

* These topics will be discussed later.

there are neutrons in all nuclei except hydrogen. The structure of the nucleus is shown explicitly for H, He, and Li.

In any discussion of the electrons in an atom, attention is focused on those in the outermost shell, as distinct from the collective number of those electrons in the inner shells. The outer shell is the occupied shell farthest away from the nucleus (also highest in energy). For example, phosphorus contains five electrons in the outer shell (shell number three) and ten electrons in the inner shells (numbers one and two). For hydrogen and helium shell number one is the outer shell.

In addition to the schematic diagrams shown in Figure 2–16, there is an additional shorthand method for representing atoms. This method is called the Lewis electron dot diagram, or simply, dot diagram. To construct the dot diagram, one determines the number of outer-shell electrons of an atom, and then arranges an equal number of dots around the atom's symbol. The dots are placed in four positions around the atom's symbol (above, below, left, and right) and are arranged so that as few dots as possible are paired.

For instance, the dot diagram for phosphorus is $\cdot \overset{\cdot\cdot}{\underset{\cdot}{P}} \cdot$, not $\cdot \overset{\cdot\cdot}{P} \overset{\cdot\cdot}{\cdot}$. See Figure 2–16 for examples of dot diagrams. The chemical and physical properties of elements are closely related to the number of electrons in their outer shell.

A table compiling all of the elements is widely used in chemistry. This table is called the periodic table, and a copy appears as Figure 2–17. Each square in the table represents one element in all its isotopic forms, that is, all the atoms with a given number of protons (same atomic number). So element number 1 is hydrogen and includes all the isotopes with atomic number 1 (hydrogen, deuterium, and tritium). The number above the atomic symbol in each square is the atomic number. The number below the atomic symbol is the atomic weight, the average weight of the atoms of the element found in nature.° Hydrogen's atomic weight is 1.008. In general the atomic weights of elements increase with increasing atomic number. A table similar to the modern periodic table was constructed in 1869 by Mendeleev, a Russian, on the basis of increasing atomic weights and similar chemical behavior.

The periodic table does more than just list the elements. It is also a gold mine of information concerning the chemical and physical properties of the elements. A vertical column of the table is called a **group**; a horizontal row is called a **period**. The elements in a group exhibit similar chemical and physical properties.

---

**Figure 2-16**  Electronic configuration of atoms. The size of the nucleus is greatly exaggerated; on the scale shown, the nucleus would be 0.0002 in. in diameter. The schematic diagram of each atom appears on the left; the dot or Lewis diagram appears on the right.

---

° See Appendix A, Section 3(a), for a further discussion of atomic weights.

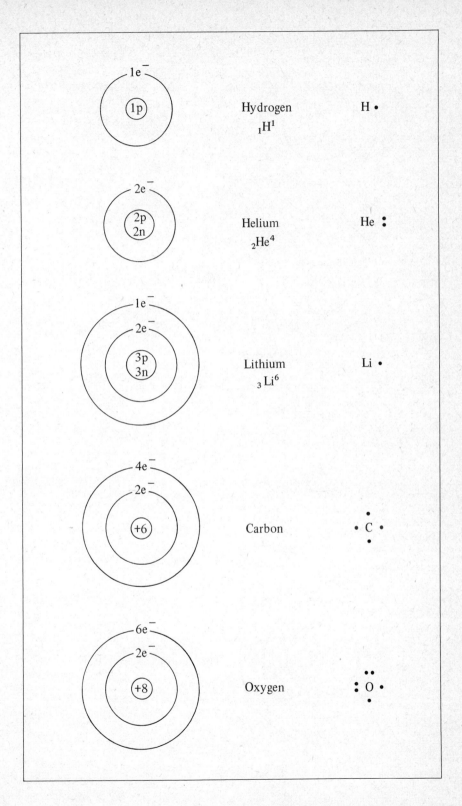

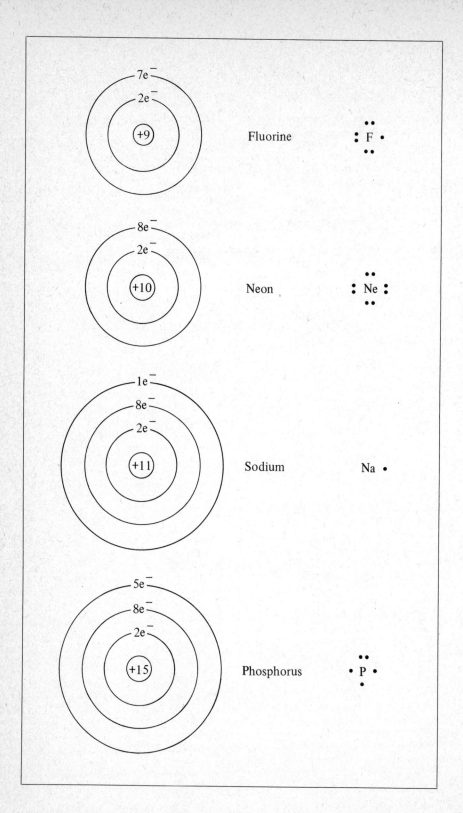

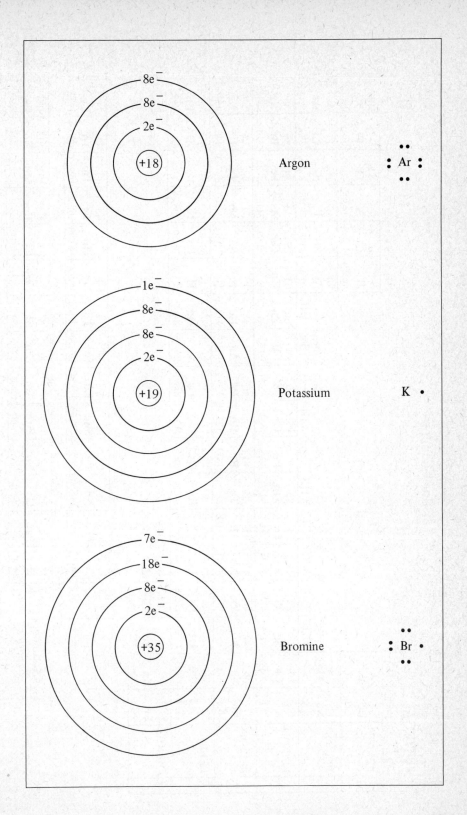

# Periodic Table of the Elements

| | IA | IIA | IIIB | IVB | VB | VIB | VIIB | VIII | VIII | VIII | IB | IIB | IIIA | IVA | VA | VIA | VIIA | Noble Gases |
|---|---|---|---|---|---|---|---|---|---|---|---|---|---|---|---|---|---|---|
| PERIOD 1 | 1 **H** 1.008 | | | | | | | | | | | | | | | | | 2 **He** 4.00 |
| PERIOD 2 | 3 **Li** 6.94 | 4 **Be** 9.01 | | | | | | | | | | | 5 **B** 10.81 | 6 **C** 12.01 | 7 **N** 14.01 | 8 **O** 16.00 | 9 **F** 19.00 | 10 **Ne** 20.18 |
| PERIOD 3 | 11 **Na** 22.99 | 12 **Mg** 24.31 | | | | | | | | | | | 13 **Al** 26.98 | 14 **Si** 28.09 | 15 **P** 30.97 | 16 **S** 32.06 | 17 **Cl** 35.45 | 18 **Ar** 39.95 |
| PERIOD 4 | 19 **K** 39.10 | 20 **Ca** 40.08 | 21 **Sc** 44.96 | 22 **Ti** 47.90 | 23 **V** 50.94 | 24 **Cr** 52.00 | 25 **Mn** 54.94 | 26 **Fe** 55.85 | 27 **Co** 58.93 | 28 **Ni** 58.71 | 29 **Cu** 63.55 | 30 **Zn** 65.37 | 31 **Ga** 69.72 | 32 **Ge** 72.59 | 33 **As** 74.92 | 34 **Se** 78.96 | 35 **Br** 79.90 | 36 **Kr** 83.80 |
| PERIOD 5 | 37 **Rb** 85.47 | 38 **Sr** 87.62 | 39 **Y** 88.91 | 40 **Zr** 91.22 | 41 **Nb** 92.91 | 42 **Mo** 95.94 | 43 **Tc**× (99) | 44 **Ru** 101.07 | 45 **Rh** 102.91 | 46 **Pd** 106.4 | 47 **Ag** 107.87 | 48 **Cd** 112.40 | 49 **In** 114.82 | 50 **Sn** 118.69 | 51 **Sb** 121.75 | 52 **Te** 127.60 | 53 **I** 126.90 | 54 **Xe** 131.30 |
| PERIOD 6 | 55 **Cs** 132.91 | 56 **Ba** 137.34 | 57 **La** 138.91 | 72 **Hf** 178.49 | 73 **Ta** 180.95 | 74 **W** 183.85 | 75 **Re** 186.2 | 76 **Os** 190.2 | 77 **Ir** 192.2 | 78 **Pt** 195.09 | 79 **Au** 196.97 | 80 **Hg** 200.59 | 81 **Tl** 204.37 | 82 **Pb** 207.19 | 83 **Bi** 208.98 | 84 **Po**× (210) | 85 **At**× (210) | 86 **Rn**× (222) |
| PERIOD 7 | 87 **Fr**× (223) | 88 **Ra**× (226) | 89 **Ac**× (227) | 104 **Ku**× (260) | 105 **Hn**× | | | | | | | | | | | | | |

Electron shell counts (left margin):
- Period 1: 2
- Period 2: 2 8
- Period 3: 2 8 8
- Period 4: 2 8 8
- Period 5: 2 8 18 8
- Period 6: 2 8 18 18 8
- Period 7: 2 8 18 32 18 8

## Lanthanide Series

| 58 **Ce** 140.12 | 59 **Pr** 140.91 | 60 **Nd** 144.24 | 61 **Pm**× (147) | 62 **Sm** 150.35 | 63 **Eu** 151.96 | 64 **Gd** 157.25 | 65 **Tb** 158.92 | 66 **Dy** 162.50 | 67 **Ho** 164.93 | 68 **Er** 167.26 | 69 **Tm** 168.93 | 70 **Yb** 173.04 | 71 **Lu** 174.97 |
|---|---|---|---|---|---|---|---|---|---|---|---|---|---|

## Actinide Series

| 90 **Th**× 232.04 | 91 **Pa**× (231) | 92 **U**× 238.03 | 93 **Np**× (237) | 94 **Pu**× (242) | 95 **Am**× (243) | 96 **Cm**× (247) | 97 **Bk**× (247) | 98 **Cf**× (251) | 99 **Es**× (254) | 100 **Fm**× (253) | 101 **Md**× (256) | 102 **No**× (254) | 103 **Lw**× (257) |
|---|---|---|---|---|---|---|---|---|---|---|---|---|---|

32

The elements (except hydrogen) in the first column, Group IA elements, are soft metals, conduct electricity, and react rapidly with water to produce alkalis, so they are sometimes called alkali metals. The second column, elements Group IIA, are hard metals that react with water, but less rapidly than Group IA metals; Group IIA elements are called alkaline earths. The elements in Group VIIA are called halogens and react with Groups IA and IIA quite easily. The inert or noble gases, the last column, are noted for their chemical stability; only Xe and Kr have been found to react and these reactions are with fluorine under specific experimental conditions. The elements in Groups IIIB, IVB, VB, VIB, VIIB, VIII, IB, and IIB are called transition metals and have similar chemical and physical properties. The lanthanides (or rare earths), elements 58 to 71, belong in the table between elements 57 and 72, but are placed below to make the table more compact; similarly the actinides fit in after element 89, actinium.

Elements with atomic numbers greater than 92 are artificially made using nuclear reactions and do not occur naturally on the earth. Groups IA, IIA, IIIA, IVA, VA, VIA, and the halogens (VIIA) are sometimes called representative elements. Remember, columns in the table represent elements with similar chemical properties.

To further illustrate the similarity of chemical and physical properties within a group, consider the following examples. Calcium (Ca), strontium (Sr), and barium (Ba) are in Group IIA. During the fallout from nuclear weapons testing experienced during the 1950s, strontium 90 was one radioactive element closely watched. Since strontium reacts like calcium, wherever one is found the other will also be present. This meant that strontium 90, like Ca, was being concentrated in the bones and teeth of growing children. Also because of its chemical similarity to calcium, the radioactive isotope of barium is used as a radioactive tracer in medical tests. The barium is partially incorporated into all areas that incorporate calcium. The radioactive label allows the medical scientist to determine where buildups of calcium are occurring and how quickly.

Silver (Ag), copper (Cu), and gold (Au), all in Group IB, are extremely good electrical conductors. Germanium (Ge) and silicon (Si), in Group IVA, are used in solid state electrical devices. While mercury (Hg) is a widely publicized environmental pollutant, its Group IIB partner, cadmium (Cd) is just beginning to get notice. Another Group IIA example is calcium and magnesium (Mg), both of which cause hard-water problems because of similar chemical properties. Table salt, NaCl (sodium chloride), is often augmented with NaI (sodium iodide) to yield iodized salt; but NaF (sodium fluoride) is never added, as excess fluoride causes a disease, fluorosis. The similarities of these sodium salts (see Group VIIA) allow them to enter the body but only

---

**Figure 2-17** Periodic table of the elements, 1973. Superscript $^\times$ indicates that all isotopes are radioactive; () indicates mass number of longest known half-life. Small figures at far left show electron distribution in preceding rare gas.

fluoride causes a problem. Sodium fluoride is used to fluoridate public water supplies, but its concentration is held below the danger level, if such a level exists. The balance between sodium and potassium in the blood is easily disturbed because these two Group IA members behave so similarly.

The table also reflects the electron arrangement or electron configuration of the atoms. This is emphasized in the table by the small numbers at the far left giving the number of electrons in each occupied shell. See Figure 2–17. Period 1 has electrons in the first shell only. Elements in Period 2 add electrons to shell 2, the first shell being full. The first element in Period 3 is sodium (Na). Sodium contains one more electron than neon (Ne), whose electronic configuration ($1^2 2^8$) appears at the beginning of Period 3. The eleventh electron of sodium is added to the third shell, since sodium is in the third period. This process repeats itself, adding one more electron and nuclear proton for each additional element until argon is reached. Argon's configuration is Ar ($1^2 2^8 3^8$); note that this is the configuration found at the far left of Period 4 (2,8,8). Potassium (K) and calcium (Ca) contain one and two electrons respectively in shell four; their electronic configurations are K ($1^2 2^8 3^8 4^1$) and Ca ($1^2 2^8 3^8 4^2$). The next element in Period 4 is scandium (Sc); its additional electron is added into the *third shell*. Titanium (Ti) also adds its additional electron into the third shell. So from scandium (Sc) to zinc (Zn), ten additional electrons are added to the third shell. This is the reason the number 10 was placed in parentheses for the third shell in Table 2–5. The numbers in parentheses correspond to electrons added for the transition metals (Group IIIB to Group IIB), the lanthanides and actinides. Not until gallium (Ga) is reached are additional electrons added into shell four. Gallium has the configuration Ga ($1^2 2^8 3^{18} 4^3$); notice that the number of electrons is 31. Bromine has the configuration Br ($1^2 2^8 3^{18} 4^7$).

These specific considerations can be generalized. All the alkali metals (Group IA) have one electron in their outer shell. Group IIA atoms have two electrons in the outer shell. Group VIIA have seven electrons in the outer shell. The inert gases have eight electrons in their outer shell with the exception of He, which has the stable electronic configuration $1^2$: a completely filled first-level shell. Neon has eight electrons in the second shell; argon, eight electrons in the third shell, and so on.

The conclusion that can be reached from the above discussion is that chemical and physical properties of elements are closely related to the electronic configurations. For instance, all Group IA elements react the same chemically and all have one electron in their outer shells. All noble gas atoms have eight electrons in their outer shells and are almost chemically inert. Since the noble gases are essentially inert chemically, eight electrons (two electrons for He) in the outer shell must be a particularly stable electronic configuration; there is little tendency for t˙e inert gas to share, lose, or gain electrons. Since an electronic configuration of eight electrons in the outer shell is particularly stable, atoms other than those of the noble gases try to gain, lose, or share electrons so that they may obtain eight electrons in the outer shell. Atoms in Groups IA and IIA readily give up electrons. Notice that $Li^+$ has the same electronic configuration as He does, $1^2$. Also note the similarity of $Na^+$ and $Mg^{+2}$ to Ne: $1^2 2^8$. So Groups IA and IIA give up electrons in order to look like stable noble gas atoms. Atoms in Groups VIA and VIIA tend to take on an electron:

$$: \ddot{\underset{..}{F}} \cdot \; + 1e^- \longrightarrow \; : \ddot{\underset{..}{F}} : ^- \; (1^2 2^8),$$

$$: \ddot{\underset{.}{O}} \cdot \; + 2e^- \longrightarrow \; : \ddot{\underset{..}{O}} : ^{2-} \; (1^2 2^8).$$

Both $F^{-1}$ and $O^{-2}$ have electronic configurations similar to Ne. Therefore atoms like to have eight electrons in an outer shell and will readily accept or lose electrons to obtain a noble gas electronic configuration.

The tendency of atoms with less than eight electrons in their outer shells to be unstable and to try to reach the stable eight-electron, or inert gas, configuration is the driving force for the formation of molecules. By combining with other atoms, those atoms that do not have eight electrons in their outer shell can obtain a stable configuration.

## MOLECULAR STRUCTURE

Our study of the structure of atoms has indicated their tendencies for reactivity. The interaction and combination of atoms leads to the formation of **molecules,** which may be defined as units of matter that contain two or more atoms. Water consists of two atoms of hydrogen and one atom of oxygen; its chemical formula is $H_2 O_1$, or $H_2 O$. Carbon monoxide, the poisonous gas released during incomplete burning of gasoline, has one carbon atom and one oxygen atom and is written $C_1 O_1$, or CO. The subscripts to the right of the atom denote the number of atoms of that particular kind in the molecule. Subscripts equal to one are omitted. Cane sugar (sucrose) has twelve carbon atoms, twenty-two hydrogen atoms, and eleven oxygen atoms, and is written $C_{12} H_{22} O_{11}$.

How do atoms in a molecule stick together? What are the driving forces involved? Why is water $H_2 O$ instead of $H_3 O$ or $H_4 O$? To answer these questions let's look at the simplest molecule, $H_2$, a molecule containing two hydrogen atoms, a total of two protons, and two electrons. The forces that hold negative electrons to positive nuclei are electrostatic forces; they are many times more powerful than gravitational forces. Therefore electrostatic forces are involved when two hydrogen atoms are brought together to form a hydrogen molecule, $H_2$.

These atoms (which individually consist of a proton with an electron whirling about it) approach one another. The electrons are moving so fast that these atoms appear to be negative clouds with positive centers. As the atoms get closer, the clouds of negative charge meet and repel each other a little. But as the atoms get still closer, the proton of one atom starts attracting the electron of the other atom, and vice versa. As the atoms get still closer together, the protons start repelling one another. There is an optimum proton-proton distance at which proton-proton and electron-electron repulsions are small and electron-proton attractions are large, and the two atoms stick together. Electrons are the "glue" that holds the two nuclei together. The electrons are free to whirl about both nuclei, sometimes about one, sometimes about the other, and much of the time between the two nuclei.

A Lewis diagram for the formation of $H_2$, using dots for each electron, is

$$H\cdot + H\cdot \longrightarrow H:H$$

This symbolically indicates that the protons are held together by a pair of electrons. *The pair of electrons forms a chemical bond* between the two protons. Sometimes $H:H$ is written $H—H$, where the dash has replaced the pair of electrons.

A few questions can be posed:

1. Do the paired electrons remain still between the two nuclei? No, each electron moves about both nuclei. The nuclei share their respective electrons and the sharing leads to the chemical bond. The electrons move freely throughout the whole molecule.
2. Do the electrons, in flying throughout the molecule, spend more time around one nucleus than the other? No, both hydrogen nuclei are the same, a proton with $+1$ charge, so there would be no distinction between one end of the chemical bond and the other.
3. Where do the electrons spend most of their time? Between the two hydrogen nuclei. Hence, the pair of electrons pulls the two protons toward each other.
4. Does the fact that each hydrogen atom has essentially achieved a noble-gas electronic configuration have anything to do with bond formation? Yes. In $H:H$ each hydrogen atom has obtained an $He:$ electronic configuration—

Each hydrogen atom can be thought of as having two electrons in its outer shell $(1^2)$, a filled outer shell. Since each hydrogen atom is sharing the pair of electrons, it has that *pair* of electrons a good part of the time and hence has a very stable electronic configuration.
5. Do the protons move? Yes, but much more slowly than the electrons. They vibrate toward and away from each other as if connected by a spring. Also the whole molecule rotates and moves through space. See Figure 2–18. Of course the electrons follow the protons. So we have found that the chemical bond consists of a pair of shared electrons which hold the two nuclei together.

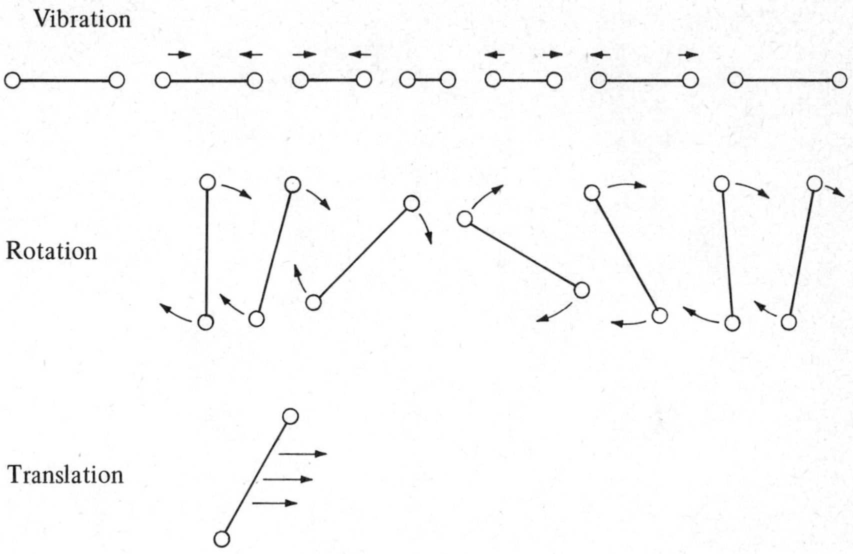

Vibration

Rotation

Translation

**Figure 2-18** Motions of the nuclei in $H_2$.

Next consider the case of a sodium atom and a chlorine atom being brought together. Their Lewis diagrams are

$$\text{Na}\cdot \quad \text{and} \quad :\overset{..}{\underset{.}{\text{Cl}}}:$$

Sodium is in Group IA and chlorine is in Group VIIA. This indicates that sodium gives up an electron easily and chlorine takes up an electron readily. So let sodium give up its electron to chlorine:

$$\text{Na}\cdot \longrightarrow \text{Na}^+ + 1e^-$$

$$1e^- + :\overset{..}{\underset{.}{\text{Cl}}}: \longrightarrow :\overset{..}{\underset{.\square}{\text{Cl}}}:^- \longrightarrow :\overset{..}{\underset{..}{\text{Cl}}}:^-$$

A "□" has been used to label sodium's electron, but it would be indistinguishable from the chlorine's electrons, once the electron transfer has occurred. The sodium is now positively charged and the chlorine negatively charged, and hence they attract each other. Neutral atoms that have lost or gained an electron are called ions. Hence $\text{Na}^+\text{Cl}^-$ is a pair of ions, usually written NaCl. Notice that chlorine now has eight electrons in its outer shell, $\text{Cl}^-$ ($1^22^83^8$), and so does $\text{Na}^+$ ($1^22^8$). Each atom has achieved a stable, inert-gas electronic configuration. Just as in the $H_2$ case, the atoms involved in chemical bonding strive to achieve an *inert-gas configuration* because of its stability.

There is a distinct difference between the chemical bond in $H_2$ and the chemical bond in NaCl. In $H_2$, the nuclei *share* a pair of electrons and this pair of electrons holds the nuclei together. In NaCl an electron is completely lost by sodium and completely taken up by chlorine; it is the attraction

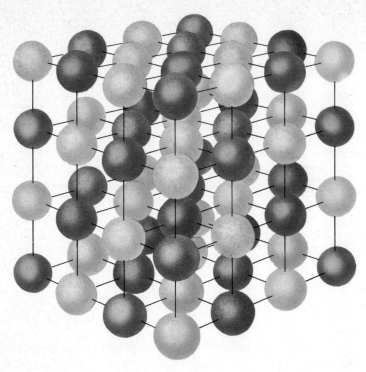

**Figure 2-19** Ball-and-stick representation of crystalline sodium chloride. Dark atoms represent $Na^+$ ions and light atoms represent $Cl^-$ ions. Each ion (except those on the surface) is surrounded by six ions of opposite charge. This continual lattice arrangement represents one type of crystal formation. All crystalline salts have a lattice network of ions, not necessarily the type shown above.

of two different particles with opposite charge ($Na^+$ and $Cl^-$) that holds the sodium and chloride ions together. These ions pack together in a specific manner to form a crystalline salt, whose structure is shown in Figure 2–19. The bond in NaCl is called an **ionic bond.** The chemical bond in $H_2$ is called a **covalent chemical bond.**

Now look at hydrogen fluoride, HF. Write Lewis structures for hydrogen and fluorine atoms,

$$H \cdot \qquad and \qquad : \overset{\cdot\cdot}{\underset{\cdot\cdot}{F}} \cdot ,$$

and note that each atom needs one electron to obtain an inert-gas electronic configuration. Hydrogen now has two electrons, the electron configuration of He ($1^2$) and fluorine has the electron configuration of neon ($1^2 2^8$). So a bond is formed between hydrogen and fluorine and a molecule, written H—F for short, is formed. Is this bond covalent or ionic? It is not ionic, since a pair of electrons is shared. But it is not the same bond as was formed in $H_2$, either.

The reason for this is that the pair of electrons involved in bonding are not equally shared between hydrogen and fluorine. Experiment shows that these electrons spend more time around fluorine than around hydrogen. The physical result is that the hydrogen fluoride molecule looks like a small sausage with the hydrogen end a little positively charged:

$$\overset{+}{\delta}\,\boxed{\text{H—F}}\,\overset{-}{\delta}.$$

The Greek $\delta$ (delta) indicates a small fractional plus or minus charge. The HF bond is still considered to be covalent, but not purely covalent in the sense that $H_2$ is, that is, the electron pair is shared but not equally. A bond in which the electrons are not shared equally is often called a polar bond because it has a separation of charge, one end of the bond is slightly positive, the other end slightly negative. Most chemical bonds lie between pure ionic and pure covalent. Table 2–6 summarizes the types of bonds. Note that inner-shell electrons play essentially no role in forming these chemical bonds.

When the question of why water is $H_2O$ instead of $H_4O$ is considered, the answer can be reached following the general rule that atoms form molecules in order to obtain the stable, eight-electrons-in-the-outer-shell configuration. Similarly, the following question can be posed. Why does the compound calcium chloride have the chemical formula, $CaCl_2$? Water is a covalent molecule while $CaCl_2$ is an ionic compound or salt. The same considerations will answer the questions posed.

First consider the water molecule. It contains hydrogen and oxygen; but how many atoms of each is the question. Write the dot formulas for oxygen and hydrogen,

$$:\!\overset{\cdot\cdot}{\underset{\cdot}{O}}\!\cdot \quad \text{and} \quad H\cdot$$

Notice that oxygen needs two electrons to obtain eight and the hydrogen needs one to look like He. See Figure 2–20. If each hydrogen shares its electron with oxygen and oxygen shares one electron with each hydrogen, then all atoms will have a stable configuration. So if another $H\cdot$ were to happen upon the scene, there would be no place for it to go. Notice that two pairs of electrons in the water molecule are not involved in bonding. These two pairs

**Table 2–6**
**Summary of Chemical Bonds**

| Type | Pure Covalent | In-Between, Polar | Pure Ionic |
|---|---|---|---|
| Description | Electron pair shared equally | Electrons not shared equally | Electron(s) transferred |
| Example | $H_2$ | HF | NaCl |

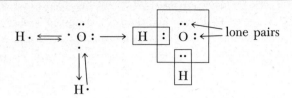

$$H\cdot \rightleftharpoons \cdot \overset{..}{O}: \longrightarrow \boxed{H} : \overset{..}{O}: \leftarrow \text{lone pairs}$$

**Figure 2-20**  Water molecule. Could this be written $H:\overset{..}{\underset{..}{O}}:H$?

are called nonbonding electrons, or **lone pairs** of electrons. Similarly, for ammonia, $NH_3$, write the Lewis diagrams,

$$H\cdot + H\cdot + H\cdot + \cdot \overset{..}{\underset{.}{N}}\cdot$$

Note that nitrogen needs three electrons and has three electrons to share. Ammonia has one nonbonded, or lone, pair of electrons.

For calcium chloride the Lewis diagram is

$$\cdot Ca\cdot + :\overset{..}{\underset{.}{Cl}}: + :\overset{..}{\underset{.}{Cl}}:$$

The calcium can lose two electrons, one to each chlorine. Each chlorine then has eight electrons about it and has the electronic configuration of argon ($1^22^83^8$), and calcium, having lost two electrons, has a double positive charge and has the electronic configuration of argon, with eight electrons in its outer shell ($1^22^83^8$). See Figure 2–21.

Finally consider methane, $CH_4$. Again write the Lewis diagrams,

$$H\cdot + H\cdot + H\cdot + H\cdot + \cdot \overset{.}{\underset{.}{C}}\cdot$$

Carbon needs four electrons and will share four electrons; four hydrogens need

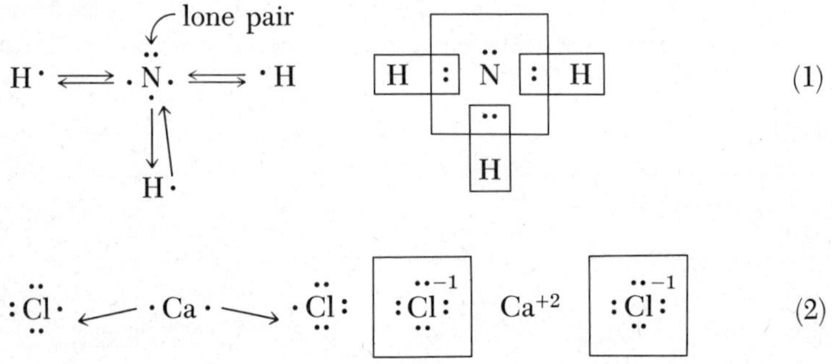

lone pair

$$H\cdot \rightleftharpoons \cdot \overset{.}{\underset{.}{N}}\cdot \rightleftharpoons \cdot H \qquad \boxed{H} : \overset{..}{N} : \boxed{H} \qquad (1)$$

$$:\overset{..}{\underset{..}{Cl}}\cdot \longleftarrow \cdot Ca\cdot \longrightarrow \cdot \overset{..}{\underset{..}{Cl}}: \qquad \boxed{:\overset{..-1}{\underset{..}{Cl}}:} \quad Ca^{+2} \quad \boxed{:\overset{..-1}{\underset{..}{Cl}}:} \qquad (2)$$

**Figure 2-21**  (1) Ammonia and (2) calcium chloride.

one electron each and will share an electron. Hence the dot diagram for methane is

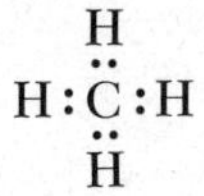

$$
\begin{array}{c}
\text{H} \\
\text{\Large H:\overset{\cdot\cdot}{\underset{\cdot\cdot}{C}}:H} \\
\text{H}
\end{array}
$$

The concept of **valence** refers to the number of chemical bonds an atom can form when it is part of a molecule. The atoms considered above have the valences shown in Table 2-7. The concept of valence is verified by a consideration of the stable atomic electronic configurations. Knowing the valence of atoms allows one to predict the molecular formula of molecules involving these atoms. In the compound containing carbon and chlorine, for instance, carbon has a valence of four, chlorine a valence of one. So it takes four chlorines to make four bonds to carbon, $CCl_4$.

The tendency for atoms to form molecules has been explained in terms of the atoms' attempts to obtain eight electrons in their outer shells. Since atoms within molecules that have this inert-gas-like electronic structure are stable, why should two molecules react to form new molecules? In fact, most chemical reactions involve molecules, not atoms. Most reactions between molecules occur because of the uneven sharing of electrons in a chemical bond(s) somewhere in the molecules involved. For instance the molecule methanol, $CH_3OH$, has the structure

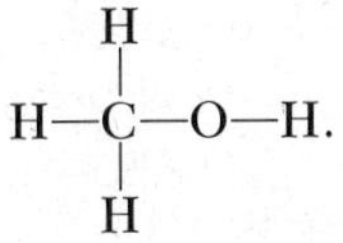

$$
\begin{array}{c}
\text{H} \\
| \\
\text{H--C--O--H.} \\
| \\
\text{H}
\end{array}
$$

The electrons involved in the C—O bond are not shared equally, being held more closely to the oxygen atom. This means that if an ion should pass by a methanol molecule, it would be attracted and repelled by the separation of charge. A hydrogen ion, $H^+$ (just a proton), would be attracted to the slightly negative oxygen. The tendency of an atom involved in a chemical

Table 2-7
Valences of Atoms

| Atom | Number of Chemical Bonds | Valence |
|------|--------------------------|---------|
| H | 1 | 1 |
| O | 2 | 2 |
| Ca | 2 | 2 |
| Cl | 1 | 1 |
| N | 3 | 3 |
| C | 4 | 4 |

**Table 2-8
Electronegativities**

| Atom | Value |
|------|-------|
| H | 2.1 |
| C | 2.5 |
| N | 3.0 |
| O | 3.5 |
| F | 4.0 |
| Si | 1.8 |
| P | 2.1 |
| S | 2.5 |
| Cl | 3.0 |
| Br | 2.8 |
| I | 2.5 |

bond to pull the bonding electrons toward it is referred to as the **electronegativity** of the atom. Of two atoms bound together, the one with the higher electronegativity will pull the bonding electrons toward it and be slightly negative; the atoms with the lower electronegativity will be slightly positive. The electronegativities of a few of the common elements are given in Table 2-8.

A bond formed between atoms with large electronegativity differences (0.8 or bigger) would be considered a polar bond and would be subject to reaction because it would attract other polar bonds or ions. In the extreme example, the electronegativity differences between atoms are so large that the pair of electrons is not shared at all, but one atom loses an electron to the other; sodium chloride, NaCl, and calcium chloride, $CaCl_2$, are examples of such cases.

Using the periodic table, note that electronegativity usually increases as you go across a period left to right (for example, C, N, O, F) and usually decreases as you go down a group (for example, F, Cl, Br, I).

If two atoms bound together have the same electronegativity, both pull the bonding electrons equally and the bond is pure covalent. This occurs in diatomic molecules involving the same atoms ($H_2$, $F_2$, $Br_2$, $O_2$, $N_2$, . . .) and in some compounds (for example, phosphine, $PH_3$). Usually the atoms involved in a chemical bond do not have equal electronegativities. In methanol, the electronegativity of oxygen (3.5) is greater than that of carbon (2.5) and hence oxygen is slightly negative. The concept of electronegativity is most useful in locating possible separation of charges within molecules.

So far only molecules where a single pair of electrons has been shared between two atoms have been considered. The molecule carbon dioxide, $CO_2$, has another arrangement. Since carbon has a valence of four and oxygen a valence of two, if only single pairs of electrons were shared, that is, only single bonds formed, the formula $CO_2$ would not be possible. But if each oxygen forms a double bond to carbon, all valences would be satisfied. This can also be verified using the dot diagrams.

$$\ddot{O}::C::\ddot{O}$$

$$O=C=O$$

Similarly triple bonds may be formed, as in nitrogen, $N_2$, for instance. Each nitrogen has a valence of three, so if a triple bond is formed, or three pairs of electrons shared, the valence of both nitrogen atoms is fulfilled:

$$:N:::N:$$

$$N\equiv N$$

Knowing the shapes of molecules in addition to the distribution of their electrons is of value in understanding them. The shapes of molecules are believed to be of great significance in odor sensation and in the regulation of chemical reactions that occur in biological systems. The factors that determine the shape of molecules include electron-electron and nuclear-nuclear repulsions and electron-nuclear attractions; the exact role these factors play is very complicated. Molecules containing two atoms are linear, since two points determine a straight line. The shapes of water, ammonia, and methane molecules are shown in Figure 2–22. The shape of methane is particularly important. The hydrogens are located at the vertices of a regular tetrahedron with the carbon in the center.°

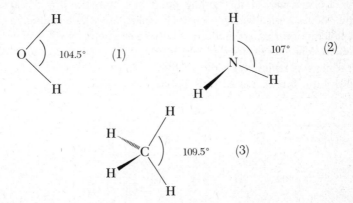

**Figure 2-22** Molecules of (1) water (O—H distance is 0.965 Å); (2) ammonia (N—H distance is 1.00 Å); and (3) methane (C—H distance is 1.10 Å). 1 Å = 1 angstrom = 0.00000001 cm. (NOTE: Solid line means bond in plane of page, wedge means bond in front, dotted line means bond behind page.)

° The shapes of other molecules will be discussed in later chapters in connection with specific topics.

Although we know how molecules are constructed, they are too small to observe. Objects large enough to see are conglomerations of molecules held together. Even though most molecules and atoms have mass, and hence experience gravitational attractions between one another, such attractions are not great enough to explain why molecules stick together. Electrostatic forces, attraction and repulsion of particles, are the answer to interactions between molecules.

Consider the Lewis diagram for water:

$$H:\overset{\cdot\cdot}{\underset{\cdot\cdot}{O}}:H$$

Like the bonding electrons in HF, the bonding electron pairs in water are found closer to oxygen than to hydrogen. So the oxygen is slightly negative and the hydrogens slightly positive.

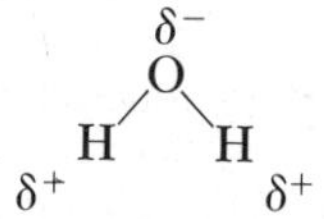

If many water molecules are thrown together, one would expect them to interact—the oxygen atom of one water molecule interacting with the hydrogen atom of another water molecule. See Figure 2–23. At room temperature water is a liquid, and our microscopic picture of water is a whole bunch of water molecules running around in the beaker attracting and repelling one another. Two water molecules repel each other if their respective oxygen atoms come close. If we start heating the water, the molecules start moving faster and faster, and at the boiling point of water, the liquid water turns to steam. Microscopically, the individual molecules, which are moving faster at higher temperatures, are no longer held to each other by the electrostatic interactions. The electrostatic forces of attraction become too small to hold the speeding molecules. A simple analogy, using gravitational forces instead of electrostatic forces, would be shooting two rockets into the sky—one a

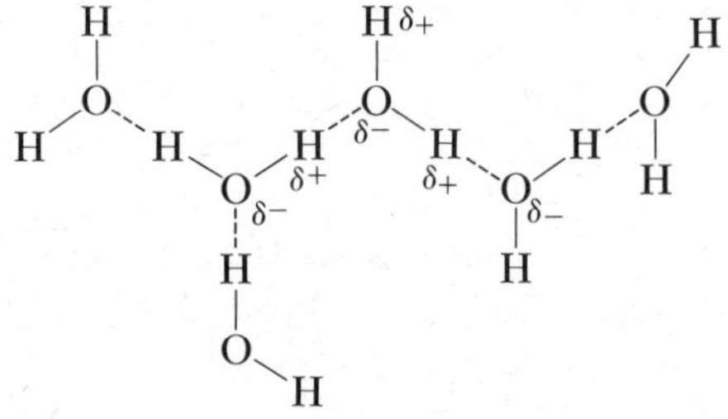

**Figure 2-23**  Interactions in liquid water. Dashed lines represent hydrogen bonds.

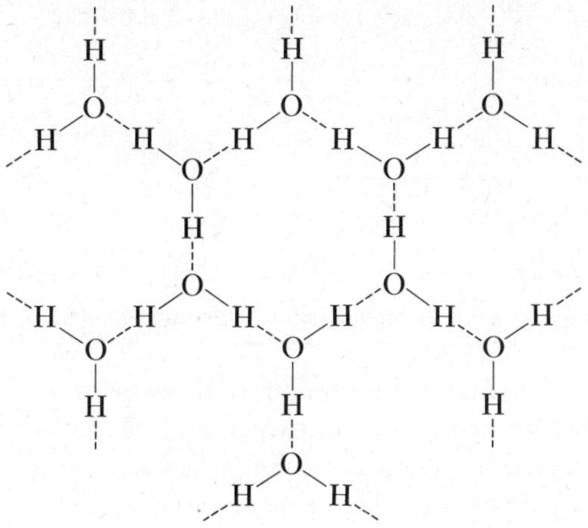

**Figure 2-24** Ice. Dashed lines represent hydrogen bonds.

small weather rocket, the other a rocket sent on a flyby of Venus. The weather rocket is shot off, burns up its fuel at 25,000 feet and falls back. The Venus rocket burns its fuel, but has enough fuel to build up enough speed to break free of the earth's gravitational pull. The same forces are present in both cases, but the Venus rocket had enough speed to break free.

If liquid water is cooled below freezing, the molecules move more and more slowly, and finally line up in a rigid structure with oxygens $(\delta-)$ next to hydrogens $(\delta+)$. This is shown schematically in Figure 2-24. In solid water (ice), the electrostatic forces are sufficient to hold the molecules together and in rigid positions. The electrostatic forces completely overcome any tendency of the molecules to move away since the molecules have very little kinetic energy.

Macroscopic matter is classified into three categories based on physical properties (*macroscopic* means large enough to see with the eye). The three categories are *solid, liquid,* and *gas.* Solids are characterized by the retention of their shapes, irrespective of the container they are in; microscopically (on the molecular scale), solids have a more or less well-defined structure. Liquids conform to the shape of their containers, but they do not try to fill the whole container. So if water is poured into a jar, it will drop to the bottom and take up the shape of the jar, regardless of the shape of its original container. Microscopically liquids are seen as individual molecules moving about randomly with little observed order. Gases also conform to the shape of their containers, but completely fill them. If a cake is being baked in the kitchen, the aroma spreads throughout the whole house. Air is found in every corner of a room, not just on the floor, as a liquid would be.

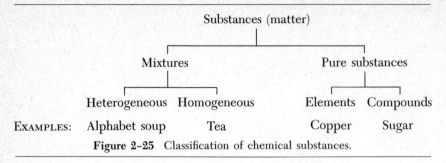

Figure 2-25  Classification of chemical substances.

Matter is also classified on the basis of *composition*. See Figure 2–25. Mixtures are substances composed of two or more different kinds of molecules that can be separated by physical means. A *heterogeneous mixture* would be a mixture of red and blue marbles, or a closed jar containing sand and water. A *homogeneous mixture* would be a mixture, such as tea, or sugar dissolved in water, where there are no discernibly different parts.

**Table 2-9**
**Atomic Weights and Atomic Numbers of the Elements**

| Element | Symbol | Atomic Number | Atomic Weight |
|---------|--------|---------------|---------------|
| Actinium | Ac | 89 | $(227)^a$ |
| Aluminum | Al | 13 | 26.9815 |
| Americium | Am | 95 | (243) |
| Antimony | Sb | 51 | 121.75 |
| Argon | Ar | 18 | 39.948 |
| Arsenic | As | 33 | 74.9216 |
| Astatine | At | 85 | (210) |
| Barium | Ba | 56 | 137.34 |
| Berkelium | Bk | 97 | (247) |
| Beryllium | Be | 4 | 9.0122 |
| Bismuth | Bi | 83 | 208.980 |
| Boron | B | 5 | 10.811 |
| Bromine | Br | 35 | 79.904 |
| Cadmium | Cd | 48 | 112.40 |
| Calcium | Ca | 20 | 40.08 |
| Californium | Cf | 98 | (249) |
| Carbon | C | 6 | 12.01115 |
| Cerium | Ce | 58 | 140.12 |
| Cesium | Cs | 55 | 132.905 |
| Chlorine | Cl | 17 | 35.453 |
| Chromium | Cr | 24 | 51.996 |
| Cobalt | Co | 27 | 58.9332 |
| Copper | Cu | 29 | 63.546 |
| Curium | Cm | 96 | (247) |

**Table 2-9 (Continued)**

| Element | Symbol | Atomic Number | Atomic Weight |
|---|---|---|---|
| Dysprosium | Dy | 66 | 162.50 |
| Einsteinium | Es | 99 | (254) |
| Erbium | Er | 68 | 167.26 |
| Europium | Eu | 63 | 151.96 |
| Fermium | Fm | 100 | (253) |
| Fluorine | F | 9 | 18.9984 |
| Francium | Fr | 87 | (223) |
| Gadolinium | Gd | 64 | 157.25 |
| Gallium | Ga | 31 | 69.72 |
| Germanium | Ge | 32 | 72.59 |
| Gold | Au | 79 | 196.967 |
| Hafnium | Hf | 72 | 178.49 |
| Hahnium | Hn | 105 | ( )[b] |
| Helium | He | 2 | 4.0026 |
| Holmium | Ho | 67 | 164.930 |
| Hydrogen | H | 1 | 1.00797 |
| Indium | In | 49 | 114.82 |
| Iodine | I | 53 | 126.9044 |
| Iridium | Ir | 77 | 192.2 |
| Iron | Fe | 26 | 55.847 |
| Krypton | Kr | 36 | 83.80 |
| Kurchatovium | Ku | 104 | (260) |
| Lanthanum | La | 57 | 138.91 |
| Lawrencium | Lw | 103 | (257) |
| Lead | Pb | 82 | 207.19 |
| Lithium | Li | 3 | 6.939 |
| Lutetium | Lu | 71 | 174.97 |
| Magnesium | Mg | 12 | 24.312 |
| Manganese | Mn | 25 | 54.9380 |
| Mendelevium | Md | 101 | (256) |
| Mercury | Hg | 80 | 200.59 |
| Molybdenum | Mo | 42 | 95.94 |
| Neodymium | Nd | 60 | 144.24 |
| Neon | Ne | 10 | 20.183 |
| Neptunium | Np | 93 | (237) |
| Nickel | Ni | 28 | 58.71 |
| Niobium | Nb | 41 | 92.906 |
| Nitrogen | N | 7 | 14.0067 |
| Nobelium | No | 102 | (254) |
| Osmium | Os | 76 | 190.2 |
| Oxygen | O | 8 | 15.9994 |
| Palladium | Pd | 46 | 106.4 |
| Phosphorus | P | 15 | 30.9738 |
| Platinum | Pt | 78 | 195.09 |

### Table 2-9 (Continued)

| Element | Symbol | Atomic Number | Atomic Weight |
|---|---|---|---|
| Plutonium | Pu | 94 | (242) |
| Polonium | Po | 84 | (210) |
| Potassium | K | 19 | 39.102 |
| Praseodymium | Pr | 59 | 140.907 |
| Promethium | Pm | 61 | (147) |
| Protactinium | Pa | 91 | (231) |
| Radium | Ra | 88 | (226) |
| Radon | Rn | 86 | (222) |
| Rhenium | Re | 75 | 186.2 |
| Rhodium | Rh | 45 | 102.905 |
| Rubidium | Rb | 37 | 85.47 |
| Ruthenium | Ru | 44 | 101.07 |
| Samarium | Sm | 62 | 150.35 |
| Scandium | Sc | 21 | 44.956 |
| Selenium | Se | 34 | 78.96 |
| Silicon | Si | 14 | 28.086 |
| Silver | Ag | 47 | 107.868 |
| Sodium | Na | 11 | 22.9898 |
| Strontium | Sr | 38 | 87.62 |
| Sulfur | S | 16 | 32.064 |
| Tantalum | Ta | 73 | 180.948 |
| Technetium | Tc | 43 | (99) |
| Tellurium | Te | 52 | 127.60 |
| Terbium | Tb | 65 | 158.924 |
| Thallium | Tl | 81 | 204.37 |
| Thorium | Th | 90 | 232.038 |
| Thulium | Tm | 69 | 168.934 |
| Tin | Sn | 50 | 118.69 |
| Titanium | Ti | 22 | 47.90 |
| Tungsten | W | 74 | 183.85 |
| Uranium | U | 92 | 238.03 |
| Vanadium | V | 23 | 50.942 |
| Xenon | Xe | 54 | 131.30 |
| Ytterbium | Yb | 70 | 173.04 |
| Yttrium | Y | 39 | 88.905 |
| Zinc | Zn | 30 | 65.37 |
| Zirconium | Zr | 40 | 91.22 |

[a] Parentheses indicate the mass numbers of the most stable or the best-known isotopes.
[b] Unknown

**Table 2-10**
**Chemical Symbols Not Derived from**
**Common Name of Element**

| Common Name | Symbol | Symbol Source |
|---|---|---|
| Antimony | Sb | Stibium |
| Copper | Cu | Cuprum |
| Gold | Au | Aurum |
| Iron | Fe | Ferrum |
| Lead | Pb | Plumbum |
| Mercury | Hg | Hydrargyrum |
| Potassium | K | Kalium |
| Silver | Ag | Argentum |
| Sodium | Na | Natrium |
| Tin | Sn | Stannum |
| Tungsten | W | Wolfram |

## PROBLEMS

1. There are three forces: gravitational, electrostatic, and nuclear. What forces are involved in the following cases:
   (a) A boy jumps off a chair and falls to the ground.
   (b) Attraction of earth for sun.
   (c) The needle on a compass points to the North Pole.
   (d) The aurora borealis.
   (e) Cobalt treatment of cancer.

2. If you drop a marble and a beach ball from the same height, will they hit the floor at the same time? Why?

3. Classify the following forms of energy as potential or kinetic.
   (a) Flashlight battery.
   (b) An Arnold Palmer putt.
   (c) The Mississippi River.
   (d) An airplane flying over the Rockies.
   (e) Sunlight.
   (f) An oak tree.

4. $^{234}_{90}$Th is formed when $^{238}_{92}$U decays. No neutrons, protons, or electrons are created or destroyed in this process. So since Thorium has two fewer protons, two fewer neutrons, and two fewer electrons, these particles must be given off. We can write this as:

$$^{238}_{92}U \longrightarrow \,^{234}_{90}Th \, + \,^{4}_{2}He$$
$$\text{alpha particle}$$

Determine $a$, $b$, and $X$ in the following:

(a) $^{210}Po \longrightarrow ^{206}Pb + {}^{b}_{a}X$

(b) $^{14}C \longrightarrow ^{14}N + {}^{b}_{a}X$

(c) $^{55}Fe \longrightarrow {}^{0}_{+1}e + {}^{b}_{a}X$

5. The half-life of $^{43}Sc$ is 3.9 hours. If you start with 40 grams of $^{43}Sc$ how much would you have left after one half-life had elapsed? After 10 half-lives? Would all the $^{43}Sc$ ever decay?

6. Consider the environmental costs of producing electrical energy by nuclear power plants. List activities necessary in nuclear power production that consume natural resources or provide hazards to man. Don't figure dollars and cents.

7. What are the electronic structures for the following? (HINT: Remember the 10 electrons for transition elements.)
   (a) Aluminum
   (b) Sulfur
   (c) Silicon
   (d) Germanium
   (e) Krypton

8. Write dot diagrams for:
   (a) Rubidium
   (b) Barium
   (c) Indium
   (d) Lead
   (e) Arsenic
   (f) Sulfur
   (g) Silicon
   (h) Krypton

9. What are the Lewis diagrams and formulas for:
   (a) Potassium bromide
   (b) Strontium chloride
   (c) Hydrogen sulfide
   (d) Aluminum fluoride

10. Indicate the type of bonding, ionic or covalent, you would expect and, where possible, determine which atoms are slightly negative due to electronegativity differences in the following molecules. What is the valence of each element in these compounds?
    (a) NaBr
    (b) $MgI_2$
    (c) $SiH_4$
    (d) $PH_3$
    (e) $H_2S$
    (f) $Br_2$

11. Lithium and hydrogen react to give LiH, lithium hydride.
    (a) Describe the steps that occur in the reaction.
    (b) The electronegativity of Li is 1.0; that of H is 2.1. What type of bond is the LiH bond? Which atom is slightly negative?

12. Complete the following table:

| Isotope | No. of Protons | No. of Neutrons | No. of Electrons |
|---------|----------------|-----------------|------------------|
| $^{55}Fe$ | ... | 29 | 26 |
| $94$ ... | ... | 144 | ... |
| $O^{-2}$ | ... | 10 | ... |

13. List as much information as you can about the chemical and physical properties of the element that will be placed in the empty box.

| VIIA | | |
|---|---|---|
| 9<br>**F**<br>19.0 | | |
| 17<br>**Cl**<br>35.5 | | |
| 34<br>**Se**<br>79.0 | | 36<br>**Kr**<br>83.8 |

14. Which of the following will have similar physical and chemical properties and why?
    (a) Na, K, Mg, KBr, $Br_2$.
    (b) $BH_3$, $CH_4$, $NH_3$, $AlH_3$, $H_2S$, HCl.


## BIBLIOGRAPHY

1. Isaac Asimov. *A Short History of Chemistry.* Garden City, N.Y.: Doubleday, 1965.
2. R. C. Austin and P. Borrelli. *The Strip Mining of America.* Sierra Club, 1971.
3. Paul Ehrlich and Anne Ehrlich. *Population, Resources, Environment: Issues in Human Ecology.* San Francisco: W. H. Freeman, 1970.
4. M. Gardner. *Relativity for the Million.* New York: Macmillan Co., 1962.
5. Strip and Surface Mining Commission. *Surface Mining and Our Environment.* Washington, D.C.: U.S. Department of the Interior, 1967.

# Organic Chemistry

# 3

1500

$\mathcal{A}\!D$

Albertus Durerus
ipsum me propriis
gebam coloribus ꝫ
anno XXVIII

*or·gan'ic* (ôr·găn'ĭk), *adj.* **Chem.** *Pertaining to that branch of chemistry which treats compounds of carbon.* **Biol.** *Pertaining to, or derived from, living organisms.*

In earlier discussions on bonding between atoms, two extremes were considered: a transfer of electrons between two nuclei (ionic bonding) and the equal sharing of electrons between nuclei (covalent bonding). An example of the latter occurs in the hydrogen molecule, $H_2$. Since both nuclei in the hydrogen molecule are identical, each will have an identical "pull" on the electrons forming the bond. The element carbon, atomic number six, has the following electron configuration: $1^2 2^4$; so there are four electrons in the outer shell (four valence electrons). To achieve a noble-gas configuration, carbon can lose four electrons (becoming $C^{+4}$ and thus resembling helium), gain four electrons (becoming $C^{-4}$ and resembling neon), or share its electrons with one electron from each of four other atoms to form four covalent bonds (resulting in no charge on carbon). Of these possibilities, the latter is almost always true. Carbon has a tendency to form covalent bonds.

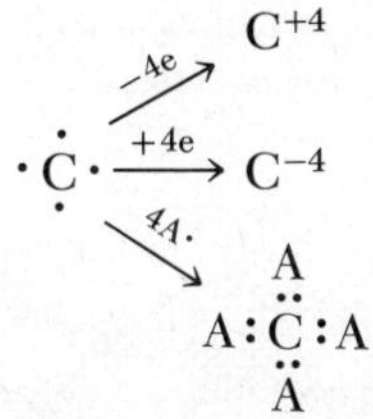

To which atoms will it tend to bond? Why not another carbon atom or even four carbon atoms as shown? The additional carbon atoms can further bond to other atoms, for example, more carbon atoms. This is one of the reasons carbon can form a disproportionately large number of compounds.

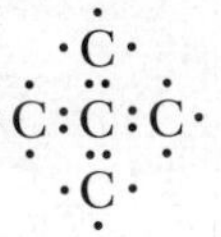

We can continue to develop a large structure of carbon bonded to carbon ad infinitum. The compound we have been describing is commonly referred to as diamond, which not only has a stable structure but is one of the hardest

---

*"I looked into a looking glass and what I saw was not the same. . . ."*

**Anonymous**

**Self-portrait of Albrecht Dürer (1500).** Dürer, a right-handed painter, executed this work while looking at his image in a mirror. The right hand we see in the painting is really his left. (*Bruckmann: Art Reference Bureau*)

materials known to man.° The ball-and-stick representation of the spatial arrangement of atoms in another form of carbon, graphite, shows the layered structure that gives graphite properties different from those of diamond. These layers can slide over one another, which accounts for the use of graphite as a lubricant.

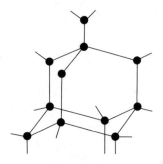

(Arrangement of atoms)
Diamond
(Black balls represent
carbon atoms.)

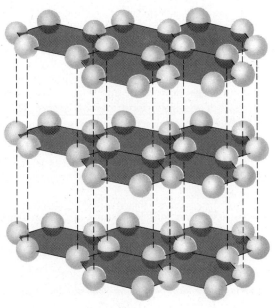

Graphite

---

°Diamond is not a true compound, since it has no specific molecular formula, but it can be represented as $C_n$.

# MOLECULAR GEOMETRY (SHAPES OF MOLECULES)

When carbon is bonded to four atoms, the geometry about the central atom is tetrahedral; that is, the bonds from the central carbon atom are directed toward the corners of a tetrahedron with the carbon positioned in the center.

Carbon also forms strong covalent bonds with hydrogen. The simplest tetrahedral carbon compound we could imagine would be one carbon atom bonded to four hydrogen atoms; it is known as methane, $CH_4$.

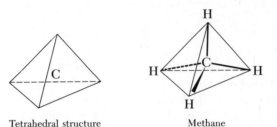

Tetrahedral structure          Methane

In order of simplicity, the next compound would be carbon bonded to three hydrogen atoms and to another carbon atom (which itself is bonded to three hydrogen atoms). This compound is ethane. If we continue in this vein, we get propane.

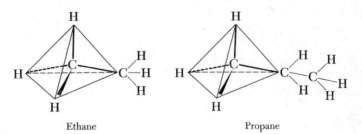

Ethane                    Propane

These compounds are composed solely of carbon and hydrogen and are called **hydrocarbons.** When each carbon is always bound to four other atoms, we say the compound is **saturated,** or has little desire to react. This entire saturated hydrocarbon series, called "paraffins," or more properly **alkanes,** was formed by starting with $CH_4$ and adding $CH_2$ units; for example, $CH_4$, $C_2H_6$, $C_3H_8$, . . . , or $C_nH_{2n+2}$. The next entry in this hydrocarbon series will cause problems. There are two individual structures that can be written for the $C_4$ compound. They are shown on page 58. Both compounds occur, and they have different physical and chemical properties. Different compounds with the same molecular formula are called **isomers.** When isomers differ in their order of attachment of atoms, they are called **structural isomers.** Both butanes have three C—C single bonds and ten C—H single

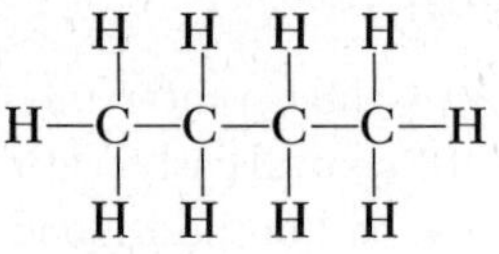

H H H H
| | | |
H—C—C—C—C—H
| | | |
H H H H

Normal butane

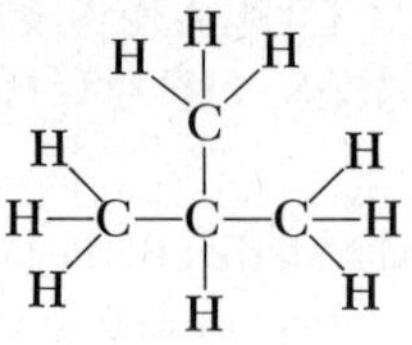

H H
H C H
| | |
H C H
| | |
H—C—C—C—H
| | |
H H H

Isobutane

bonds, yet the structures are obviously not identical. The difference is in the sequence of attachment of the atoms—which atoms are attached to which. The central carbon atom of isobutane is bonded to three carbon atoms while no carbon atom in normal butane is bonded to more than two carbon atoms.

The problem will be magnified as we go to the $C_5$ compound, for now three structural isomers are possible. Try to write them. After you write them, how do you name them? The naming of organic compounds (nomenclature) is considered in Appendix B. This knowledge is useful but not essential to the material in this text. It should be read at least once after reading this section for the first time and before starting the section on drugs.

There is free rotation about the axis of a C—C single bond. Thus, the butane molecule can exist in space with a large number of shapes. Certain spatial arrangements are more favorable than others. For a graphic representation, specific orientations are shown herewith. In Figure 3–1, structures I, II, and III concentrate on rotations about the central C—C bond. The various shapes a molecule may have by rotation about its single bonds, without breaking any bonds, are known as **conformations.** All conformations of a given molecule are interconvertible. The most favorable conformation represents the most favorable shape (lowest energy) or the way the molecule actually looks most of the time. Structure III is the most favorable conformation of butane.

Therefore, the two structures written below (A and A′) represent the same compound; they are not isomers. (See footnote page 60.)

$$CH_3—CH_2—CH_2—CH_3$$

A

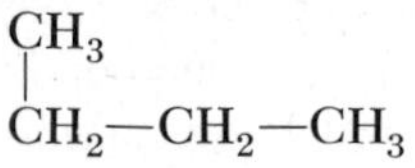

$$CH_3$$
$$|$$
$$CH_2—CH_2—CH_3$$

A′

Structural theory is extremely important because it allows the scientist to describe the physical architecture of complex molecules. The shape of molecules is often a key in understanding some of their properties, reactivity, and chemistry. One theory suggests that the individual shapes of different molecules can account for the senses of smell and taste, the chemistry of which is still not well understood. The importance of conformations will be dramatized later in this chapter.

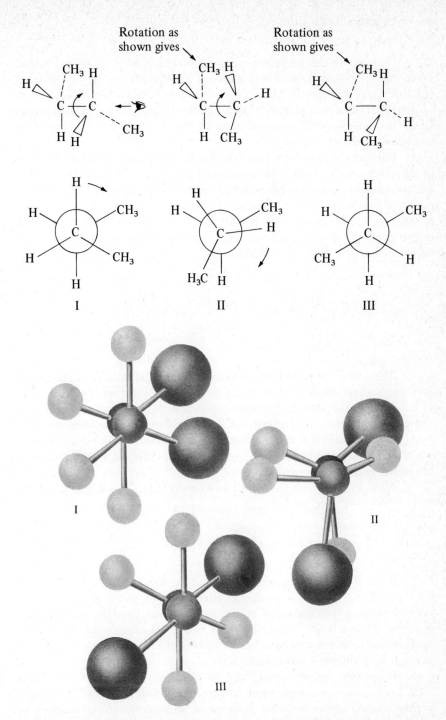

**Figure 3-1** Conformations around central C—C bond. Solid line means bond in plane of paper, wedge means bond in front, dotted line means bond behind paper. Drawings are projected as if viewer is sighting down the axis of the central C—C bond. For clarification of this concept, refer to the punch-out models provided.

Compounds that contain a carbon atom with a double bond to another carbon atom are unsaturated hydrocarbons, or **alkenes;** ethene, or ethylene, $H_2C{=}CH_2$, is an example.° Doubly (or triply) bonded carbon atoms are no longer tetrahedral, but planar. Thus all of the bonds to those carbon atoms lie in the same plane. The carbon atom bonded to four atoms (saturated) is still tetrahedral (see Figure 3–2).

Rotations around carbon-carbon double bonds are restricted since one of the bonds must be broken for such rotation to occur. Thus, *cis-* and *trans-*1,2-dichloroethene (shown herewith) are different compounds that are not readily interconvertible (it is necessary to put in enough energy to break one of the bonds). The groups attached to the doubly bonded carbons may be in two relatively different positions. Groups on the same side are called *cis*, those on alternate sides are called *trans*.

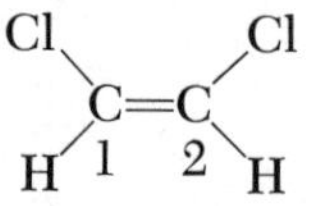

cis-1,2-Dichloroethene
(boiling pt. = 60 °C)

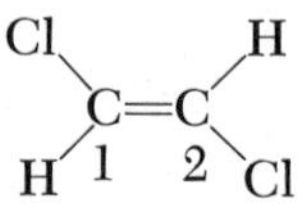

trans-1,2-Dichloroethene
(boiling pt. = 48 °C)

These two compounds have the same molecular formula ($C_2H_2Cl_2$) so they are isomers. Each compound not only has the same number and types of bonds (as in the case of the butane structural isomers) but also the same sequence of attachment of atoms. Yet the *cis* and *trans* compounds are obviously not identical. They have a different spatial geometry. Isomers that have the same sequence of attachment of atoms but a different spatial geometry are called **geometric** isomers.

A third isomer with the molecular formula $C_2H_2Cl_2$ is shown herewith.

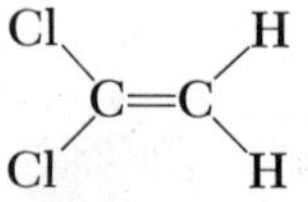

1,1-Dichloroethene

It is not a geometric isomer of the *cis* or *trans* compound because it does not have the same sequence of attachment of atoms (there are two chlorine atoms on one carbon atom). It does, however, have the same number and types of bonds (one C—C double bond, two C—H single bonds, two C—Cl single bonds) so it is a structural isomer of both the *cis* and *trans* compounds and will have different properties.

When two carbon atoms share six electrons, there is a C—C triple bond. Compounds containing triple bonds are called **alkynes.** All of the bonds to these carbon atoms lie in the same plane and are linear (see diagram of propyne in Figure 3–2).

---

° The designation $CH_2$ is a shorthand reference to a carbon atom with single bonds to two hydrogen atoms; $CH_3$ refers to a carbon atom with single bonds to three hydrogen atoms; therefore, normal butane can be written as $CH_3{-}CH_2{-}CH_2{-}CH_3$.

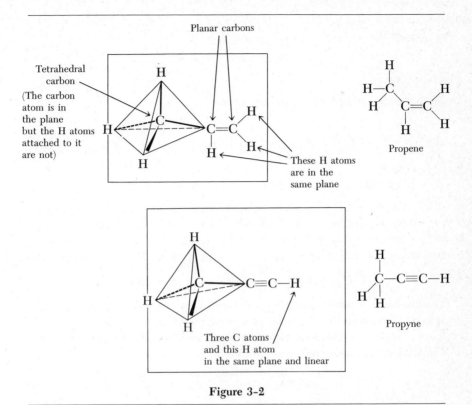

Figure 3-2

## Chemistry of Saturated and Unsaturated Hydrocarbons

Unsaturated hydrocarbons have an affinity for other atoms because not all of the carbon atoms are bonded to four other atoms. Thus, they are more reactive than the saturated hydrocarbons undergoing **addition reactions** with hydrogen gas, bromine, and other reagents; for example,

$$H_2C=C\overset{H}{\underset{CH_2-CH_3}{}} + Br_2 \longrightarrow H_2C-\overset{H}{\underset{Br}{\overset{|}{C}}}-CH_2-CH_3$$

1-Butene

$$H_2C=CH_2 + H_2O \longrightarrow H_2C-CH_2$$
$$\qquad\qquad\qquad\qquad\quad H\;\;OH$$

Ethylene          Ethanol

$$H_3C-\overset{|}{\underset{H}{C}}=CH_2 + H_2 \longrightarrow H_3C-CH_2-CH_3$$

Propene                    Propane

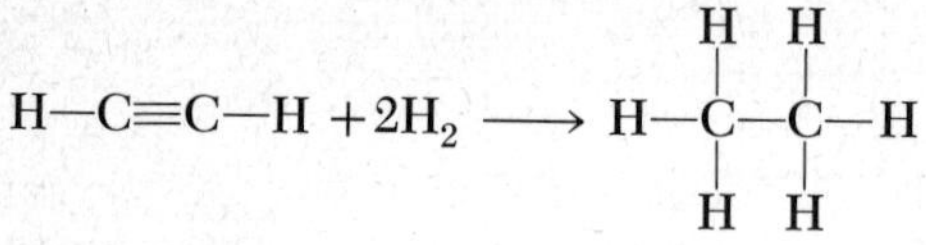

$$H-C \equiv C-H + 2H_2 \longrightarrow H-\underset{\underset{\displaystyle H}{|}}{\overset{\overset{\displaystyle H}{|}}{C}}-\underset{\underset{\displaystyle H}{|}}{\overset{\overset{\displaystyle H}{|}}{C}}-H$$

Acetylene

and being cleaved by reaction with ozone, $O_3$:

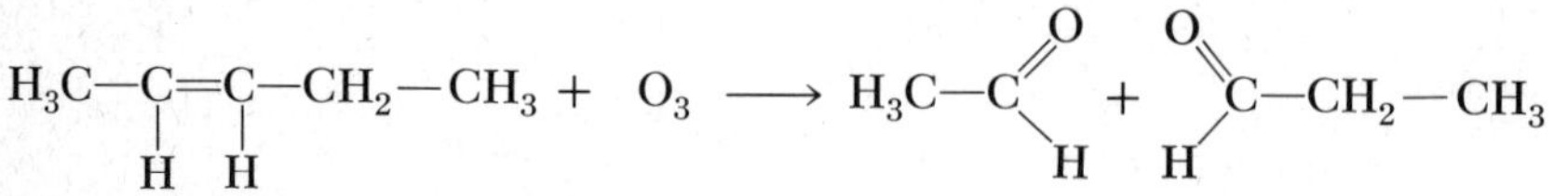

$$H_3C-\underset{\underset{\displaystyle H}{|}}{C}=\underset{\underset{\displaystyle H}{|}}{C}-CH_2-CH_3 + O_3 \longrightarrow H_3C-C\overset{\overset{\displaystyle O}{\parallel}}{\underset{\displaystyle H}{\diagup}} + \overset{\overset{\displaystyle O}{\parallel}}{\underset{\displaystyle H}{\diagdown}}C-CH_2-CH_3$$

2-Pentene                Ozone        Acetaldehyde

All saturated hydrocarbons react with oxygen at elevated temperatures to give carbon dioxide, water, and energy. You caused this **oxidation reaction** to be performed today if you drove to the university. Gasoline normally contains hydrocarbons from $C_6$ to $C_{10}$, a mixture of over 100 compounds!

$$C_7H_{16} + 11O_2 \longrightarrow 7CO_2 + 8H_2O + energy$$

The combustion reaction of acetylene with oxygen gives such an enormous amount of energy that it is used as a welder's torch.

When two ends of a molecule are joined, a cyclic compound is produced. It can be saturated or unsaturated.

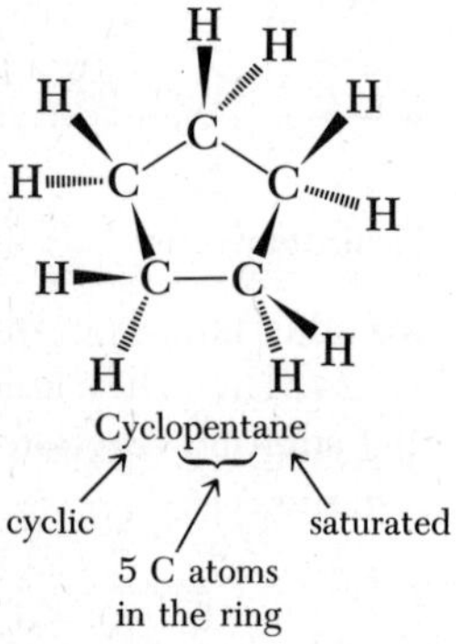

Cyclopentane

cyclic     saturated

5 C atoms
in the ring

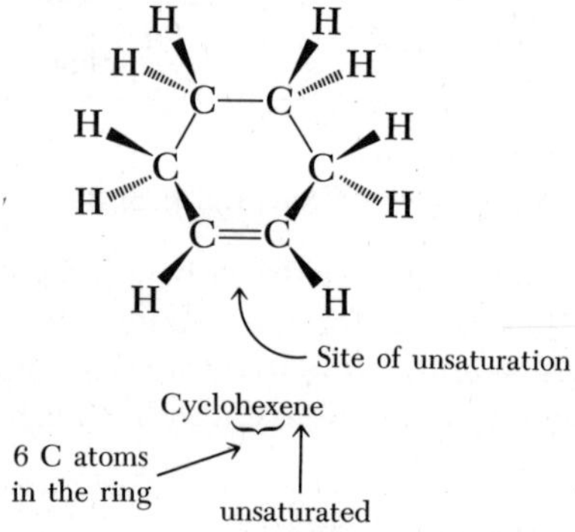

Site of unsaturation

Cyclohexene

6 C atoms
in the ring

unsaturated

This leads us to an interesting and industrially important unsaturated cyclic compound called benzene.

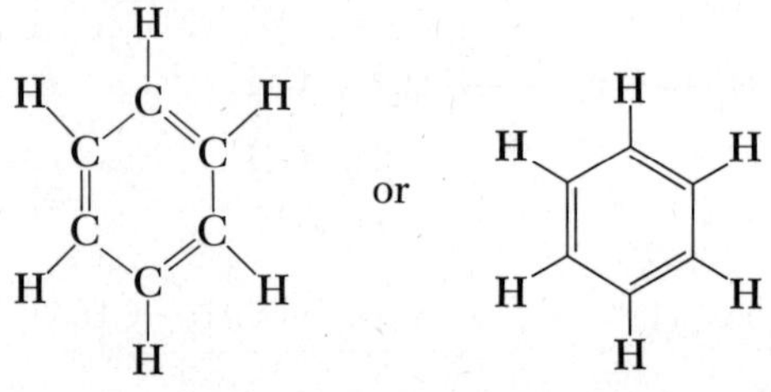

or

Benzene

(For simplicity, benzene rings are often written without showing individual

C atoms so benzene can be written as shown.) Compounds that contain this structure have unusual stability. We should expect them to undergo addition reactions as readily as other unsaturated hydrocarbons, but they do not. Compounds that contain a benzene ring are called **aromatic.** Their odors are no more repugnant that other unsaturated hydrocarbons; however, many strong-smelling compounds do contain a benzene ring. Naphthalene (mothballs) contains two benzene rings fused together.

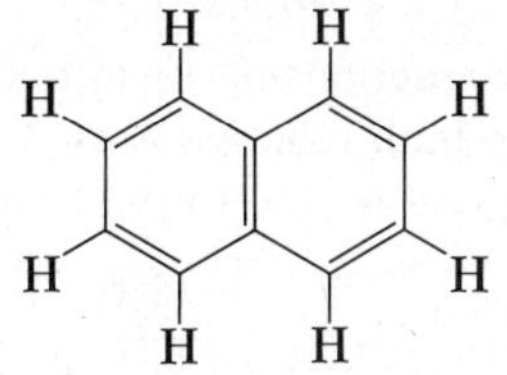

Naphthalene

Compounds are often named as substituted benzenes, for example, methylbenzene (acceptable alternative names are shown in parentheses).

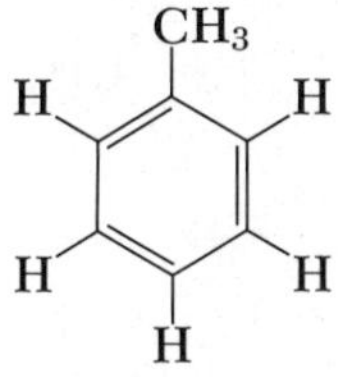

Methylbenzene
(toluene)

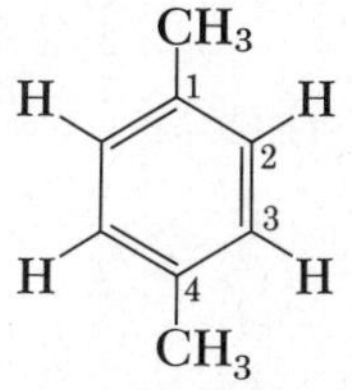

1,4-Dimethylbenzene

### Sites of Chemical Reactivity within the Molecule

The reactivity of compounds is due to charge separations within the molecule, for example, where one position is positive and another position negative. At these charged sites, reactions tend to occur. Our discussion so far has dealt with molecules that have little or no charge separation, in particular, with carbon atoms, which are covalently bonded to carbon and hydrogen. When carbon is bonded to atoms other than C and H, the electrons in the bond may not be shared equally to such an extent. The distribution will depend on the relative electronegativities of the atoms involved. Bonds within the molecule that are characterized by a large separation of charge, that is, those that are more polar, are most reactive. This is extremely important when one considers that over two million organic compounds are known and characterized and tens of thousands of new compounds are prepared annually. The fact that we can expect reactions to occur at the **reactive site** allows us to classify these compounds into a relatively small number of **classes** and thereby deduce their chemistry and properties from our experiences with other members of that class. The reactive portion of a molecule is known as the **functional group.** Any organic compound that has a particular functional group is included in the class the functional group represents; for example, any compound that contains a carbon atom bonded to an oxygen atom which is itself bonded to a hydrogen is considered to be an alcohol. Thus, the universally popular ethanol, where *-ol* is the suffix for alcohols, and also the now unpopular cholesterol are alcohols. Despite large differences in

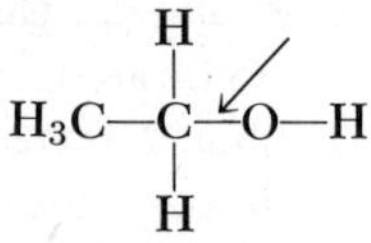

$$H_3C-\underset{\underset{H}{|}}{\overset{\overset{H}{|}}{C}}-O-H$$

Ethanol

molecular formulas, similar reactions occur in both compounds at the C—O and O—H bonds, which are both reactive sites. We might have deduced this by a consideration of the greater electronegativity of oxygen, as compared

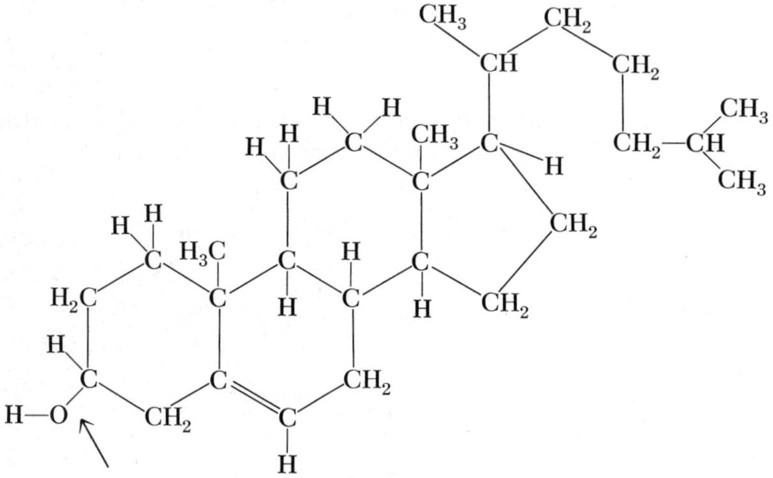

Cholesterol

with carbon and hydrogen, which causes charge separation at these sites (see Table 2–8). Chemical reactions involve the breaking and making of bonds. It is possible for the C—O and O—H bonds to be cleaved. Reactions at these sites occur preferentially to those at C—C and C—H bonds.

Another compound with the same molecular formula as ethanol ($C_2H_6O$) is dimethyl ether. Although the same number of atoms of carbon, hydrogen, and oxygen are present in both molecules, dimethyl ether has two C—O bonds but no O—H bond. This results in a chemically less reactive compound, whose boiling point is more that 100° lower (the boiling point of ethanol is 78 °C; of dimethyl ether, −24 °C). The difference in chemical reactivity of the two isomers is due to the difference in the chemical reactivities of O—H bonds as compared with C—O bonds.

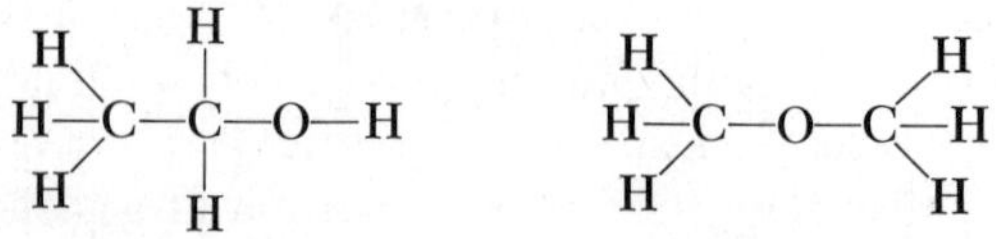

Ethanol
(boiling point 78 °C)
[intoxicant]

Dimethyl ether
(boiling point −24 °C)
[analgesic]

**Table 3-1**
**Table of Functional Groups and Classes of Compounds**

| Functional Group | Class | Example |
|---|---|---|
| 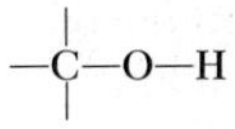 $-\overset{\mid}{\underset{\mid}{C}}-O-H$ | Alcohol | $H_3COH$ Methanol (wood alcohol)[a] |
| 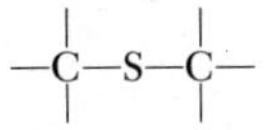 $-\overset{\mid}{\underset{\mid}{C}}-S-\overset{\mid}{\underset{\mid}{C}}-$ | Sulfide | $H_3C-S-CH_3$ Dimethylsulfide |
| 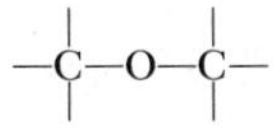 $-\overset{\mid}{\underset{\mid}{C}}-O-\overset{\mid}{\underset{\mid}{C}}-$ | Ether | 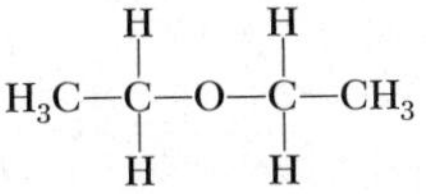 $H_3C-\overset{H}{\underset{H}{C}}-O-\overset{H}{\underset{H}{C}}-CH_3$ Diethyl ether [anesthetic][b] |
| 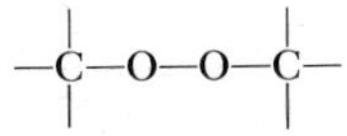 $-\overset{\mid}{\underset{\mid}{C}}-O-O-\overset{\mid}{\underset{\mid}{C}}-$ | Peroxide | 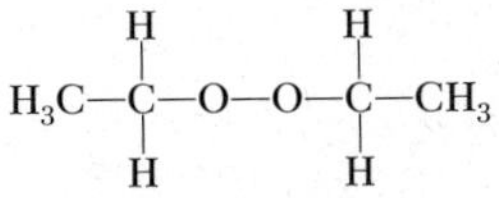 $H_3C-\overset{H}{\underset{H}{C}}-O-O-\overset{H}{\underset{H}{C}}-CH_3$ Diethyl ether peroxide |
| 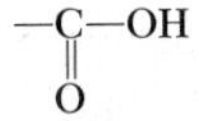 $-\overset{\mid}{\underset{\parallel O}{C}}-OH$ | Carboxylic acid | 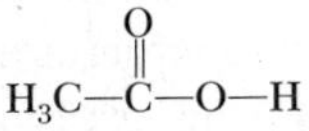 $H_3C-\overset{O}{\overset{\parallel}{C}}-O-H$ Ethanoic acid (acetic acid) [vinegar] |
| 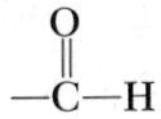 $-\overset{O}{\overset{\parallel}{C}}-H$ | Aldehyde | 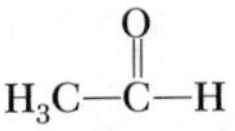 $H_3C-\overset{O}{\overset{\parallel}{C}}-H$ Ethanal (acetaldehyde) |
| 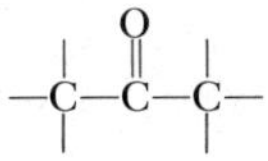 $-\overset{\mid}{\underset{\mid}{C}}-\overset{O}{\overset{\parallel}{C}}-\overset{\mid}{\underset{\mid}{C}}-$ | Ketone | 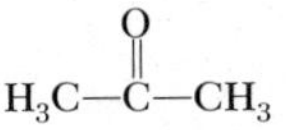 $H_3C-\overset{O}{\overset{\parallel}{C}}-CH_3$ Propanone (acetone) [solvent] |
| 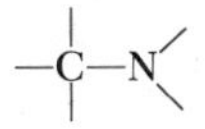 $-\overset{\mid}{\underset{\mid}{C}}-N\diagup$ | Amine | 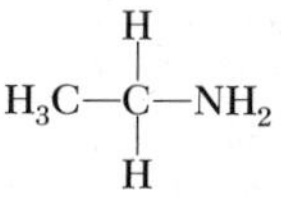 $H_3C-\overset{H}{\underset{H}{C}}-NH_2$ Ethylamine |
| 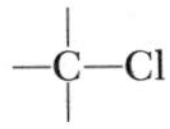 $-\overset{\mid}{\underset{\mid}{C}}-Cl$ | Organochlorides | $HCCl_3$ Trichloromethane (chloroform) |
| 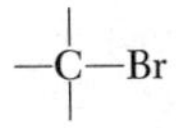 $-\overset{\mid}{\underset{\mid}{C}}-Br$ | Organobromides | $Br-CH_3$ Bromomethane |
| 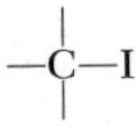 $-\overset{\mid}{\underset{\mid}{C}}-I$ | Organoiodides | 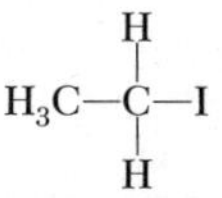 $H_3C-\overset{H}{\underset{H}{C}}-I$ Iodoethane |
| 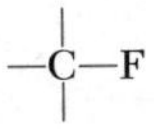 $-\overset{\mid}{\underset{\mid}{C}}-F$ | Organofluorides | $CF_4$ Tetrafluoromethane (carbon tetrafluoride) [a Freon] |

Organohalides (bracket spanning Organochlorides, Organobromides, Organoiodides, Organofluorides)

[a] Names in parentheses are alternatives.
[b] Bracketed information indicates use.

The difference in boiling point of these two molecules is due to a form of bonding that is possible between ethanol molecules but not ether molecules; it is called **hydrogen bonding.** When a hydrogen atom is covalently bonded to a highly electronegative atom (for instance, O, N, F), it is attracted to a second electronegative atom. This polar attraction, resulting from an unequal sharing of electrons in the oxygen-hydrogen bond, is hydrogen bonding.

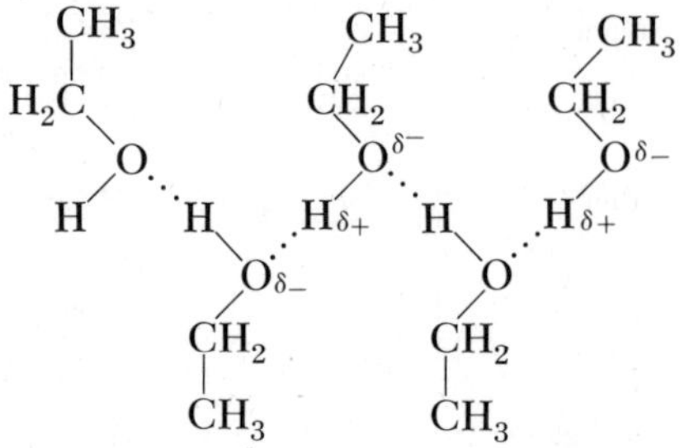

Intermolecular hydrogen bonds

There are seven isomers that have the formula $C_4H_{10}O$. Four are alcohols that are structural isomers of one another, and three are ethers that are structural isomers of one another. The four alcohols will be more similar to one another in chemical reactivity and boiling point than to the three ethers (which are themselves more similar to one another than to the alcohols). The

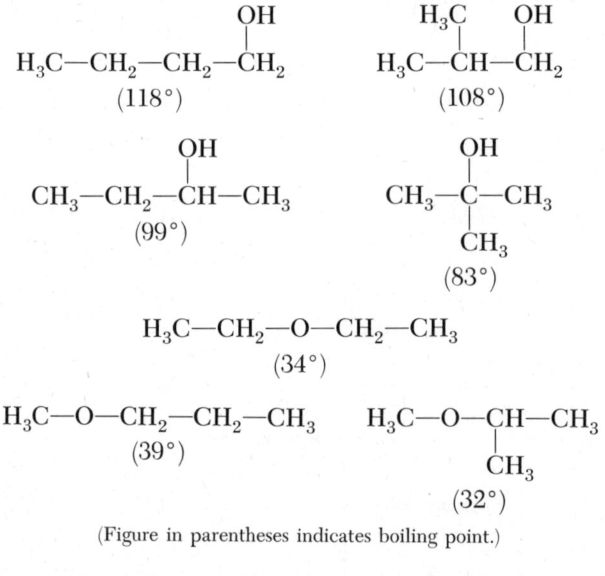

(Figure in parentheses indicates boiling point.)

differences in boiling point among the four isomeric alcohols (as much as 35 °C) allude to the dependence of molecular architecture on physical properties. The differences in boiling point between the alcohols and ethers are a direct consequence of the hydrogen bonding present in the former.

Knowledge of the chemistry of some organic compounds by class will give us insight into the reactivity of hundreds of thousands of compounds. Since the remaining portion of the molecule (the hydrocarbon portion, or nonpolar end) can be largely ignored, we shall designate it merely by a symbol

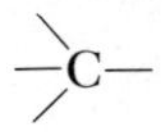

$$-\overset{\displaystyle |}{\underset{\displaystyle |}{C}}-$$

It can be seen in the structural formulae of the following compounds, among others: an aldehyde,

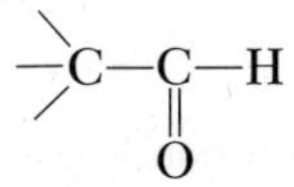

$$-\overset{\displaystyle |}{\underset{\displaystyle |}{C}}-\overset{\displaystyle \phantom{|}}{\underset{\displaystyle \parallel}{C}}-H$$
$$\phantom{--}O$$

an ether,

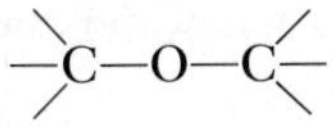

$$-\overset{\displaystyle |}{\underset{\displaystyle |}{C}}-O-\overset{\displaystyle |}{\underset{\displaystyle |}{C}}-$$

an alcohol,

$$-\overset{\displaystyle |}{\underset{\displaystyle |}{C}}-O-H$$

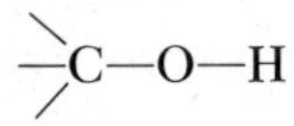

See Table 3–1 and read it through.

## Chemistry of Compounds by Class

Carboxylic acids react:

·1. With alkali bases (for example, sodium hydroxide) to give water and the corresponding salt of the acid,

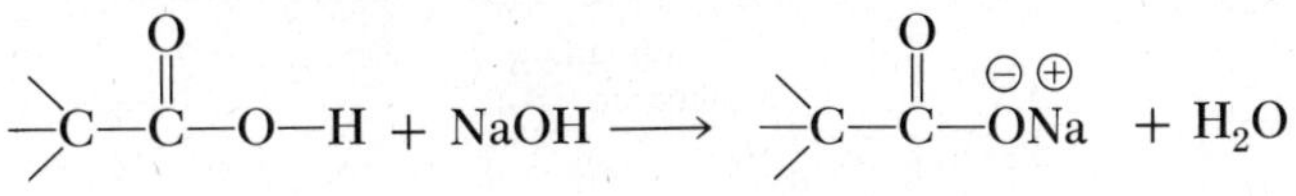

$$-\overset{|}{\underset{|}{C}}-\overset{O}{\overset{\parallel}{C}}-O-H + NaOH \longrightarrow -\overset{|}{\underset{|}{C}}-\overset{O}{\overset{\parallel}{C}}-\overset{\ominus\ \oplus}{O}Na\ + H_2O$$

A sodium carboxylate

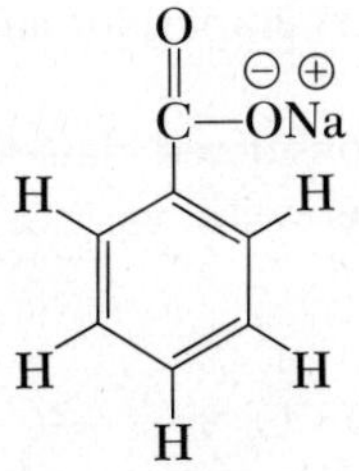

Sodium benzoate
[a preservative]

2. With alcohols to give water and compounds known as **esters.**

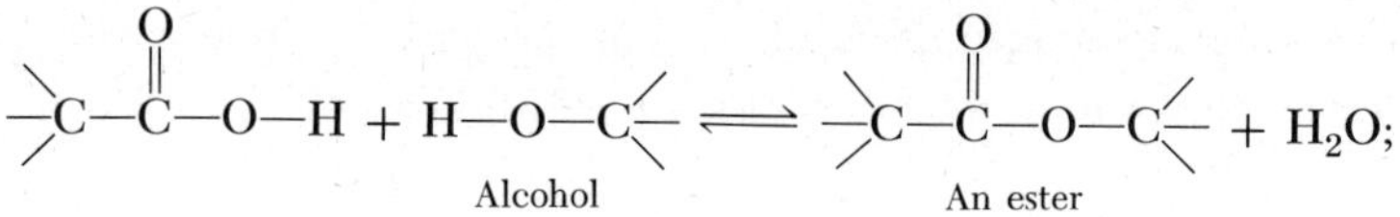

$$\overset{\displaystyle O}{\underset{\displaystyle \parallel}{}}C-C-O-H + H-O-C \rightleftharpoons C-C-O-C + H_2O;$$

Alcohol                    An ester

$$H_3C-\overset{\displaystyle O}{\underset{\displaystyle \parallel}{C}}-O-CH_2-CH_2-\overset{\displaystyle CH_3}{\underset{\displaystyle CH_3}{C-H}}$$

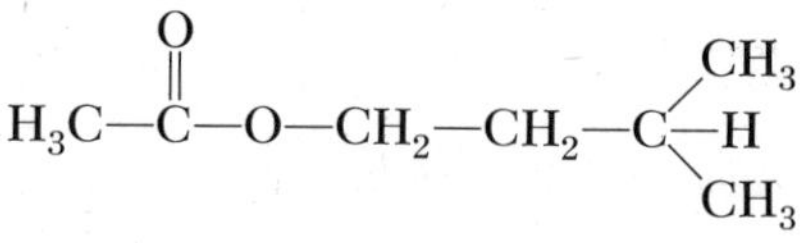

Isoamylacetate
(banana oil)
[flavoring]

Under different reaction conditions this reaction is reversible; for example, esters can be **hydrolyzed** (cleaved by reaction with water) to carboxylic acids and alcohols.

3. With amines or $NH_3$ to give water and compounds known as **amides.**

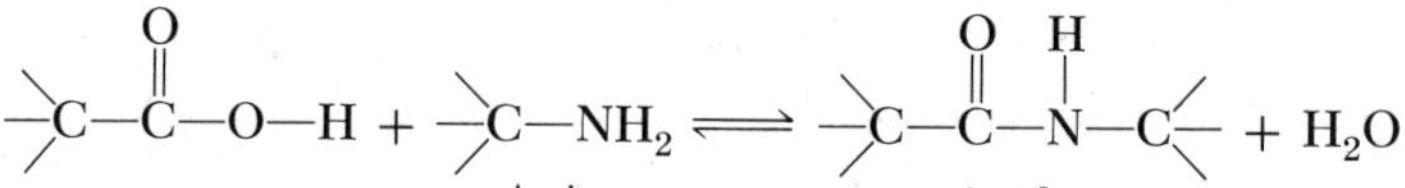

$$\overset{\displaystyle O}{\underset{\displaystyle \parallel}{}}C-C-O-H + C-NH_2 \rightleftharpoons \overset{\displaystyle O \quad H}{\underset{\displaystyle \parallel \quad |}{}}C-C-N-C + H_2O$$

Amine                    Amide

$$\overset{\displaystyle O}{\underset{\displaystyle \parallel}{}}C-C-O-H + H-NH_2 \rightleftharpoons \overset{\displaystyle O \quad H}{\underset{\displaystyle \parallel \quad |}{}}C-C-N-H + H_2O$$

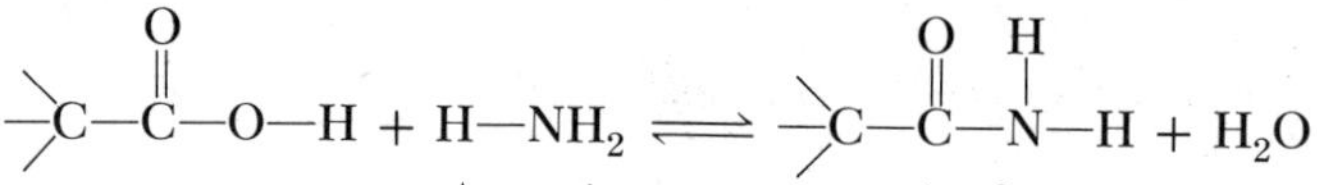

Ammonia                    Amide

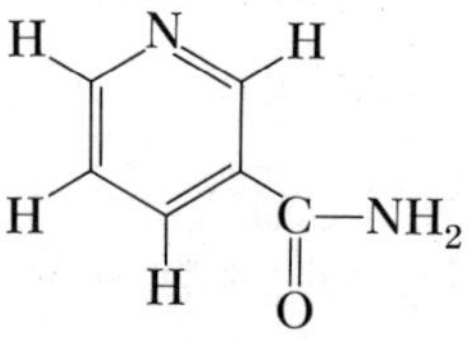

Nicotinamide
(Vitamin $B_3$)

This reaction is reversible as well. Hydrolysis of amides gives carboxylic acids and amines.
The latter two reactions are very important and will be discussed in greater detail.

Amides and esters are considered classes of compounds as well. Their functional groups are shown herewith.

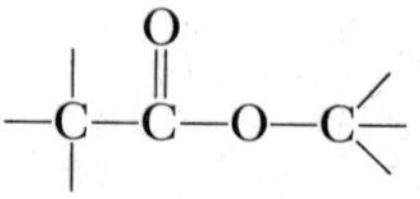

$$-\overset{\displaystyle O}{\underset{\displaystyle |}{C}}-C-O-C \qquad -\overset{\displaystyle O}{\underset{\displaystyle |}{C}}-C-N$$

Ester 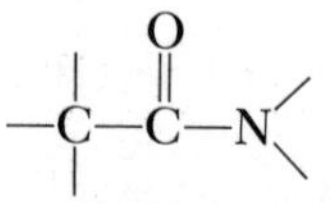                    Amide

Fats are esters formed from carboxylic acids with large hydrocarbon ends (called fatty acids) and the alcohol glycerol (a compound containing more than one alcohol functional group). The hydrocarbon portion of fatty acids can be saturated or unsaturated, leading to the familiar term *polyunsaturated fats* (or fats with many points of unsaturation). Since unsaturated sites are more reactive than saturated ones, polyunsaturated fats (viscous liquids) are broken down in the body (metabolized) thereby becoming more digestible than hydrogenated fats (in which hydrogen has been added to the carbon double and triple bonds to produce paraffinlike saturated fats). Treatment of fats with aqueous sodium hydroxide causes hydrolysis of the esters to the carboxylic acids and glycerol. The acids react with the sodium hydroxide to give salts of carboxylic acids, which can be used as soaps. Nonpolar hydrocarbons (grease) dissolve in the hydrocarbon end of the soap but the polar, salt end of the molecule is water-soluble, allowing it to be "washed" away in water.°

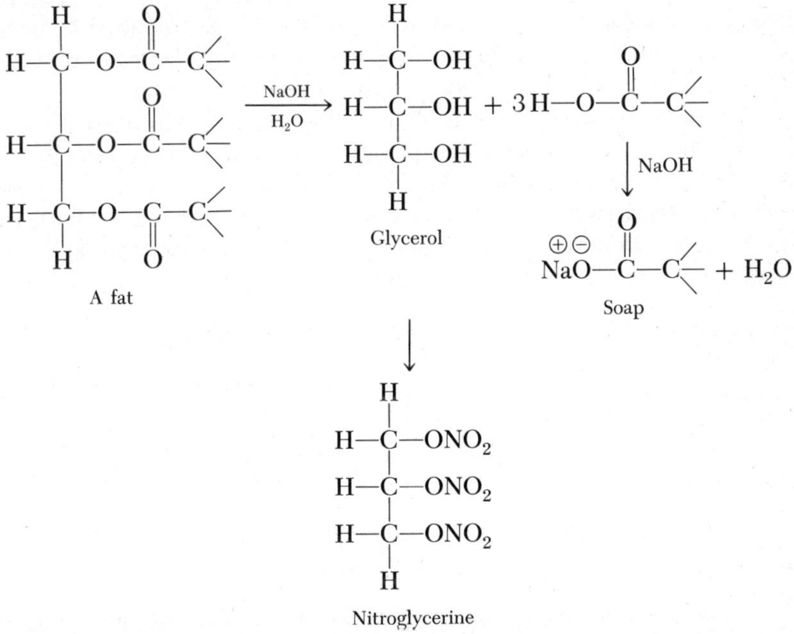

Glycerol is easily nitrated to form a compound called nitroglycerine, which is an explosive. In World War II, animal fats became extremely important as a natural source of glycerol. When supplies ran low, the Nazis resorted to human fat to produce soaps and explosives. Carboxylic acids with relatively few atoms per molecule are found in many common places. Acetic acid ($CH_3CO_2H$) is found in vinegar, and formic acid ($HCO_2H$) occurs in high

---

° This is discussed in greater detail in Chapter 5.

concentration in ants. The burning sensation and swelling from a red ant bite is the result of an injection of formic acid. Citrus fruits contain the tricarboxylic acid, citric acid, which is used in popular tangy drink mixes. It provides the tang.

$$
\begin{array}{ccc}
 & & \overset{\displaystyle O}{\overset{\displaystyle \|}{CH_2-C-OH}} \\
\overset{\displaystyle O}{\overset{\displaystyle \|}{H-C-O-H}} & H-O-\overset{|}{C}\cdots\cdots\overset{\displaystyle C}{\|}-OH \\
 & HO-\overset{\displaystyle C}{\|}-CH_2 \quad O \\
 & O
\end{array}
$$

Formic acid                    Citric acid

Organohalides have proven to be useful, dangerous, and controversial compounds. Relatively small compounds such as $CHCl_3$ (chloroform), $CCl_4$ (carbon tetrachloride), and $Cl_3C-CH_3$ (1,1,1-trichloroethane) are used as solvents and dry (nonaqueous) cleaning agents, owing to their ability to dissolve organic compounds (grease, oil, and so on) and their high volatility. Unfortunately, halogenated compounds affect the liver and in high concentration can be toxic, so that careful control must be exercised over their use.

The infamous insecticide DDT is a chlorinated hydrocarbon, one of the many undesirable synthetic compounds that all of us now have in our bodies. At the same time, some halides are essential to our normal life processes. Iodine is necessary for the production of thyroxin, which is the hormone that controls the important thyroid gland. Iodized salt is recommended when seawater fish (a natural source of iodine) is not included in the diet.

DDT

Many aromatic compounds, especially those containing nitrogen, are believed to be **carcinogenic** (inducing cancer formation). Such compounds (for example, dibenzacridine and pyrene) can be formed by high-temperature reactions of organic molecules: the type of reaction that occurs in combustion engines, furnaces, and the glowing tip of a cigarette when there is a limited supply of oxygen.

Aldehydes are reactive compounds that usually have strong odors. Historically, many have been used as perfumes. Benzaldehyde is an odoriferous constituent of almonds; and cinnamaldehyde, of cinnamon; and who can forget the odor (or commercial utility) of formaldehyde. The aldehyde group can

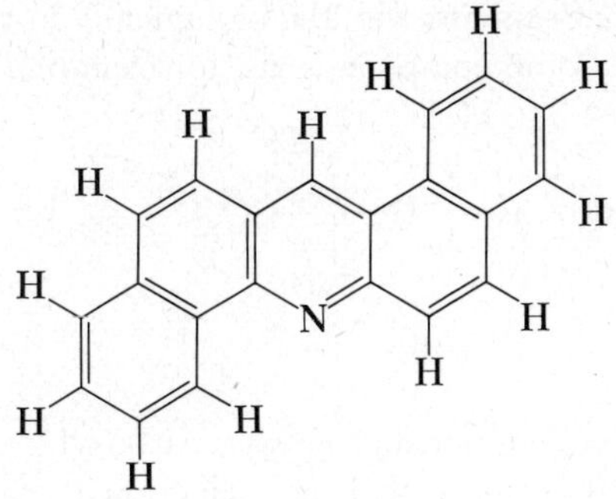

Dibenzacridine

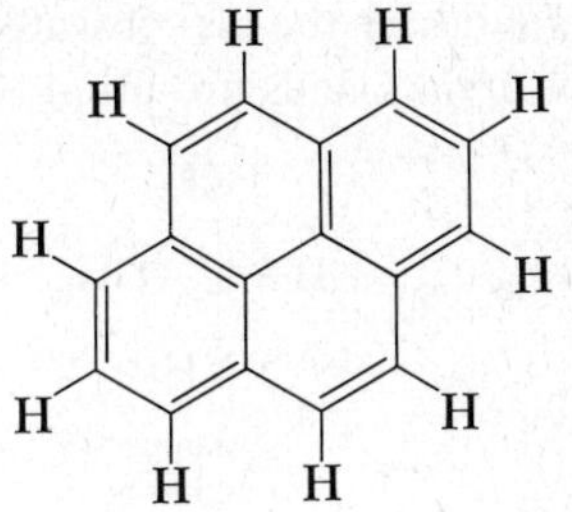

Pyrene

be easily oxidized to a carboxylic acid in the presence of oxygen. In the past few years, aldehydes such as acrolein have become notorious as air pollutants.

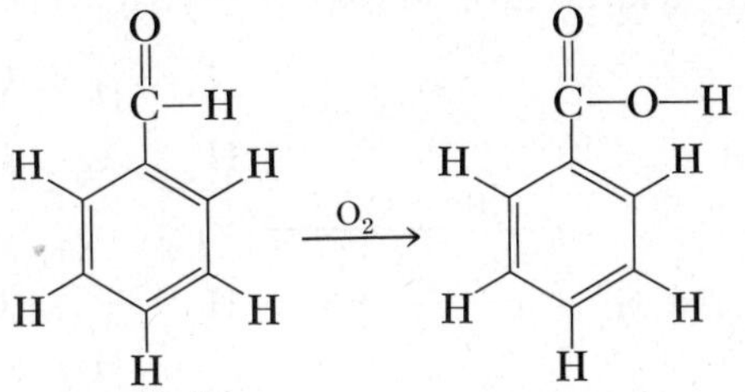

Benzaldehyde

Benzoic Acid

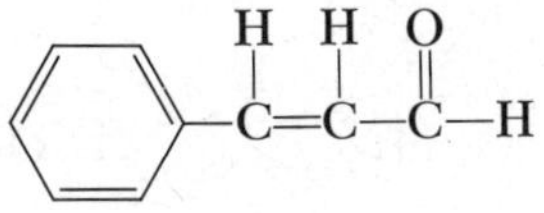

Cinnamaldehyde

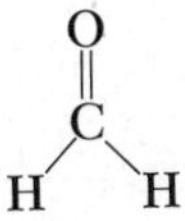

Formaldehyde

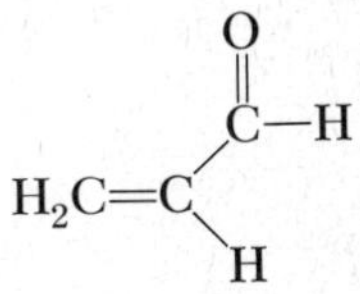

Acrolein

Amines have an unpopular fishlike odor. They can react with carboxylic acids to form water-soluble ammonium salts which can then be converted

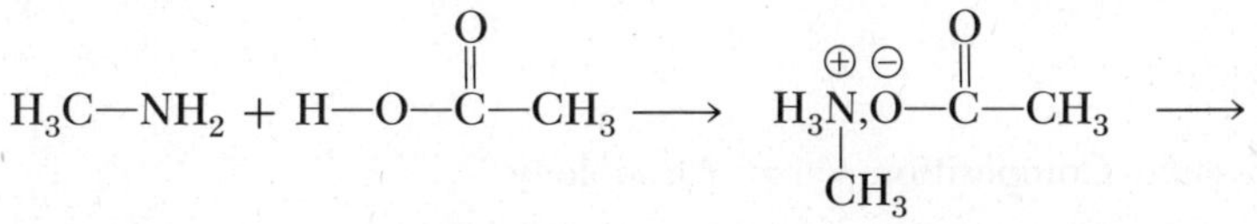

Methylamine          Acetic acid          Methylammonium acetate
(a salt)

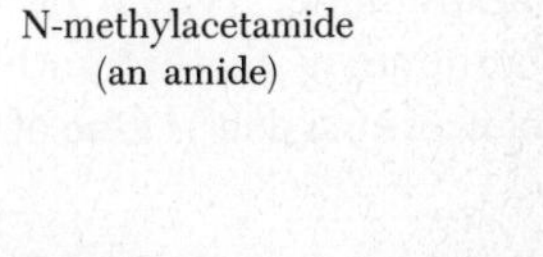

N-methylacetamide
(an amide)

to amides by the loss of water. This fact explains why the biologically important amino acids are often written as ionic compounds, as, for example,

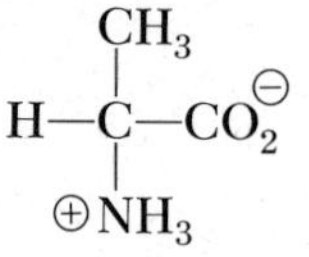

$$
\underset{\substack{| \\ \oplus NH_3}}{\overset{\substack{CH_3 \\ |}}{H-C-CO_2^{\ominus}}} \qquad \text{instead of as} \qquad \underset{\substack{| \\ NH_2}}{\overset{\substack{CH_3 \\ |}}{H-C-CO_2H}}
$$

Alanine
(an amino acid)

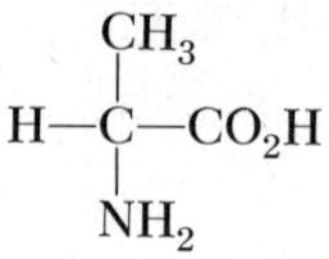

Note that the amino acid has two functional groups—a carboxylic acid
and an amine. Many compounds contain more than one functional group.
Thus, there can be more than one reactive site on a molecule. Each site may
exhibit its own chemistry. For example, salicylaldehyde is a compound with
both alcohol and aldehyde functional groups. The aldehyde can be oxidized
to form a carboxylic acid (salicylic acid); the alcohol can react with a carboxylic acid (acetic acid) to give an ester (acetylsalicylaldehyde).

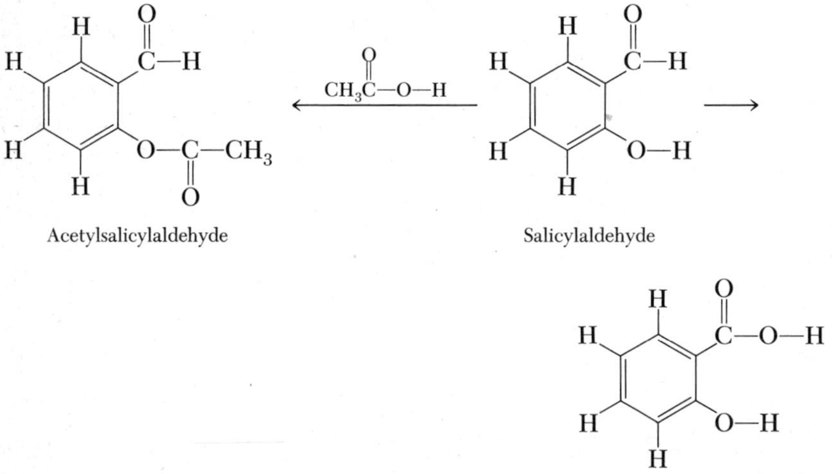

Acetylsalicylaldehyde                    Salicylaldehyde

$CH_3\overset{\overset{O}{\|}}{C}-O-H$

Salicylic acid

## DRUGS

### Molecular Composition Versus Physiology

Looking at the examples in Figure 3–3, we are tempted to try to relate
chemical structures and functional groups with physiological activity. Both
epinephrine and amphetamine have similar structures and both are stimulants.
Benzocaine and methyl anthranilate are aromatic compounds with amine and
ester functional groups. One is an anesthetic and the other a perfume. Is it
possible to predict the physiological effect that would be produced by ethyl
anthranilate? Will it be a pleasant-smelling compound suitable for use as a
perfume (as methyl anthranilate is) or will it have anesthetic properties (as
benzocaine does)? One of the most important areas of scientific research today,

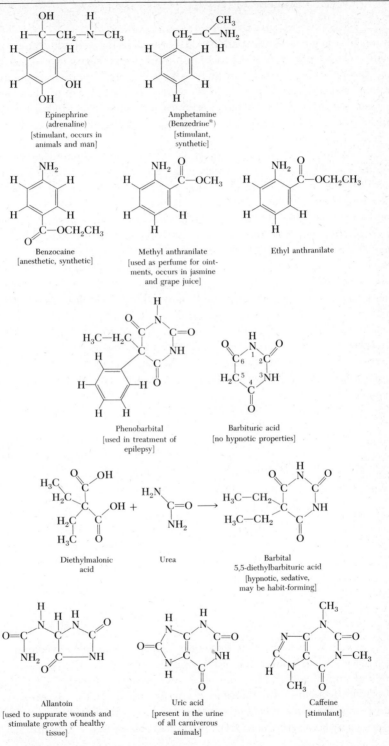

**Figure 3-3** Examples of drugs: ® = a registered trademark.

although it is incompletely understood as yet, is concerned with the development and understanding of the relationship between the structure of molecules and the physiological response they initiate. In other words, we are trying to find what physical features of a compound are necessary to produce desired responses from drugs, medicines, perfumes, flavorings, dyes, and so forth. The present state of the art necessitates the tedious preparation and testing of all related compounds.

Barbital, one of the barbiturate sedatives, is prepared as shown in Figure 3–3. The reaction is similar to the formation of an amide. Barbituric acid itself has no important physiological properties but the compound with two ethyl groups at the 5 position is extremely active. It would seem reasonable to prepare other similar compounds with different groups at position 5 to determine whether they are also active and whether they have more desirable properties, such as fewer undesirable side effects or an enhanced effect and consequent lower dosage requirement. By this reasoning, the sedative phenobarbital, which is a barbituric acid with one ethyl group and one phenyl group at position 5, and many other 5,5-disubstituted barbituric acids have been synthesized and tested. Note the similarity of uric acid (which can be easily converted to the suppurative drug allantoin) to the barbiturate sedatives and the stimulant caffeine. For simplicity and clarity, the cyclic structures will be written without showing the individual C atoms or the H atoms attached to them, except for the $CH_3$ groups. Thus, morphine is written as shown herewith.

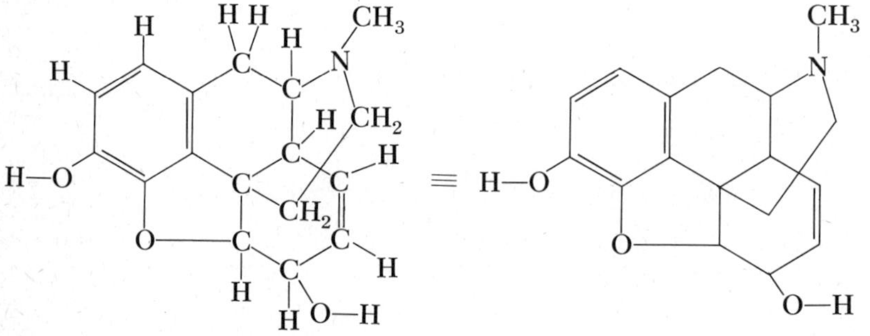

Morphine

Another example that shows the delicate balance between structure and physiology is opium, the dried resinous juice of the unripe seeds of the oriental poppy, which has been used since antiquity for the relief of pain. Morphine, codeine, and thebaine occur in opium. Heroin is prepared by converting the alcohol groups in morphine to esters of acetic acid. Codeine is formed by converting one OH group in morphine to a methyl ether ($OCH_3$);

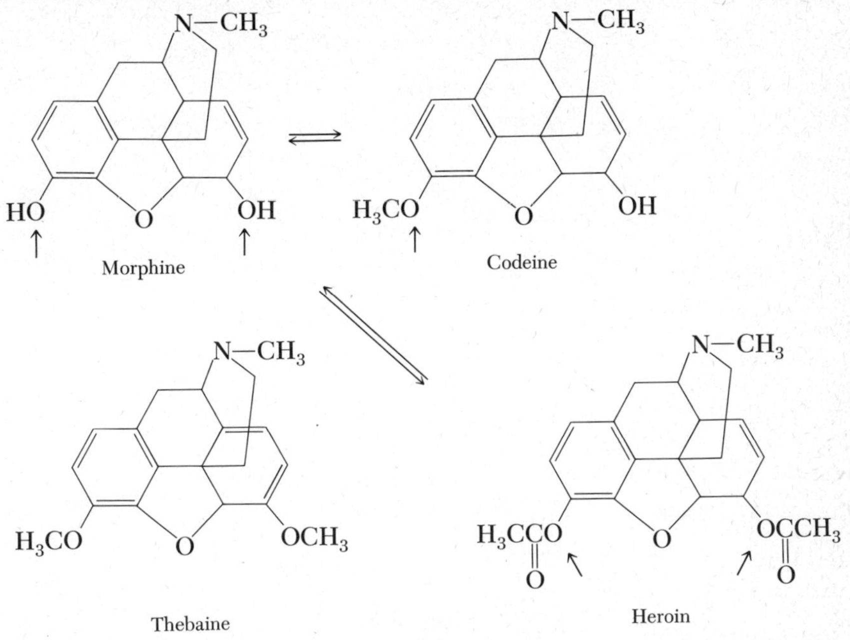

**Figure 3-4** Note the recurring structural similarity in these and many of the following structural formulas.

see Figure 3–4. Morphine, codeine, and heroin are **analgesics** (pain relievers) but they produce adverse effects in the development of physical dependence (the body cannot function properly without the drug) and increased **tolerance** (a cumulatively increasing dosage is needed to produce analgesia and **narcosis,** or sedation). Thebaine, though structurally similar to the other three compounds, is toxic and medically useless at present. For over a half-century, scientists have been attempting to prepare substitute compounds with morphinelike, pain-relieving properties and negligible side effects. Initial attempts were concentrated on minor chemical changes of the naturally occurring compounds. Current research involves totally synthetic compounds prepared in the laboratory and not known as occurring in nature. One such compound is methadone, which bears only scant structural resemblance to morphine (see Figure 3–5). It is being used experimentally in rehabilitation therapy for heroin and morphine addicts. Synthetic compounds more closely related to morphine in structure include Cyclazocine (which appears useful as a rehabilitation agent) and dextromethorphan (which is a nonnarcotic cough-relieving drug). With the development of less abusive synthetic analgesics, the problem of drug dependence of the morphine type (narcotic addiction) might be greatly alleviated.

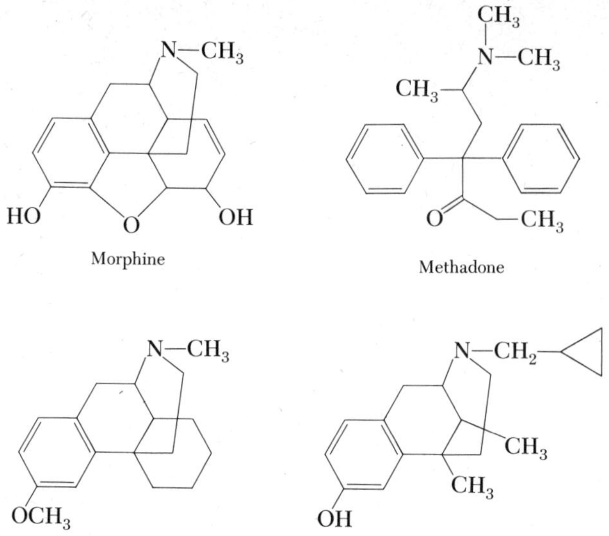

Morphine

Methadone

Dextromethorphan

Cyclazocine

**Figure 3-5**

## Everyday Drugs

*Strawberry juice and plantain water mixed with eight liters of mulberry juice, one liter of the dung of a white dog, and a little vinegar is good for ulcers of the throat if used as a gargle.*

Drug Preparation (1485 A.D.)

In the past decade, drugs have become a large part of the American scene. In many cases, they have proven to be salvation in a bottle. Used erroneously, they can be hell in tablet form. Whether we like it or not, drugs are as much a part of the American way of life as apple pie. There are probably more individual drugs in your house than apples. The consumer will purchase a bottle of tablets for ten dollars that probably contains a chemical he cannot pronounce and supposedly causes a physiological effect he does not understand. Do you remember the outdated movie showing the medicine man selling his famous elixir that cured everything for two dollars? Unbelievable? That wouldn't happen in today's modern society? Don't bet on it. One of the most highly advertised compounds on the market is acetylsalicylic acid (aspirin). A major portion of the U.S. supply is produced by the Monsanto Company. Yet under different trade names, the price of this product fluctuates wildly, and the best-selling brands are usually not the cheapest! Remember that ten-dollar bottle of tablets mentioned earlier? What if it contained aspirin worth approximately ten cents? Would you know the difference? It is not surprising that an expanding illegal business in the United States is counterfeit drugs when one considers that the U.S. legitimate drug industry has over a $7-billion-a-year global business. The understaffed Food and Drug Administration does what it can to protect the consumer but lack of public concern has prevented adequate funding of their programs. The medicine show continues while the ethical scientist continues in pursuit of drugs that heal. The slow but continuous stream of wonder drugs that are patiently developed to combat ageless crippling diseases is a testimonial to his resourcefulness and creativity.

Among the problems related to the drug industry are:

1. Duplicate drugs produced by minor modifications of the molecular structure of existing drugs, which offer no significant therapeutic advantage but result in additional marketable products.
2. Side effects. According to a National Institutes of Health survey, 1.5 million people are hospitalized each year because of adverse drug reactions.
3. Misleading, deceptive, or fraudulent advertising for over-the-counter drugs.
4. Reckless use of prescription drugs.

To correct these costly and dangerous problems, suggestions have been made to:

1. Establish a national drug testing and evaluation center for the controlled clinical testing of new drugs by the government, to obtain uniform, reliable, and unbiased testing.
2. Create a federal drug compendium that would summarize available information for all drugs on the market.
3. Establish a uniform drug coding system that would permit ready identification of the drug.
4. Establish a system for the review of advertising claims.
5. Improve the inspection and quality control of all drug production.

Tougher federal laws requiring "substantial evidence" of therapeutic efficacy for unapproved drugs to be allowed on the market have resulted in a reduction of new single-chemical entity drugs (as opposed to combination drugs). Only 73 were marketed in the period of 1966–70 and the total number of new drug products was only 416. The Food and Drug Administration reviewed drugs approved between 1938 and 1962, requiring manufacturers to show that clinical experience with a drug justifies its use. By 1969, more than 16,500 claims of therapeutic efficacy for about 4,300 formulations of some 2,800 drugs had been reviewed. About 11% of the 4,300 drug formulations were judged ineffective and about 15%, effective. Another 47% were found to be possibly

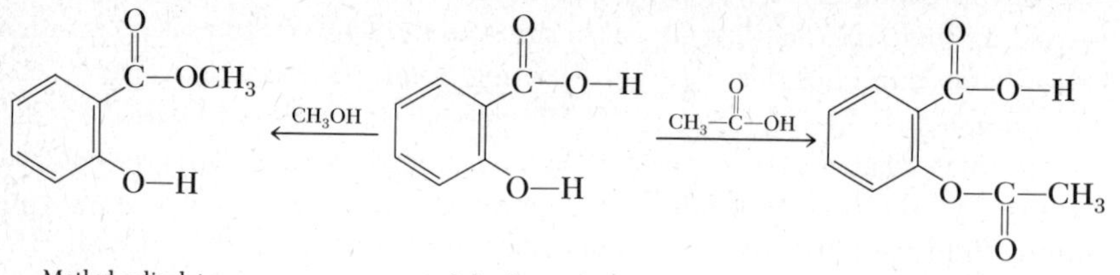

Figure 3-6 shows the structures:

Methyl salicylate (oil of wintergreen) [used as flavoring and counterirritant] — structure with $C-OCH_3$ and $O-H$ groups.

Salicylic acid — structure with $C-O-H$ and $O-H$ groups, with $CH_3OH$ reaction arrow on left and $CH_3-C-OH$ (acetic acid) reaction arrow on right.

Acetylsalicylic acid (aspirin) [analgesic and antipyretic, or fever depressant] — structure with $C-O-H$ and $O-C-CH_3$ groups.

**Figure 3-6** Two derivatives of salicylic acid. Aspirin is the acetic acid ester of salicylic acid, and oil of wintergreen in the methyl ester.

effective and 27%, probably ineffective. In 1972, FDA announced plans to start a new review of all over-the-counter drugs sold in the U.S. There are approximately 100,000 products but still only some 200 significant active ingredients. Figure 3–6 shows two common derivatives of salicylic acid.

Controlled clinical testing is expensive and may be misleading. If aspirin were a new drug to be introduced to the consumer market it would probably be banned owing to evidence of its being teratogenic (causing birth defects) in mice. However, extensive use in human systems has shown it to be relatively safe. Drug manufacturers argue that drugs that have been used extensively and shown to be beneficial by experience should not have to be reexamined, that the expense of such testing could be better used to investigate new drugs.

There will continue to be much discussion on the cost of ethical drugs in the U.S.° Drug manufacturers argue that the returns from the small number of profitable drugs that are sold must pay for the enormous costs of research and testing not only for the successful drugs but for those that could not be utilized. This is over and above the cost of manufacturing and promoting the drug.

One important aspect of a drug is to have it get to the right place as soon as possible. Thus, in some cases it is necessary to inject the compounds directly into the blood stream. In other cases, slow ingestion via typical digestive processes is adequate. The practicality of the brandy snifter is that ethanol proceeds through the nasal membrane to do its thing faster than through the stomach wall. You can be wiped out by sniffing without drinking an ounce. One aspirin manufacturer points with pride to the fact that his tablet

---

° An ethical drug is a specific preparation of a physiologically active compound or compounds usually protected by a trademark. There has been much debate about the fact that doctors' prescriptions often specify the ethical drug by name (referring to a specific formulation) rather than the generic name (the name of the compound itself), which would allow the consumer a choice of several available formulations (or allow him to practice competitive buying).

dissolves faster in the stomach and thus begins to work more rapidly. If you want to beat his time, crush the tablet.

A very powerful inexpensive solvent, dimethyl sulfoxide,

$$H_3C-\overset{\overset{\textstyle O}{\|}}{S}-CH_3,$$

has been used recently for treatment of bursitis and other painful muscular diseases, under a physician's direction. Its healing properties were discovered accidentally, when a scientist with a headache rubbed his temple with his hand, which inadvertently contained a small amount of this material. The headache disappeared. Further informal testing showed that rubbing it into sore spots relieved the ache rapidly. Newspapers picked up the story of this inexpensive wonder drug. The fear of mass usage of this untested compound led to rapid controlled examination, which showed that high concentrations produced harmful effects. In this case the public was warned in time.

The use of hallucinogens has become a controversial issue today. You may be surprised to learn that it was so controversial in the 1930s that a big study on the effects of marijuana was initiated by Mayor La Guardia of New York. The carefully documented report gave little cause to ban the drug, nor did the 1972 report of the National Committee on Marijuana and Drug Abuse. Although marijuana cannot be classified as a narcotic, it has been suggested (also refuted) that marijuana use significantly impairs automobile driving ability. The extensive use of this controversial drug and its increasing popularity necessitates further research on its physiological effects.

How do we know when any drug is safe? Should we carefully follow the effects on users for one year, five years, a generation? While we prohibit the use of a new drug, many people who might safely benefit from its therapeutic properties may suffer. This is a delicate question, which has been only partially resolved. In the United States, new drugs are tested for several years before they are released for general consumption. Other countries are not so careful, with such occasional consequent horrors as the thalidomide disaster, in which the widespread use of thalidomide as a sedative by pregnant women led to thousands of disfigured babies. A case in point is the recent approval to market a drug called L-dopa, for therapy of patients with Parkinson's disease. Considerable dangerous side effects accompany the high doses that are needed for effective use. Yet, it was felt that the need outweighed the danger. Research continues for an effective substitute requiring lower dosage.

A new set of initials has become familiar in the past decade—LSD, lysergic acid diethylamide—10,000 times more powerful a hallucinogen than mescaline, from peyote. Its unregulated use and the apparent danger therefrom led the United Nations Narcotic Commission to issue a call for immediate international controls to limit its use to medical and scientific purposes. Its physiology is still not well understood, and disastrous results to users are well

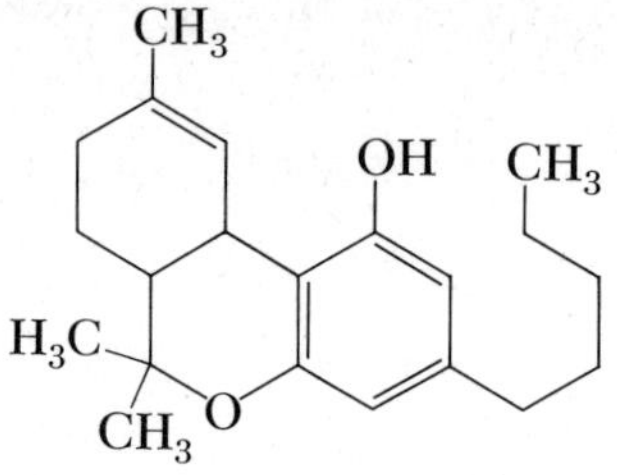

Tetrahydrocannabinol
[active constituent of marijuana (hashish)]

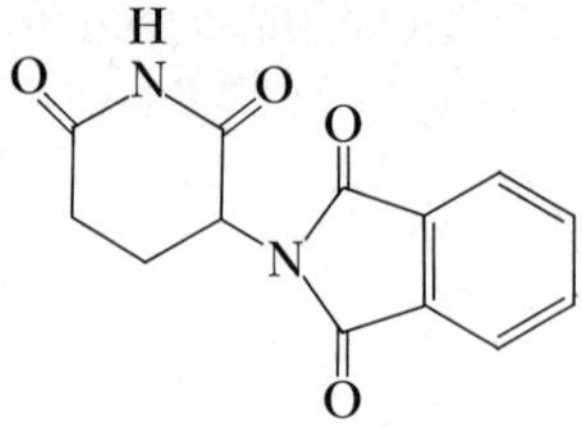

Thalidomide
[sedative associated with
fetal abnormalities]

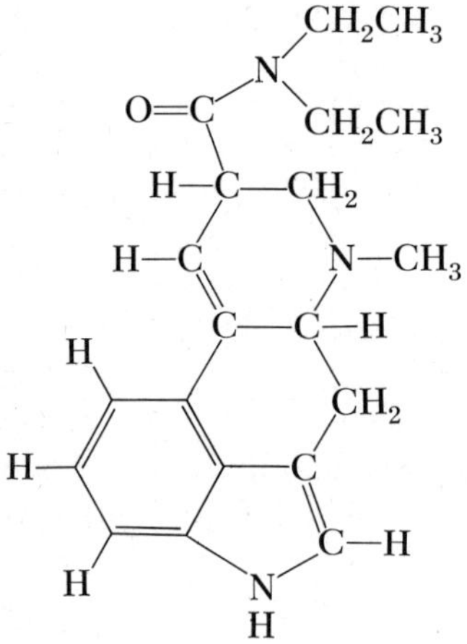

LSD

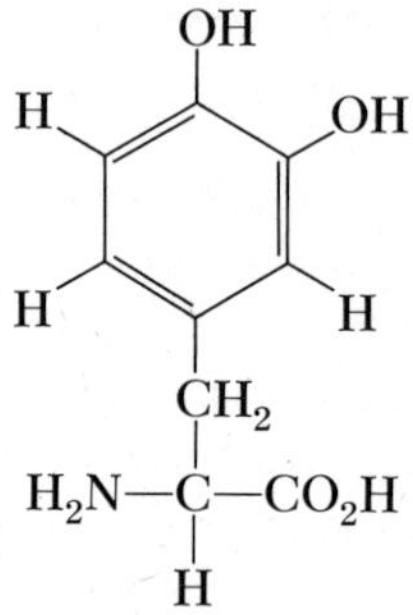

3,4-Dihydroxy-L-phenylalanine
(L-dopa)
("L" signifies one of
the two possible mirror-
image forms, as explained
in text)

documented. It has recently been suggested that LSD may affect the chromosomes of the cell, thereby causing mutant progeny; such an effect would be a horrible legacy from a parent who may have felt lucky in not experiencing a "bad trip."

Each of us will buy many drugs in our lifetime. We will not always be in a position to assess completely the gaily colored substances that will alter our body chemistry, but until a satisfactory, unbiased regulating agency is established, caution would seem to be in order—*caveat emptor.*

### A Compendium of Drugs

*TERMINOLOGY*

1. *Abuse, misuse*    Use of a substance beyond the generally accepted limits of medical therapy or limitations imposed by current laws.
2. *Soft drugs*    Drugs considered arbitrarily not to lead to physical dependency.

3. *Hard drugs*    Drugs that lead to physical dependency.
4. *Psychoactive drugs*    Those which have a profound tendency to alter mood, perception, cognition, consciousness, and behavior.
5. *Psychedelic*    Rough synonym for psychoactive, usually applied to those drugs, such as LSD, that powerfully alter mood and perception.
6. *Psychomimetic*    Mimicking a psychosis, or mental disease.
7. *Drug dependence*    State of psychic or physical dependence, or both, on a drug, arising in a person following administration of that drug on a periodic or continuing basis.
8. *Psychic (psychological) dependence*    Requirement of periodic or continual administration of a drug to produce pleasure or avoid discomfort related to a psychic drive.
9. *Physical dependence*    Adaptive state that manifests itself by intense physical disturbances when the administration of a drug is suspended or when its action is affected by a specific antagonist.
10. *Withdrawal (abstinence) syndromes*    Intense physical disturbances accompanying withdrawal of a drug on which a person is physically dependent.
11. *Tolerance*    Condition in which the user has progressively to increase the dose of a given drug to obtain the same intensity of effect.

## SURVEY OF SOME MISUSED DRUGS

*Stimulants (antidepressants):* The structures of several of the well-known stimulant drugs are represented in the accompanying figures. In general, these drugs are central-nervous-system stimulants. For example they elevate blood pressure, suppress appetite, produce a decreased sense of fatigue, and increase confidence and alertness. Repeated administration produces tolerance (except for cocaine) and drug dependence. Side effects include nervousness, insomnia, and headache. Large doses of amphetamines lead to amphetamine psychosis (in which the person loses contact with reality and is convinced that others are out to harm him); this condition persists several weeks after withdrawal.

Amphetamines have been used for many years by truck drivers, airline pilots, hospital personnel, and students, and more recently, athletes in competition, primarily because they delay fatigue and sustain effort. They are prescribed for fatigue and mild depression, and to suppress appetite, but now are so misused that many countries are following Sweden's lead and completely banning prescription or sale. Many housewives originally taking these drugs for weight loss have become amphetamine-dependent. The World Health Organization regards stimulant misuse as a worldwide epidemic.

Amphetamines show a tolerance effect; from a normal oral dose of 0.01–0.02 gram per day, this may build up to 0.2 gram and even 1 gram per day. Withdrawal symptoms of amphetamine dependency include deep sleep, marked hunger, and depression that may reach suicidal proportions.

The milder stimulants—phenmetrazine, pipradol, and methylpheni-date—have effects similar to amphetamines and are used for appetite sup-pression. There is considerably less misuse of these stimulants.

Cocaine, derived from cocoa leaves, has been regarded as a narcotic, and is prescribed only as a topical anesthetic. It has recently become quite popular among young drug users because it does not produce tolerance. However, its effects are similar to those from amphetamines, in a more marked degree. With high doses, it produces hallucinations and leads to pronounced psychological dependence. Possible side effects include nausea and coma.

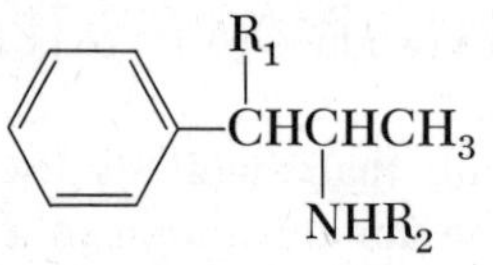

| $R_1$ | $R_2$ | |
|-------|-------|--------------------|
| H | H | Amphetamine |
| H | $CH_3$ | Methamphetamine |
| OH | $CH_3$ | Ephedrine |

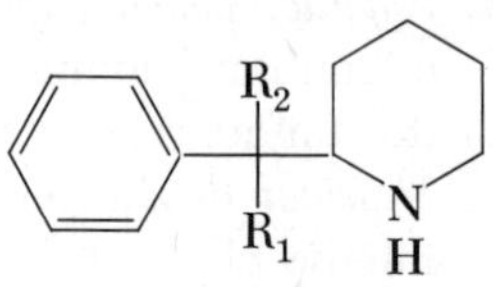

| $R_1$ | $R_2$ | |
|-------|---------|----------------|
| OH | H | Pipradol |
| H | $COOCH_3$ | Methylphenidate |

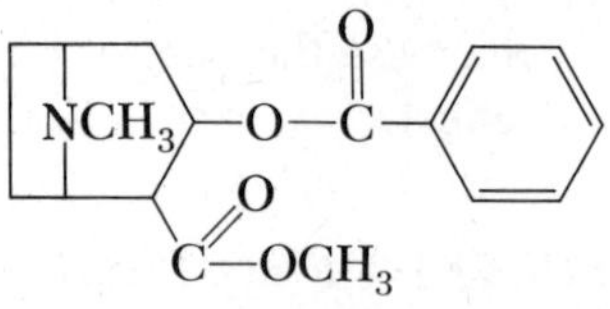

Cocaine

*Sedatives and Depressants:* Barbiturates are thought to exert an inhibi-tory action at nerve response centers. They are sedatives used to induce sleep. Many nonbarbiturate sedatives, such as chloral hydrate, can, like the barbitu-rates, produce psychological and physical dependence. Antihistamines often produce an appreciable degree of sedation. Ethyl alcohol (alcohol) is a de-pressant because it depresses the functions of the brain that control thinking and coordination. In high doses, it produces drowsiness and sleep. Alcohol is an addictive drug since it can cause withdrawal symptoms at least as serious as the other addictive drugs. It is estimated that alcohol intoxication is respon-sible for approximately half of the traffic fatalities in the United States (about 25,000 per year).

Barbiturates are probably the most misused drugs other than alcohol and nicotine. Accidental overdose is quite common since anything over a therapeutic dose impairs memory and judgment; thus the user can unwittingly and repeatedly ingest more and more tablets. Barbiturates taken in overdose (usually in a suicide attempt) cause more deaths than any other simple sub-stance. Barbiturates must be withdrawn with great caution from chronically dependent individuals because of the convulsions and delirium that abstinence can produce.

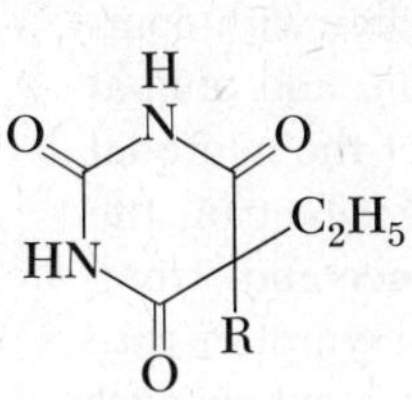

R

$C_2H_5$        Barbital
$C_6H_5$        Phenobarbital
$CHCH_2CH_2CH_3$    Pentobarbital
|
$CH_3$

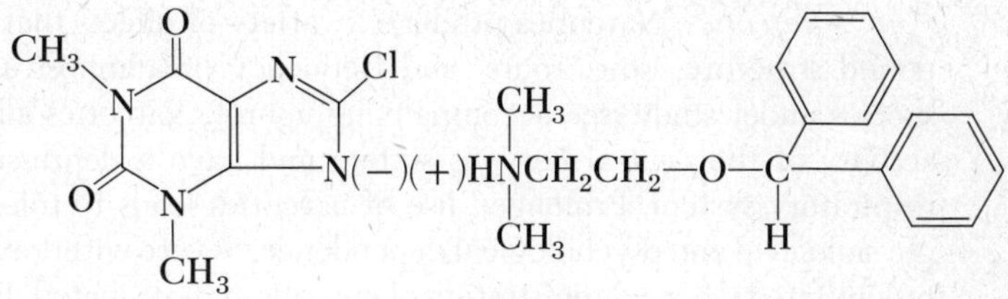

Dimenhydrinate
(Dramamine)

$Cl_3CCH(OH)_2$        $H_3C—CH_2OH$

Chloral hydrate        Alcohol
(Ethanol)

*Tranquilizers:* Tranquilizers calm tension and anxiety but do not cause sleep. These drugs can produce physical dependence and a withdrawal syndrome similar to that of barbiturates. Some of the minor tranquilizers are widely overprescribed by physicians, often when psychotherapy is what is indicated. Tranquilizers are sometimes used to combat unpleasant psychedelic intoxication. They are also a popular vehicle for suicide attempts. Some of the tranquilizers have muscle-relaxant properties and should not be taken with alcoholic beverages. It is also dangerous to drive when taking these medications.

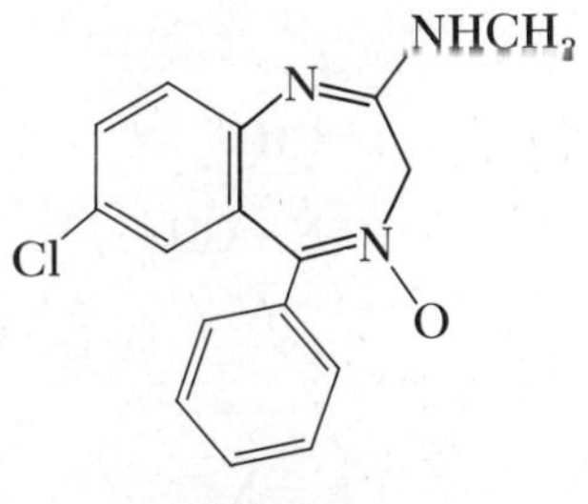

Chlordiazepoxide
(Librium)

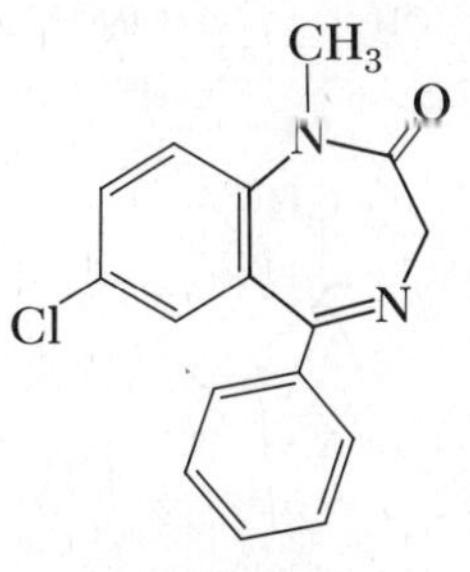

Diazepam
(Valium)

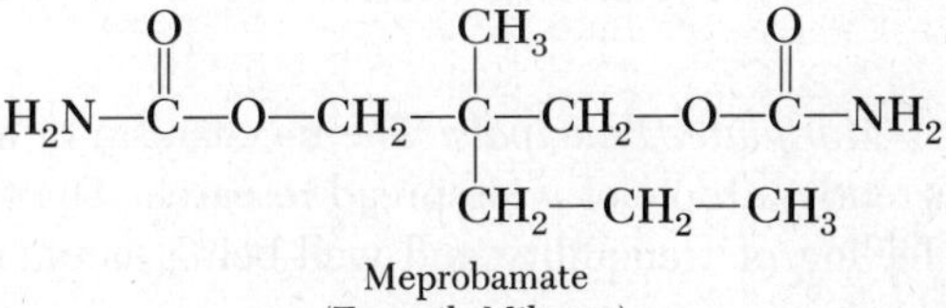

$H_2N—\overset{O}{\overset{\|}{C}}—O—CH_2—\overset{CH_3}{\underset{CH_2—CH_2—CH_3}{\overset{|}{\underset{|}{C}}}}—CH_2—O—\overset{O}{\overset{\|}{C}}—NH_2$

Meprobamate
(Equanil, Miltown)

*Narcotics:* Narcotics produce a variety of effects that differ with compound structure, dose, route, and frequency of administration, and animal species under study (see accompanying figure). Narcotics affect the neuronal activity of the central nervous system and have a depressing effect on the respiratory system. Prolonged use of narcotics leads to tolerance and strong physiological and psychological dependence. Severe withdrawal symptoms are precipitated when administration of narcotics is terminated. Persistent changes in narcotic drug tolerance are observed long after acute withdrawal symptoms disappear. Addiction among military personnel returning to the United States from Southeast Asia and Europe and among young drug users graduating from soft to hard drugs has increased tremendously in the past decade. Recently, through political pressure, the United States has influenced Turkey, currently the primary source of raw opium, to ban cultivation of the opium poppy.

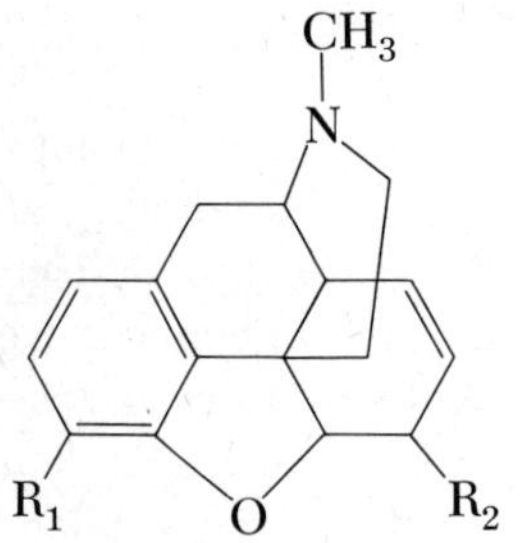

| $R_1$ | $R_2$ | |
|-------|-------|--|
| OH | OH | Morphine |
| OCOCH$_3$ | OCOCH$_3$ | Heroin |
| OCH$_3$ | OH | Codeine |

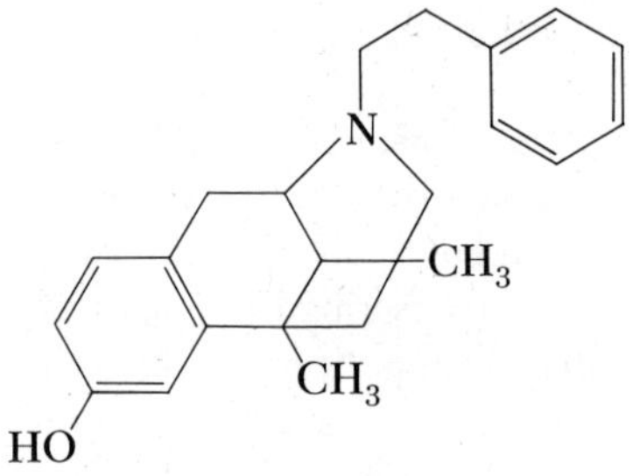

Phenazocine

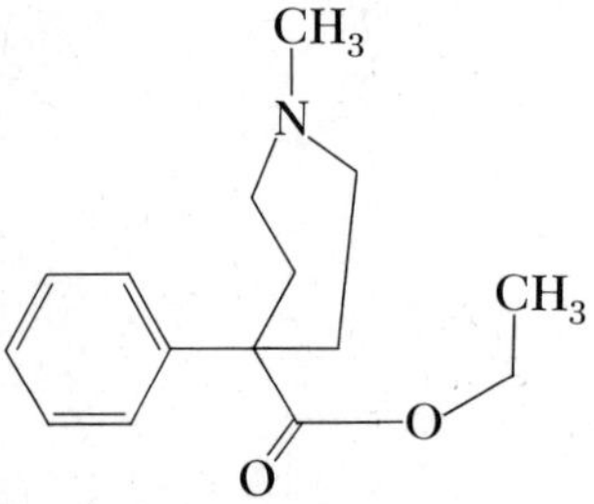

Pethidine
(Demerol)

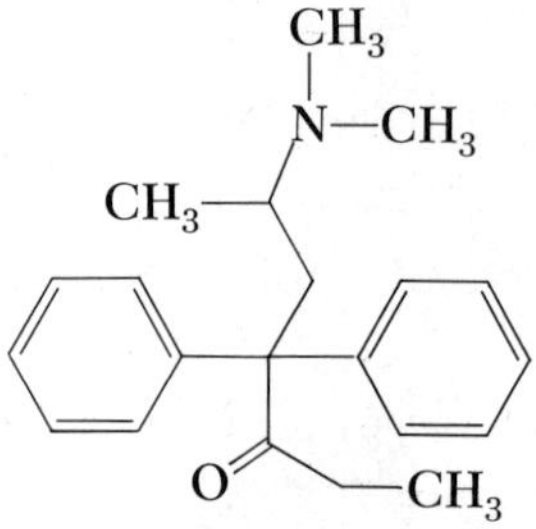

Methadone

*Marijuana Principals:* The mechanism of action of cannabis products is currently a topic of widespread research. The typical marijuana syndrome is a feeling of tranquility and well-being, euphoria, a sensation of floating, changes in perception and mood, and an altered experience of time. The pupils do not dilate; the reported dilation may have been due to the fact that most users smoke in darkened rooms. The cannabis products, marijuana, hashish, and tetrahydrocannabinol, are constituents of *Cannabis sativa* L. Var. *indica*, which grows wild, ideally in warm, dry climates. The exuded resinous matter of the tops of the female plant contain the highest concentration of active ingredients.

Marijuana refers to a preparation from the dried, mature flowering tops of the female plant and is usually smoked as a cigarette. It can be smoked in a pipe, eaten in various recipes, and drunk as a "tea." The effects of one or two "joints" reach a maximum in about thirty minutes and last up to three hours. Hashish is produced from the resin of the tops of the plant and is some five times more potent than marijuana. The effects are similar and more pronounced. Tetrahydrocannabinol is the active ingredient of marijuana and hashish. It is not very stable and is difficult to synthesize; thus, it is not found on the "street" market.

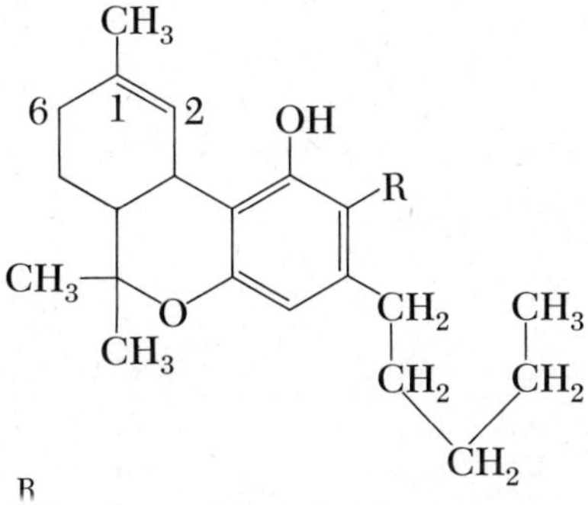

| R | |
|---|---|
| H | $\Delta^1$-Tetrahydrocannabinol |
| $CO_2H$ | $\Delta^1$-Tetrahydrocannabinol acid |

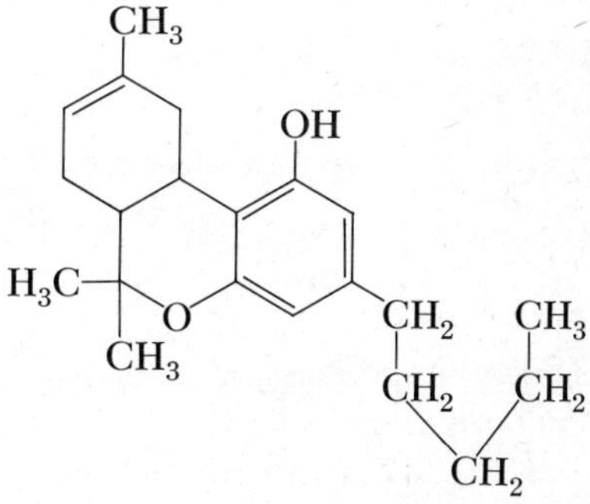

$\Delta^6$-Tetrahydrocannabinol

*Psychedelic Drugs:* Psychedelic drugs are mysterious compounds that change a person's perception of his surroundings. They induce hallucinations and delusions, impair intellectual activity, and lead to wide disparities between the judgments a person makes of his own activities compared with what others make of them. Their mechanism of action on the brain is not well understood. Many compounds in this category structurally resemble compounds in the body that affect the brain. With intake of LSD, one experiences time and space distortions, brighter colors, vivid sounds, and a sense of beauty in common objects, as well as fear and panic. Psychedelic drugs are the foundation of a subculture that has strongly affected popular art, music, and fashions. They have no proven therapeutic use. Before the discovery of LSD in 1944, mescaline was the main hallucinogen in use. Listed in order of

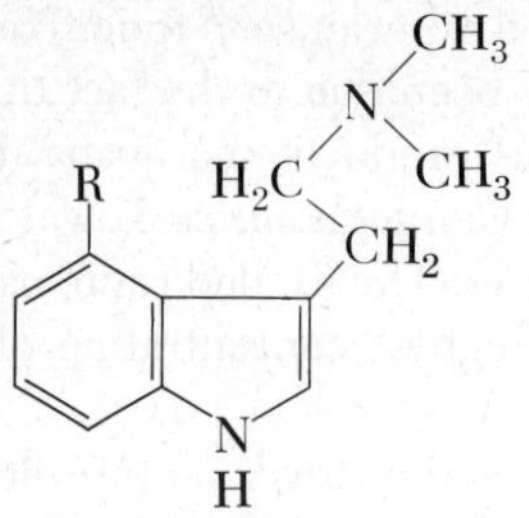

| R | |
|---|---|
| H | DMT |
| PO₄H₂ | Psilocybin |

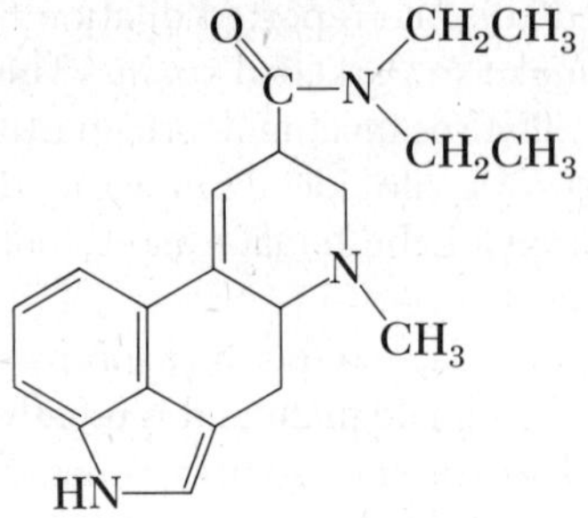

LSD

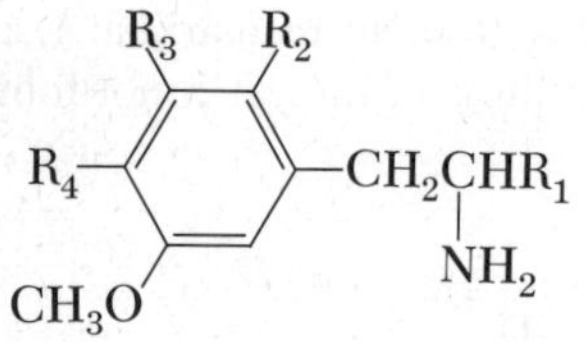

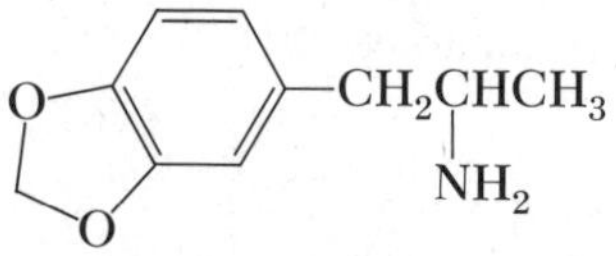

MDA

| R₁ | R₂ | R₃ | R₄ | |
|----|----|----|----|---|
| H | H | OCH₃ | OCH₃ | Mescaline |
| CH₃ | OCH₃ | H | CH₃ | STP |

increasing potency, some of the more popular psychedelic drugs are DMT, psilocybin, mescaline, LSD, MDA, and STP.

DMT is the active ingredient in snuffs used by South American Indians. DMT intoxicates for short periods (30–45 minutes). It is inactive orally and must be inhaled. Although weaker than LSD, it gives a greater number and variety of visual illusions and sometimes induces panic when taken intramuscularly. Psilocybin is active orally but is not presently in wide use. Mescaline, obtained from the peyote cactus, first causes nausea, tremors, and perspiration, and then a dreamlike intoxicated state. The trancelike state is accompanied by vivid kaleidoscopic visions and finally deep sleep.

LSD is one of the most potent psychochemicals known; one ounce could "turn on" a city of 300,000 inhabitants. LSD is about 4,000 times as potent as psilocybin and mescaline. A dose of 0.0001 gram can induce a psychedelic experience. Effects begin about 30 minutes after injection and last 9–12 hours. The experience is complex and unpredictable. LSD complications include psychotic disorders (chronic depressions); nonpsychotic disorders (chronic anxiety, acute panic states, antisocial behavior); and neurological reactions (convulsions, permanent brain damage). This drug produces neither a physical dependence nor an abstinence syndrome; tolerance develops and disappears rapidly.

MDA and STP (for the words *serenity, tranquility, peace*) are among the newer psychedelics alleged to produce a particularly tranquil psychedelic experience. STP first appeared on the illicit market in 1967 and was considered a highly potent and dangerous substance. Poisonings and death were attributed to its misuse. Subsequent investigations of poisonings suggest the symptoms were due to other compounds in the mixtures rather than to pure STP.

## MIRROR IMAGE ISOMERISM

*"I looked into a looking glass and what I saw was not the same. . . ."*

**Anonymous**

Take a look at your hands. Are they identical? No; your two hands are not identical but they are mirror images of one another (Figure 3–7). It is not possible for you to superimpose one hand over the other. If there are four different substituents attached to a carbon atom, there are two geometrical arrangements of these substituents about the carbon atom. The relationship of these two configurations is such that they are mirror images of one another (just as your two hands are mirror images), which are not superimposable (just as your two hands are not superimposable). The carbon atom is said to be **asymmetric** (without symmetry). Assembling the punch-out models will dramatize this point. (See Problem 12.)

Consider as in three-dimensional form the compound shown in Figure 3–8(a) and its mirror image in (b). Their central C atoms have four different

**Figure 3-7** The left hand and right hand are not identical but are mirror images of one another. If they were identical, it would be possible to overlap one hand on the other. (*Photograph by Kenneth Henderson. Reprinted by permission.*)

substituents, so they are asymmetric. It is not possible to slide (a) over (b) so that they coincide exactly. Now consider compound (c), whose central C atom does not have four different substituents, and its mirror image (d). By turning (c) to the right 60° and sliding it over (d), a perfect overlap is obtained. The reason this is possible for (c) but not for (a) is that (c) is not without symmetry. We can see this by passing a vertical plane through the middle of (c) and noticing that the left half is identical (symmetrical) with the right half (see Figure 3–8). Thus, (c) and (d) are identical structures whereas (a) and (b) are not. Structures (a) and (b) bear a special relationship to one another; they are nonsuperimposable mirror-image structures of the same molecule. The two mirror-image forms of a compound (for example, (a) and (b)) differ in the way they interact with light; other than that they have identical physical properties.

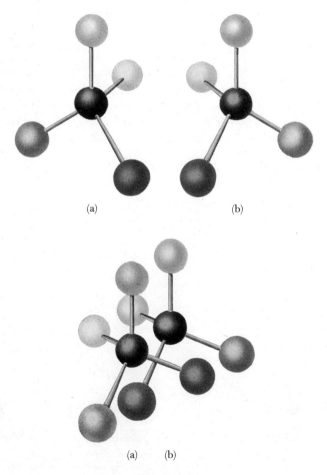

(a)                    (b)

(a)     (b)

**Figure 3–8**

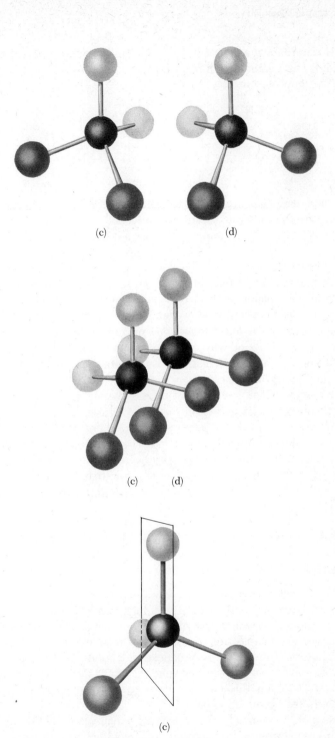

(c)          (d)

(c)     (d)

(c)

Figure 3-8 (Continued)

To determine if a compound can have mirror-image isomer forms, first find out if it has a carbon atom bonded to four *different* substituents (a single atom or a group of atoms).

$$\begin{array}{ccc}
& CH_3 & \\
& | & \\
H-C-CH_2-CH_3 & \\
& | & \\
& H-O &
\end{array}
\qquad\qquad
\begin{array}{ccc}
H & CH_3 & \\
| & | & \\
H-C-C-CH_3 \\
| & | & \\
H-O & H &
\end{array}$$

A                                        B

Compound *A* can exist in two forms that are mirror-image isomers since there are four different substituents attached to the central carbon atom (H, OH, $CH_3$, $CH_2CH_3$). Compound B does not have an asymmetric carbon atom.

In the laboratory preparation of an asymmetric compound, unless there is a particular reason for one mirror-image form to predominate, both are produced in equal amounts. Conventional techniques will not separate the individual mirror-image forms. However, for many different classes of organic compounds found in living systems, one mirror-image form predominates, often exclusively. How or why Nature developed such specificity is a fact which somehow must be rationalized in any theory concerning the origin of life. The physiology of living systems is so highly specialized that often only one mirror-image form is utilized while the "wrong mirror-image form" is completely unreactive. A pair of mirror-image forms is labeled D and L. Only the L form of the drug "dopa" is therapeutic. One mirror-image form of LSD is hallucinogenic; the other is not. It has recently been suggested that only the nonhallucinogenic form of LSD causes chromosome damage, the physiologically active form may not. Unfortunately, for "trippers," LSD from illegal sources normally contains an equal amount of both forms.

## NATURAL AND UNNATURAL POLYMERS

### Addition Polymers

The repeated combination of small molecules of relatively low mass to form very large molecules of relatively high mass is known as **polymerization.** The products are called **polymers.** Those that occur in nature (starch and cellulose, for example) are known to have been in existence since earliest recorded history. Those that are produced in laboratories have been known for less than thirty years, but in that time they have incredibly enhanced the standard of living on this planet. However, the ecological price of that prosperity is just becoming apparent.

Most of the commercially important polymers are **addition polymers** formed from one unsaturated starting compound called a **monomer.** When the gas ethylene is used as the monomer and polymerized, one obtains polyethylene, a white waxy-looking solid. It is flexible, begins to soften at 80 °C, melts near 115 °C and has excellent electrical insulating properties. Polyethylene is an example of an addition polymer where a big molecule is formed by the continual addition of smaller molecules. A polyethylene molecule is

$$H_2C{=}CH_2 + H_2C{=}CH_2 + H_2C{=}CH_2 + \cdots \longrightarrow$$

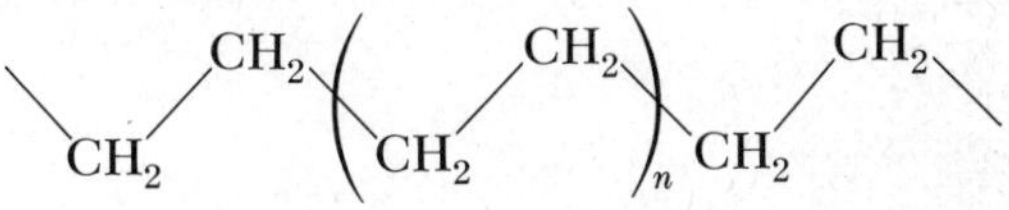

or

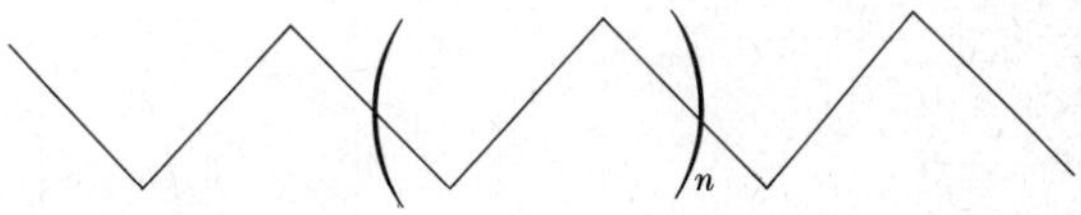

(Repeating units are shown in parentheses; $n$ represents a large number.)

a long chain of repeating $-CH_2-CH_2-$ units, which can be repeated thousands of times. An estimated 5.7 billion pounds of polyethylene were produced in the United States in 1970.

Alteration of the monomer can result in a polymer with dramatically different properties. Replacing one of the hydrogens of ethylene with a chlorine atom gives a compound called vinylchloride (see Figure 3–9). Polymerization gives polyvinylchloride (PVC), a tough, colorless, rigid, transparent solid used for containers, plastic films, and wrappings. The substitution of a phenyl group for a hydrogen gives the compound styrene. Its polymer, polystyrene, is a clear glasslike brittle material. Replacing all the hydrogens of ethylene with fluorine atoms gives tetrafluoroethylene (Teflon), whose polymer withstands temperatures of above 350° and has an extremely low coefficient of friction. For these reasons, it has been used in bearings and bushings and as a surface coating in nonstick frying pans. Other examples of important addition polymers are shown in Table 3–2.

Two important variations in structure affect the properties of the polymer:

1. **Size of the molecules**    Each polymer molecule (for example, in polyethylene) will not be the same length, so the term **average length** is used. Fortunately, it is possible to vary the average length of the polymer by technological methods, which include control of the reaction temperature. Polymers used for fibers must have a good resilience, which is obtained only at specific minimum molecular lengths.
2. **Isomerism**    The C atoms in the polymer may be linked in a straight linear chain (A), in a linear chain backbone with short branches (B), or a linear chain backbone with larger branches (C),

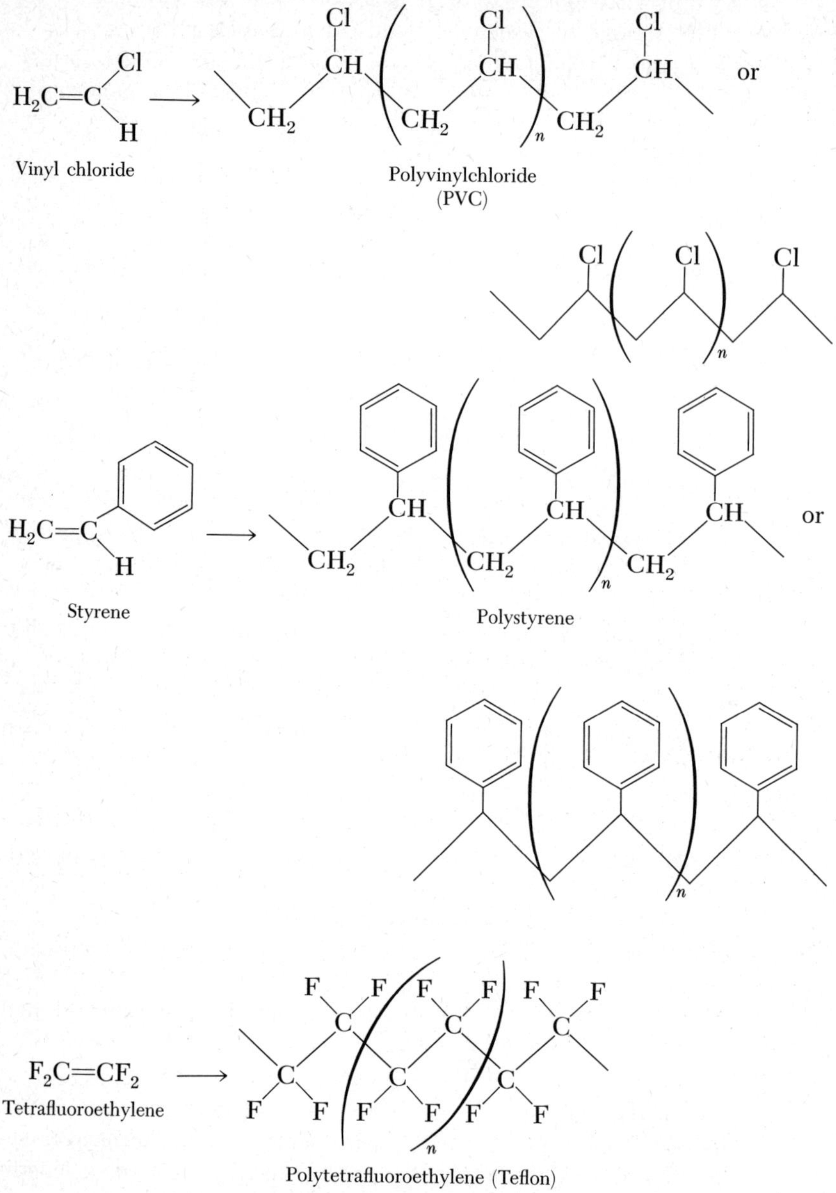

**Figure 3-9** The formation of addition polymers from monomers.

as shown in Figure 3–10. This structural isomerism is identical to that discussed for the smaller pentane molecule, but it occurs on a much larger scale. The linear polyethylene can assume a more ordered structure than the branched polyethylene and will have a

**Table 3-2**
**Addition Polymers**

| Monomer | Polymer | Trade Name | Use |
|---------|---------|------------|-----|
| 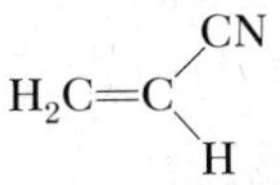 $H_2C\!=\!C\overset{CN}{\underset{H}{}}$  Acrylonitrile | 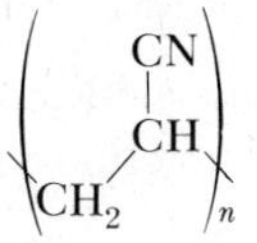 $\left(\!\!\begin{array}{c} CN \\ CH \\ CH_2 \end{array}\!\!\right)_n$ | Orlon | Fibers for clothing. |
| 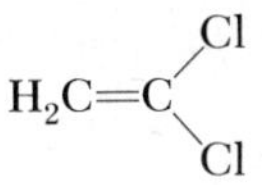 $H_2C\!=\!C\overset{Cl}{\underset{Cl}{}}$  Vinylidene chloride | 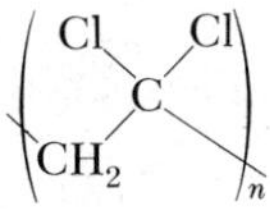 $\left(\!\!\begin{array}{c} Cl \quad Cl \\ C \\ CH_2 \end{array}\!\!\right)_n$ | Saran | Self-adhering food wrapper. |
| 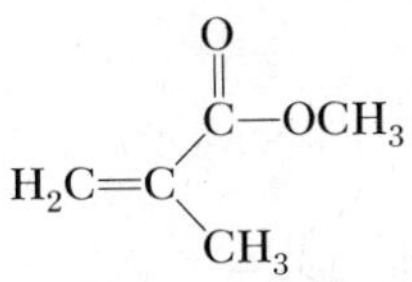 $H_2C\!=\!C\overset{\overset{O}{\parallel}\;C\!-\!OCH_3}{\underset{CH_3}{}}$  Methyl methacrylate | 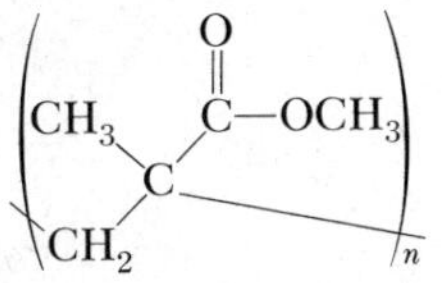 $\left(\!\!\begin{array}{c} CH_3 \quad \overset{O}{\overset{\parallel}{C}}\!-\!OCH_3 \\ C \\ CH_2 \end{array}\!\!\right)_n$ | Plexiglas | Unbreakable substitute for glass; used in paints. |

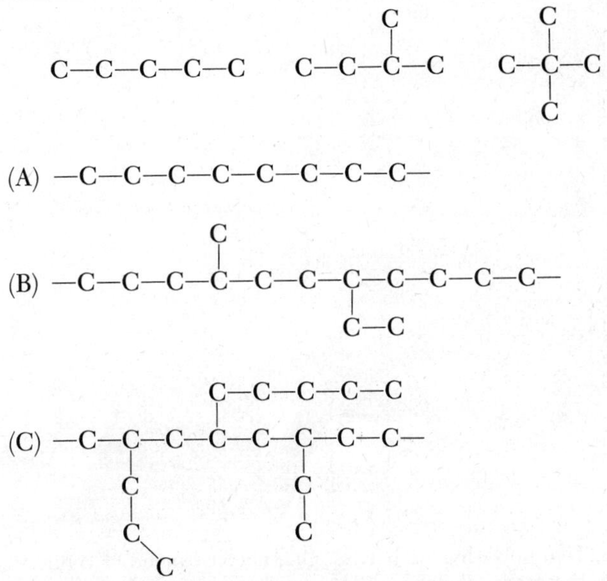

Figure 3-10 Structural isomerism in polymers.

"high density" and greater rigidity than the "low-density" material, whose flexibility allows it to be used for completely different applications. (See Figure 3–11.)

### Peptides and Condensation Polymers

The reaction of amines and carboxylic acids to form amides was mentioned earlier. Amino acids contain both functional groups. Thus, both the amino end and acid end of one molecule can form a bond by reacting with the acid end and amino end of two other amino acid molecules. The resultant molecule is called a **peptide.** The bonds that link the amino acid units are called **peptide bonds** (see Figure 3–12).

We can continue this process, using amino acids to form long peptide

**Figure 3–11** Two polyethylene bottles. One is constructed of rigid "high-density" polyethylene; the other, of flexible "low-density" polyethylene. (*Photograph by Kenneth Henderson. Reprinted by permission.*)

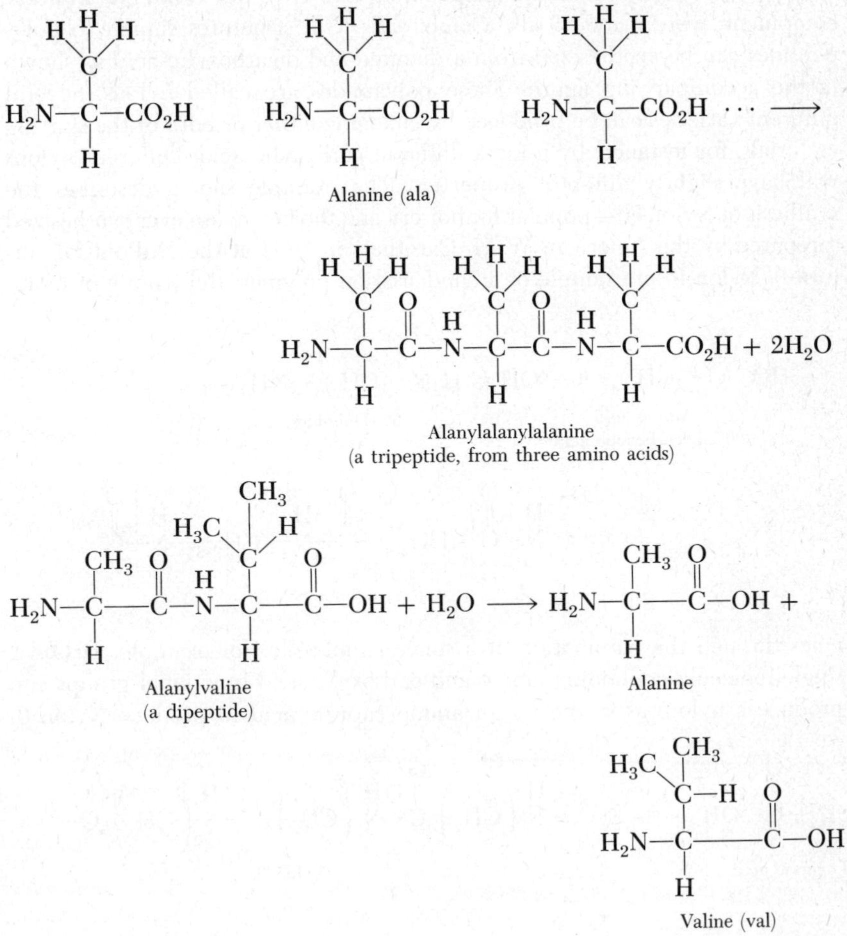

$$H_2N-\overset{\overset{\displaystyle H}{|}}{\underset{\underset{\displaystyle H}{|}}{C}}-CO_2H \qquad H_2N-\overset{\overset{\displaystyle H}{|}}{\underset{\underset{\displaystyle H}{|}}{C}}-CO_2H \qquad H_2N-\overset{\overset{\displaystyle H}{|}}{\underset{\underset{\displaystyle H}{|}}{C}}-CO_2H \cdots \longrightarrow$$

Alanine (ala)

$$H_2N-\overset{H}{\underset{H}{C}}-\overset{O}{C}-N-\overset{H}{\underset{H}{C}}-\overset{O}{C}-N-\overset{H}{\underset{H}{C}}-CO_2H + 2H_2O$$

Alanylalanylalanine
(a tripeptide, from three amino acids)

$$H_2N-\overset{CH_3}{\underset{H}{C}}-\overset{O}{C}-N-\overset{H_3C\quad CH_3}{\underset{H}{C}}-\overset{O}{C}-OH + H_2O \longrightarrow H_2N-\overset{CH_3}{\underset{H}{C}}-\overset{O}{C}-OH +$$

Alanylvaline
(a dipeptide)

Alanine

$$H_2N-\overset{H_3C\quad CH_3}{\underset{H}{C}}-\overset{O}{C}-OH$$

Valine (val)

**Figure 3-12** Peptides and amino acids. Top formula: Formation of a tripeptide from the condensation of three amino acids. Bottom formula: Hydrolysis of a dipeptide into two amino acids.

chains (polypeptides). Proteins are polypeptides that are essential to the life process (see Chapter 6). The digestion of protein from other sources (animal, fish, vegetable) is the reverse reaction, namely breaking down the food protein into individual amino acids, which the organism then uses to build its own chain (see Figure 3–12). Commercial silk fibroin is a polypeptide containing a proportionately large number of alanine units produced by the larvae of mulberry silk moths. A strong, lightweight material, its availability became limited when the United States and Japan (a primary source of silk) were engaged in conflict. It was then necessary to produce large quantities of material that had similar properties, not merely because the fairer sex needed more stockings, but because parachutes, among other strategic military equipment, were made of silk. Compounds with structures similar to polypeptides can be synthesized from a diamine and dicarboxylic acid as shown in the accompanying figure. These polyamides are called nylons. Several different varieties can be produced by changing either or both of the starting materials, for instance, by using a different carboxylic acid. Different nylons will have slightly different properties. The example shown describes the synthesis of Nylon 66—popular for hoisery and the first nylon ever synthesized (prepared by the American W. H. Carothers in 1934 at the DuPont laboratories). Nylon is an example of a condensation polymer: the joining of mole-

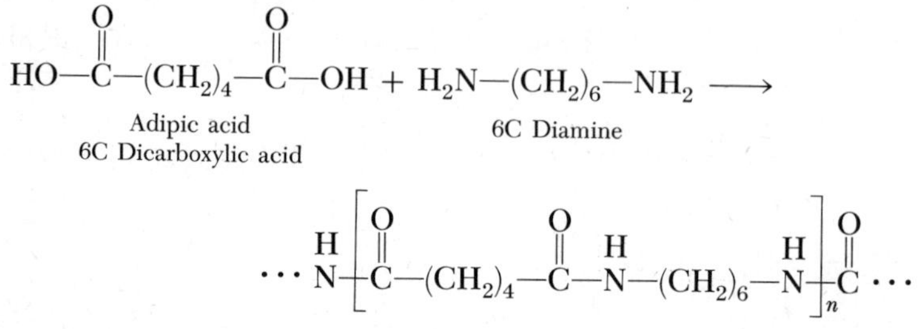

$$HO-\underset{\underset{\displaystyle O}{\|}}{C}-(CH_2)_4-\underset{\underset{\displaystyle O}{\|}}{C}-OH \; + \; H_2N-(CH_2)_6-NH_2 \longrightarrow$$

Adipic acid<br>6C Dicarboxylic acid         6C Diamine

$$\cdots \; \underset{\displaystyle H}{N}\left[\underset{\underset{\displaystyle O}{\|}}{C}-(CH_2)_4-\underset{\underset{\displaystyle O}{\|}}{C}-\underset{\displaystyle H}{N}-(CH_2)_6-\underset{\displaystyle H}{N}\right]_n \underset{\underset{\displaystyle O}{\|}}{C} \; \cdots$$

Nylon 66

cules through the elimination of a smaller molecule (for example, $H_2O$). A single molecule containing amine and carboxylic acid functional groups can produce a nylon, as in the use of amino caproic acid to produce Nylon 6.

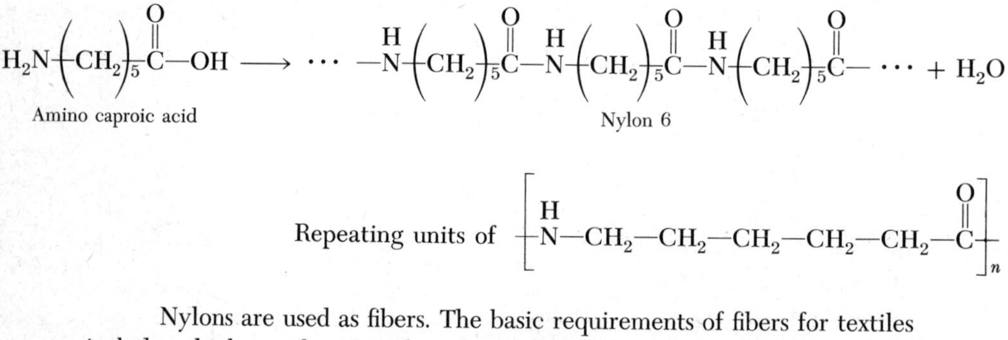

$$H_2N\left(CH_2\right)_5\underset{\underset{\displaystyle O}{\|}}{C}-OH \longrightarrow \cdots -\underset{\displaystyle H}{N}\left(CH_2\right)_5\underset{\underset{\displaystyle O}{\|}}{C}-\underset{\displaystyle H}{N}\left(CH_2\right)_5\underset{\underset{\displaystyle O}{\|}}{C}-\underset{\displaystyle H}{N}\left(CH_2\right)_5\underset{\underset{\displaystyle O}{\|}}{C}- \cdots + H_2O$$

Amino caproic acid                           Nylon 6

$$\text{Repeating units of} \; \left[\underset{\displaystyle H}{N}-CH_2-CH_2-CH_2-CH_2-CH_2-\underset{\underset{\displaystyle O}{\|}}{C}\right]_n$$

Nylons are used as fibers. The basic requirements of fibers for textiles include a high tensile strength and a melting point that is high enough to permit being ironed but not too high for the manufacturing process (between 200 °C and 300 °C). For tensile strength in fibers, chain length and forces between the chains are important so that the chains are held together in some orderly fashion. These forces should not be permanent bonds in fibers; they are usually attractive forces. Models of the structure of Nylon 66 show that hydrogen bonding is possible between the chains (see Figure 3–13 and p. 66), allowing stronger attractive forces. The importance of the hydrogen bonding is dramatized by comparing the commercially unimportant "methyl nylon" with Nylon 66. Methyl nylon has $CH_3$ groups on N, so no hydrogen bonding

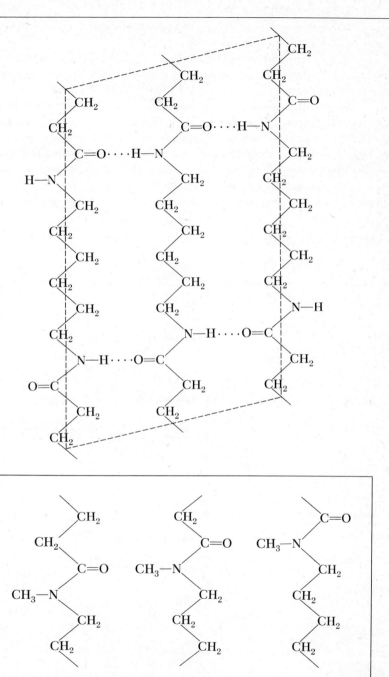

**Figure 3-13** Hydrogen bonding in Nylon 66 (top), which is not possible in methyl nylon (bottom).

can occur (see Figure 3–13). The melting point of methyl nylon is 145 °C as compared with 265 °C for Nylon 66.

By utilizing a dicarboxylic acid and a diol, one can make an ester polymer (polyester). The accompanying example describes the synthesis of a particular polyester commonly called Dacron. If the isomeric dicarboxylic acid is used, another polyester is formed with significantly different chemical and physical properties (see Figure 3–14). If a triol (glycerol) is used instead of a diol, the resulting polymer still contains alcohol groups (see Figure 3–15). On vigorous heating, the OH groups can react to form permanent bridges between the chains (called **cross-links**). The heating has now set the material (thermosetting) and it cannot be made to flow. Contrast this with the polyethylene, which softens whenever heated and then rehardens on cooling

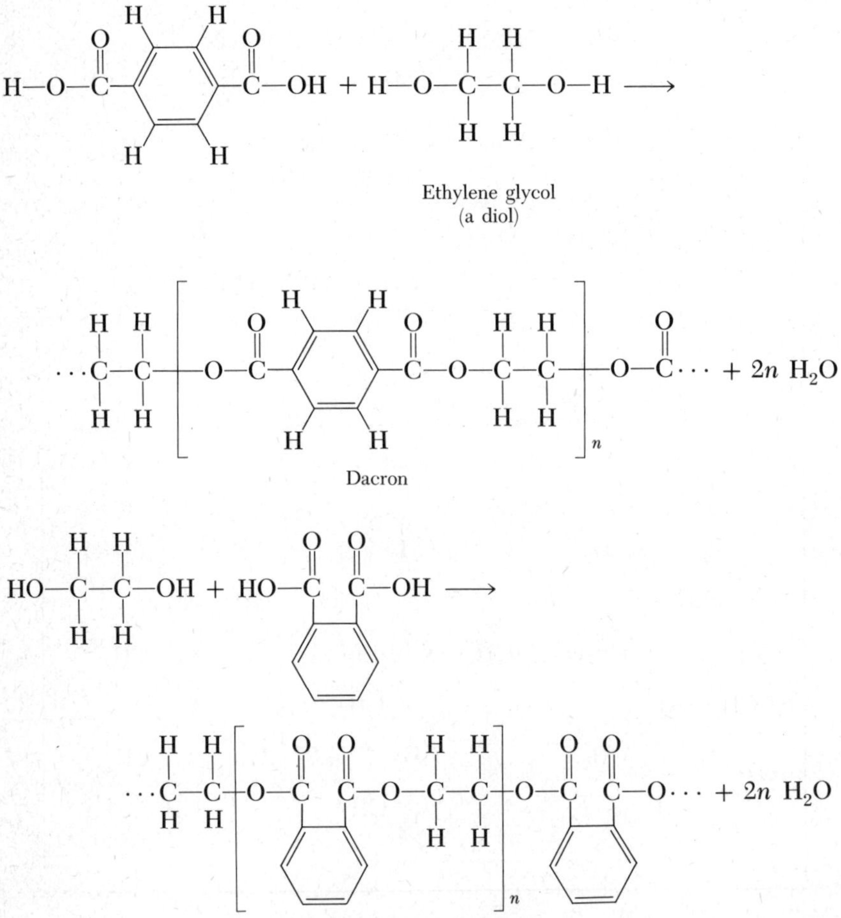

Ethylene glycol
(a diol)

Dacron

Figure 3–14  The formation of polyesters.

(thermoplastic). The technological uses of thermosetting materials include paints and adhesives (some epoxy resins), which set because cross-links are formed.

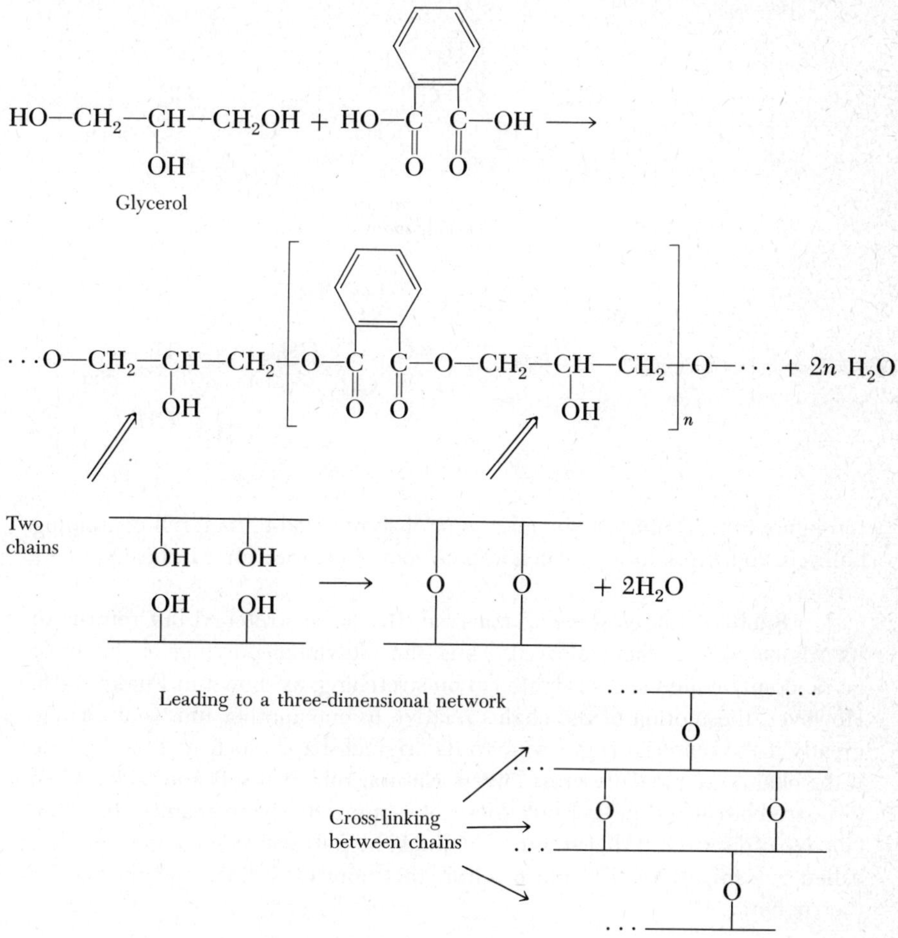

---

**Figure 3-15**   Cross-linking of polymers.

---

### Rubber

An interesting example of an attempt to duplicate a naturally occurring polymer is the case of rubber. It can be thought of as being formed from isoprene units. In the formation of the chain, a double bond is left in each original monomer unit. Thus, two stereoisomers are possible, one with the $CH_3$ and H groups on the double bond *cis* to one another, and one with those groups *trans* to one another. The rubber tree makes the *cis* polymer. The synthesis of rubber from inexpensive isoprene has been the goal of scientists

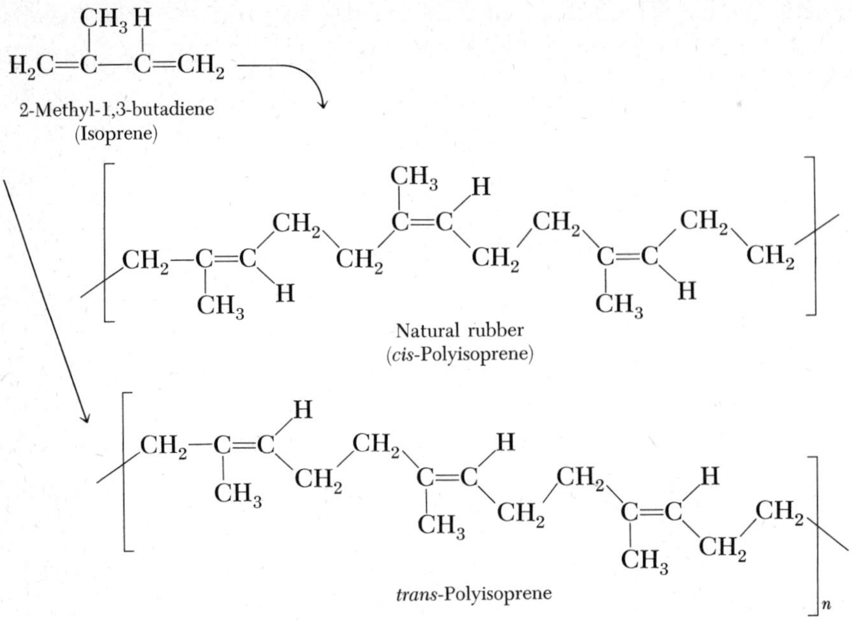

$$CH_3 \quad H$$
$$H_2C=C—C=CH_2$$

2-Methyl-1,3-butadiene
(Isoprene)

Natural rubber
(*cis*-Polyisoprene)

*trans*-Polyisoprene

throughout this century. Polymerization of styrene led to material containing both *cis* and *trans* forms, which was a poor substitute for Mother Nature's product.

Rubber is an *elastomeric* material. It can be stretched but returns to its original shape when released. Thus, the individual polymer chains must be randomly coiled and straightened on stretching, as shown in Figure 3–16. However, the motion of the chains relative to one another must not change greatly if the material is to revert to its original shape. Such will be the case if the chains are partially cross-linked. Natural rubber is soft and tacky when hot, stretches when pulled but does not resume its shape readily. In 1839, Goodyear discovered that sulfur cross-links the polyisoprene chain (a process called vulcanization). The more sulfur, the more cross-links and the harder the rubber.

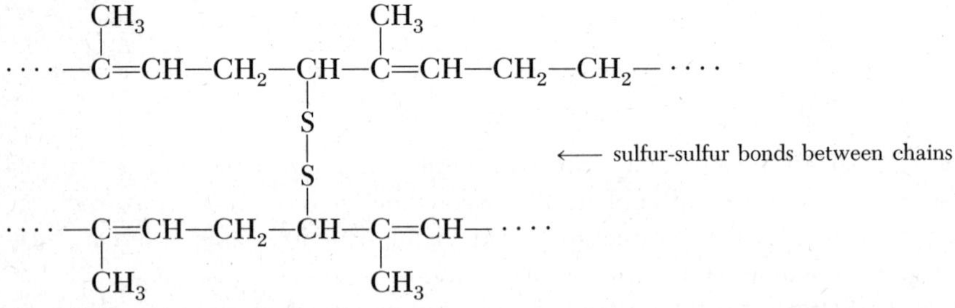

$$CH_3 \qquad\qquad CH_3$$
$$\cdots\cdots—C=CH—CH_2—CH—C=CH—CH_2—CH_2—\cdots\cdots$$
$$\qquad\qquad\qquad\qquad S$$
$$\qquad\qquad\qquad\qquad |$$
$$\qquad\qquad\qquad\qquad S \qquad\qquad \longleftarrow \text{ sulfur-sulfur bonds between chains}$$
$$\qquad\qquad\qquad\qquad |$$
$$\cdots\cdots—C=CH—CH_2—CH—C=CH—\cdots\cdots$$
$$\qquad CH_3 \qquad\qquad CH_3$$

Example of cross-linking in rubber

It is now not only possible to prepare synthetic rubber that is virtually

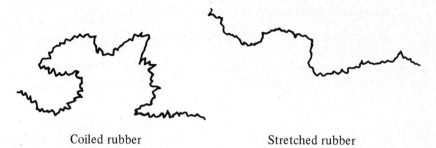

Coiled rubber                    Stretched rubber

**Figure 3-16** Elastomeric materials such as rubber can be stretched but return to their original shape when released.

identical with Nature's product, but rubbers that Nature hasn't deemed necessary. For example, replacing the $CH_3$ of isoprene with a Cl atom gives neoprene, which is not affected by hydrocarbon solvents the way other elastomers are.

$$H_2C=\overset{\underset{|}{Cl}}{C}-CH=CH_2 \longrightarrow \left[ CH_2-\overset{\underset{|}{Cl}}{C}=CH-CH_2 \right]_n CH_2-$$

Neoprene

The unsaturated sites in rubbers are readily attacked by oxygen, causing the polymer to become brittle. A stabilizer called an antioxidant is added to the rubber to prevent this. It will react with the oxygen faster than the rubber and thus forestall attack on the rubber. By an unfortunate choice, $\beta$-naphthylamine had been used as an antioxidant for rubber. It is highly carcinogenic, inducing cancer of the bladder in people who come in contact with it. The apparently similar $\alpha$-naphthylamine is harmless.

$\beta$-Naphthylamine                    $\alpha$-Naphthylamine

A possible shortage of rubber during World War II prompted research in the area of elastomers. In fact, World War II produced a dramatic change in all areas of American science and technology. The rapid growth of science in this country was favored by the importation of scientific ideas and laboratory designs from Western Europe. It was aided enormously by the influx of distinguished foreign scientists, many of whom were refugees from a Europe financially and politically troubled and racially distraught. Financial support of academic research prior to World War II was almost nonexistent but is now essential for further progress.

### Practical Considerations of Polymer Technology

At this point, after discussing the fantastic progress in an industry that has been developed since World War II (polymers), before we continue with the elucidation of scientific achievement, it is appropriate to include a quotation from an address by Albert Einstein to the student body at Caltech in 1931:

> *It is not enough that you should understand about applied science in order that your work may increase man's blessings. Concern for man himself and his fate must always form the chief interest of all technical endeavors . . . in order that the creations of our mind shall be a blessing and not a curse to mankind. Never forget this in the midst of your diagrams and equations.*

Synthesized polymers do not often occur naturally. Most of them have been known for less than thirty years. Unlike the natural polymers which have been in our ecological system for millions of years, they do not fit in. They are not degraded by organisms or natural surroundings—in fact, we designed them that way! How do we get rid of them? Scientists are working on that problem and new degradable polymers are being sought. Can we readily give up our synthetic polymers? It would mean dependence on natural sources for clothing, building and structural materials, furniture, floor coverings, tires, etc. Moreover, many synthetic polymers have unique properties not found in naturally occurring materials, as illustrated by abrasion-resistant fabrics, lightweight insulating materials, nonmarring surface material, chemically inert surgical dressings, and contact lenses and dentures. We have reached a stage of sophistication where it is possible to consider a particular need and design a polymer specifically for that purpose. These needs may be as diverse as material for bathing suits, which must be able to be woven (thermoplastic), stretchable (elastomeric), and quick-drying (no attraction for $H_2O$ molecules), and synthetic material for the nose cones of space ships that is able to be molded but resistant to high temperatures and shock.

### Coal and Petroleum: Raw Materials for Polymers

Raw materials for polymers come from substances found in nature: petroleum, coal, and vegetation (for example, cellulose for rayon). When coal is heated to 1,000 °C in the absence of air, coke (a form of carbon), coal tar, coal gas, and $NH_3$ are produced. The coke is used to produce methanol and formaldehyde via the initial formation of CO and $H_2$:

$$C + H_2O \longrightarrow CO + H_2;$$
Coke

$$CO + 2H_2 \longrightarrow CH_3OH.$$
Methanol

Treatment of coke with lime gives $CaC_2$, which reacts with water to form acetylene:

$$3C \ + CaO \rightarrow \ CaC_2 \ + CO;$$

Coke     Lime     Calcium carbide

$$CaC_2 + H_2O \rightarrow HC\equiv CH + CaO.$$

Acetylene reacts with HCl to give vinyl chloride, used for making polyvinyl-chloride:

$$HC\equiv CH + HCl \longrightarrow H_2C\!=\!\underset{\underset{Cl}{|}}{CH} \longrightarrow \text{polyvinylchloride.}$$

Crude oil is very competitive as a source of organic compounds. Heating petroleum in the absence of air breaks down the large molecules to smaller more useful ones, a process called cracking. It is possible to separate and collect compounds or mixtures of compounds containing various numbers of C atoms since the smaller molecules will be more volatile (lower boiling point). Thus, the $C_1$ to $C_4$ hydrocarbons can be obtained for use as fuel (like methane or propane) or as starting materials for polymers. Ethylene, available from petroleum, is now used instead of acetylene for making vinylchloride:

$$H_2C\!=\!CH_2 + Cl_2 \longrightarrow ClCH_2\!-\!CH_2Cl;$$
$$ClCH_2\!-\!CH_2Cl \longrightarrow H_2C\!=\!CHCl + HCl.$$

It is also used as a source of ethanol:

$$H_2C\!=\!CH_2 + H_2O \longrightarrow H_3C\!-\!CH_2OH.$$

Ethanol

But only 3 percent of processed petroleum is used for making chemicals. Most of it is used for making gasoline ($C_5$—$C_{12}$), kerosene ($C_{12}$—$C_{16}$), and oils ($C_{15}$—up). Thus, the $C_1$—$C_4$ gases are often discarded by burning, lighting the sky near refineries at night and burning with even less utility by day.

## WHAT PRICE DUNG?
## THE PROBLEM OF ORGANIC WASTES

*There are more chickens in the U.S. than there are people in the world.*

**Dr. G. Alex Mills, U.S. Bureau of Mines**

Two billion tons of animal waste are produced in the United States each year. That is ten tons per person per year or enough dung to cover the Chicago metropolitan area ten feet deep! It is substantially greater than the 250 million tons of household, commercial, and municipal solid wastes generated yearly.

Manure is no longer valuable as a fertilizer because it is difficult or unpleasant to handle and spread and may have an undesirable composition. The amount of potassium and other minerals that are available in one ton of manure can be obtained in a 100-pound bag of fertilizer. Grain farmers prefer fertilizers with a lower nitrogen content than found in manure, because they keep the grain stalk from growing high. High stalks are more susceptible to damage from wind, rain, and so on. Chicken manure can no longer be used for many crops because chicken feed contains many antibiotics that are passed into the feces and in turn into the agricultural crop. Contaminated water from feed-lot runoff can enter municipal waterways; a dangerous situation when one considers that over a hundred animal diseases can be passed to humans. What can be done with this abundant natural resource?

The U.S. Bureau of Mines has recently discovered that if you heat manure in a reaction vessel with carbon monoxide at 1,200 pounds per square inch pressure at 380 °C for 20 minutes you get oil. The oil can be used for fuel. Manure will not solve the energy problem but using it as energy could solve the manure problem. Why not? Manure has been used for fuel in undeveloped countries for centuries.

## PROBLEMS

1. List all of the functional groups in the following compounds, whose structures appear in the text:
   (a) citric acid; (b) LSD; (c) morphine.

2. Draw the structures of the compound(s) that would be used to make the following polymers and determine if they are addition or condensation polymers.

(a)

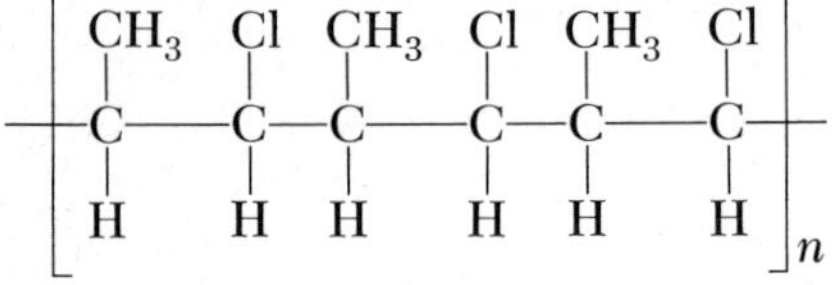

$$\left[ \begin{array}{cccccc} CH_3 & Cl & CH_3 & Cl & CH_3 & Cl \\ | & | & | & | & | & | \\ C & C & C & C & C & C \\ | & | & | & | & | & | \\ H & H & H & H & H & H \end{array} \right]_n$$

(b)

$$\left[ \begin{array}{c} \\ O=C \quad C=O \qquad O=C \quad C=O \\ -O-CH_2-CH_2-O \quad O-CH_2-CH_2-O \quad O-CH_2-CH_2-O- \end{array} \right]_n$$

(c)

$$\left[ \begin{array}{c} O \quad CH_3 \qquad CH_3 \quad O \qquad O \quad CH_3 \qquad CH_3 \quad O \qquad O \\ \| \quad | \qquad | \quad \| \qquad \| \quad | \qquad | \quad \| \qquad \| \\ -C-N-CH_2-N-C-CH_2-C-N-CH_2-N-C-CH_2-C- \end{array} \right]_n$$

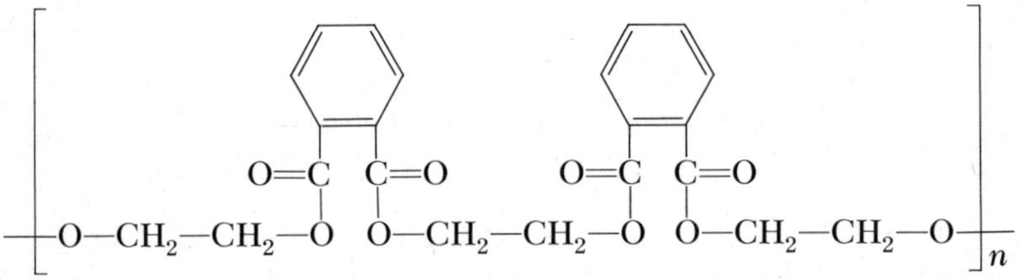

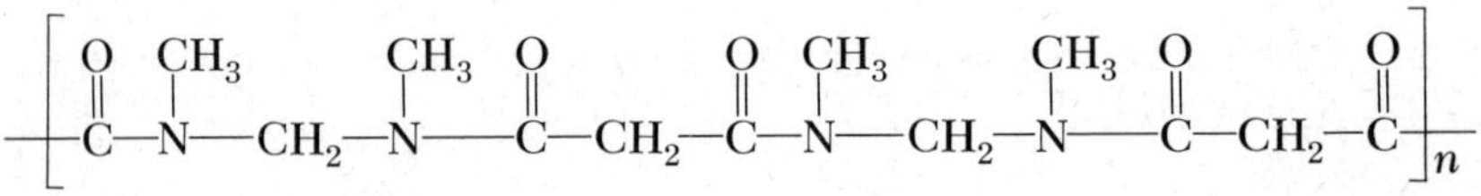

(d)

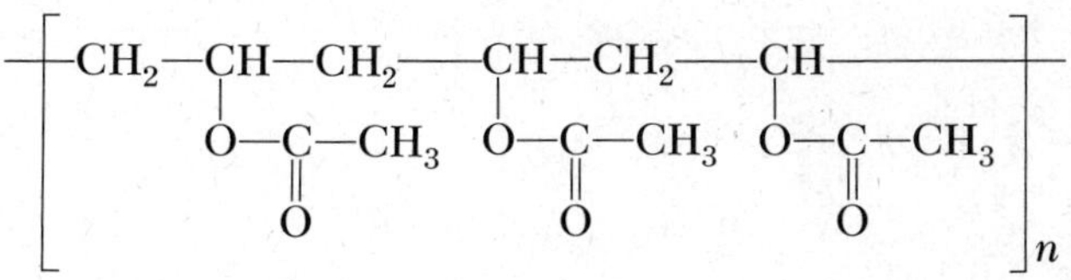

$$\left[ -CH_2-\underset{\underset{\underset{O}{\parallel}}{\underset{|}{O-C-CH_3}}}{CH}-CH_2-\underset{\underset{\underset{O}{\parallel}}{\underset{|}{O-C-CH_3}}}{CH}-CH_2-\underset{\underset{\underset{O}{\parallel}}{\underset{|}{O-C-CH_3}}}{CH}- \right]_n$$

3. Which of the compounds below will have mirror-image isomer forms? First determine if the compound has a carbon atom bonded to four different groups.

(a)

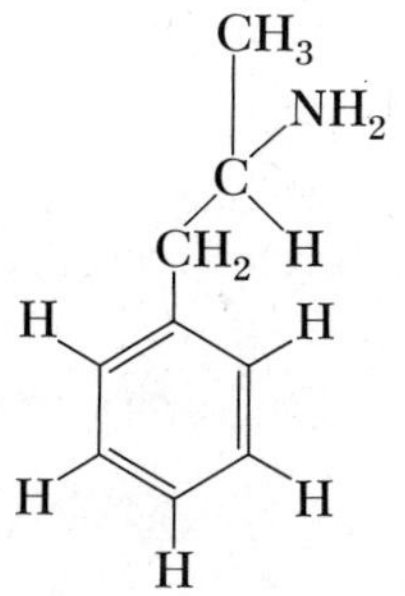

(b)

$$H_3C-\underset{\underset{HO}{|}}{\overset{\overset{H}{|}}{C}}-\underset{\underset{CH_3}{|}}{\overset{\overset{CH_3}{|}}{C}}-CH_3$$

Benzedrine

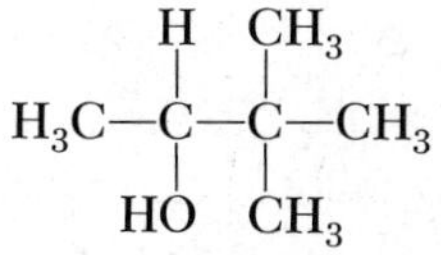

(c) DDT

4. Write and label the structures of the *cis* and *trans* isomers of

$$\underset{Br}{\overset{H}{\diagdown}}C=C\underset{CH_3}{\overset{H}{\diagup}}$$

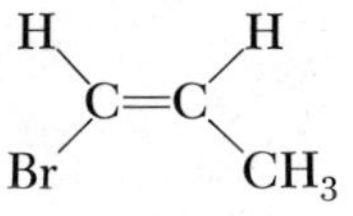

Why do you *not* have *cis* and *trans* isomers of vinyl chloride?

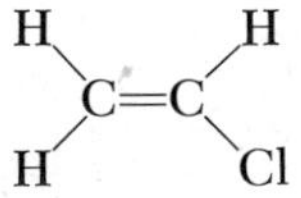

$$\underset{H}{\overset{H}{\diagdown}}C=C\underset{Cl}{\overset{H}{\diagup}} \quad \text{(HINT: Try to write them.)}$$

5. How many asymmetric carbon atoms are there in alanylalanylalanine? (See Figure 3–12.)

6. What products will be produced by hydrolysis of the following compounds?

(a)

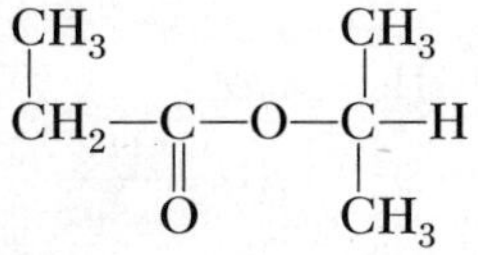

$$\underset{\underset{O}{\parallel}}{\overset{\overset{CH_3}{|}}{CH_2-C}}-O-\underset{\underset{CH_3}{|}}{\overset{\overset{CH_3}{|}}{C}}-H$$

(b)

$$-CH_2-\overset{\overset{O}{\parallel}}{C}-N\underset{CH_3}{\overset{CH_3}{\diagup}}$$

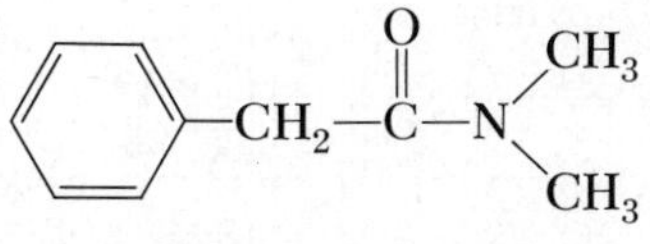

7. What is the difference between science and technology?

8. Are the following compounds *cis* or *trans?*

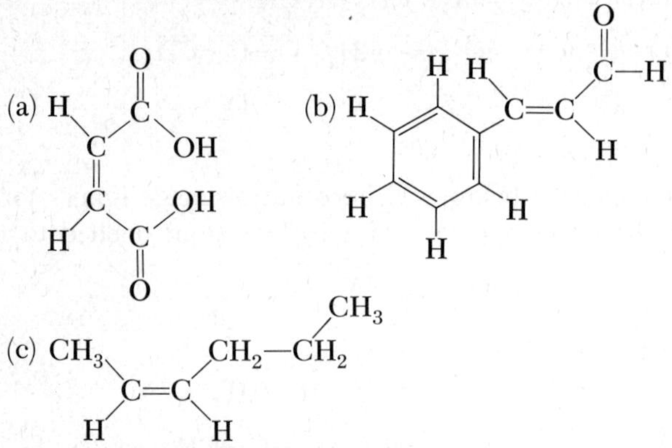

(a) H
C
C
OH
C
OH
H
C
O

(b) H
H H
C=C
C—H
O
H
H
H
H

(c) CH₃
CH₂—CH₂
CH₃
C=C
H
H

9. Which of the following compounds are identical? Which are structural isomers? Which are geometric isomers? (Identical structures cannot be isomeric.)

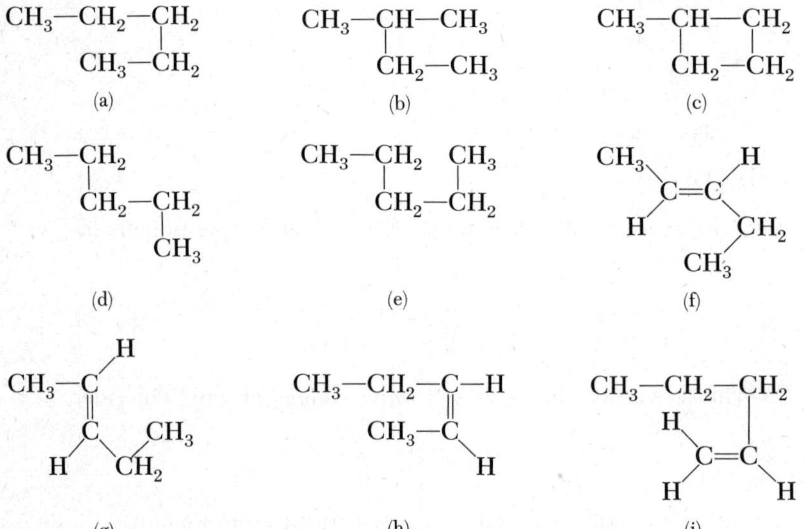

CH₃—CH₂—CH₂
CH₃—CH₂
(a)

CH₃—CH—CH₃
CH₂—CH₃
(b)

CH₃—CH—CH₂
CH₂—CH₂
(c)

CH₃—CH₂
CH₂—CH₂
CH₃
(d)

CH₃—CH₂   CH₃
CH₂—CH₂
(e)

CH₃          H
C=C
H          CH₂
CH₃
(f)

CH₃—C
H
C   CH₃
H   CH₂
(g)

CH₃—CH₂—C—H
CH₃—C
H
(h)

CH₃—CH₂—CH₂
H
C=C
H          H
(i)

10. All naturally occurring amino acids except one have an asymmetric carbon. If the following tripeptide is hydrolyzed, three amino acids are obtained. Write their structures. Which of the three cannot have mirror-image isomer forms?

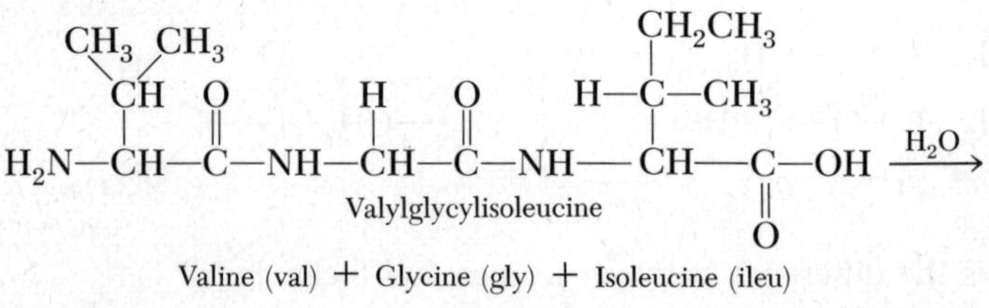

CH₃  CH₃                          CH₂CH₃
CH   O          H   O     H—C—CH₃
H₂N—CH—C—NH—CH—C—NH— CH—C—OH  $\xrightarrow{\text{H}_2\text{O}}$
O
Valylglycylisoleucine

Valine (val) + Glycine (gly) + Isoleucine (ileu)

11. A statement often made is that one plant while growing will produce enough oxygen in one year to support $X$ rabbits. Attack this statement by using the following two chemical equations and assuming that the rabbit doesn't eat the plant.

$$\left(\begin{array}{c}\text{Approximate}\\\text{composition}\\\text{of plants}\end{array}\right) C_6H_{12}O_6 \ + 6O_2 \ \xrightarrow{\text{bacteria}} \ 6CO_2 + 6H_2O.$$

$$\text{Photosynthesis: } 6CO_2 + 6H_2O \ \xrightarrow{\text{light, chlorophyll}} \ 6O_2 + C_6H_{12}O_6.$$

12. (a) Using the set of models provided, construct a model with four different colored groups. Set it aside and then without referring to it, construct a second model. See if the second model is superimposable on the first or if it is a mirror image. Disassemble the second model, and at random, reconstruct it. Repeat this approximately ten times noting the relative number of times you obtain a mirror image or a superimposable structure. What is the percentage of each isomer obtained randomly?

(b) Construct two models, carbon atoms with four different substituents, in such a way that they are superimposable. Now change the relative positions of two of the four groups by reversing anyone of the crescent-shaped pieces. Are the two models now superimposable (identical) or mirror images? Repeat several times and note if the models are identical or mirror images after each exchange.

(c) Repeat part (a) but use only three different color groups. Are all models you have now constructed identical? Why?

# BIBLIOGRAPHY

1. N. L. Allinger, M. P. Cava, D. C. DeJongh, C. R. Johnson, N. A. Lebel, and C. L. Stevens. *Organic Chemistry*. New York: Worth, 1971.

2. P. L. Cook and J. W. Crump. *Organic Chemistry: A Contemporary View*. Lexington, Mass.: D. C. Heath, 1969.

3. The Editors, *Consumer Reports*. *The Medicine Show*. Mount Vernon, N.Y.: Consumers Union, 1970.

4. "Consumerism." Special report on the drug industry. *Chemical and Engineering News*, July 26, 1971, pp. 24–28.

5. M. Kaufman. *Giant Molecules*. Garden City, N.Y.: Doubleday, 1968.

6. *The Merck Index*, 8th ed. Rahway, N.J.: Merck, 1968.

7. E. L. May. "The Evolution of Totally Synthetic, Strong Analgesics." *Aldrichimica acta*. Milwaukee: Aldrich Chemical Co., 1969.

8. *Chemical and Engineering News*, August 15, 1971, p. 43.

9. "Drugs." Special report. *Chemistry in Britain* 8, no. 3 (1972):98–130. An excellent concentrated review of various aspects of this subject.

10. J. R. Unwin. "Non-Medical Use of Drugs." *Journal of the Canadian Medical Association* 101 (1969):72–88 (cumulative pages 804–20).

# The Atmosphere

# 4

*". . . this most excellent canopy, the air, look you, this brave o'erhanging firmament, this majestical roof fretted with golden fire—why, it appears no other thing to me than a foul and pestilent congregation of vapours."*

*Hamlet,* act II, sc. ii.

Imagine that we are viewing the earth from outer space. We see a blue disk with white swirls that largely cover the disk and an occasional brown area below the swirls. It's quite a breathtaking view, and brings us to the realization that the earth is a small planet in a very large universe. If we come in for a closer look, we encounter the atmosphere: a layer of gases surrounding the earth, without which life, as we know it, could not survive. In addition to providing oxygen and carbon dioxide to living organisms, the atmosphere regulates the amounts and kinds of energy that reach the earth's surface from the sun. Through transportation of weather systems, it also provides a medium for the distribution of water over land areas.

What properties of the atmosphere allow it to carry out these functions? To answer these questions and to discuss problems associated with man's use of the atmosphere, we need to know the composition of the atmosphere and the general properties of gases.

## ATMOSPHERIC PROFILE

The atmosphere is a sea of many gases that surrounds the earth. The density of this mixture is greatest near the earth's surface and the air gets "thinner" as the altitude increases. The atmosphere is usually divided into four layers: troposphere, stratosphere, mesosphere, and thermosphere; see Table 4–1. The troposphere is the layer nearest the earth and the only region in which life exists. However, the other layers are also essential to life. These layers perform two functions: filtering out damaging radiation from outer space, originating primarily from the sun, and preventing the large-scale loss of heat due to infrared radiation reflected from the surface of the earth.

The filtering of high-energy radiation is accomplished by a series of chemical reactions that occur in the stratosphere. Here molecular oxygen ($O_2$) decomposes into oxygen atoms by absorbing ultraviolet radiation. This reaction, called a photochemical reaction because it is initiated by light, is the first step in a cycle that continuously filters ultraviolet radiation, preventing it from reaching the lower regions of the atmosphere. Without this process,

---

**United States Steel's Duquesne plant** used to pour 150 tons of particulates and gas per day into the skies. New equipment now washes the smoke and disposes of the particulates. (*Photograph by John R. Shrader, EPA.*)

**Table 4-1**
**Regions of the Atmosphere**

| Region | Height (miles) | Temperature, °F | Chemical Species |
|--------|----------------|-----------------|------------------|
| | Space | $-340$ | |
| Thermosphere | 59 | $-160$ | $NO^+$, $O_2^+$, $O^+$ |
| Mesosphere | | | $NO^+$, $O_2^+$ |
| | 31 | $+32$ | |
| Stratosphere | | | $O_3$ |
| | 7.4 | $-60$ | |
| Troposphere | 0 | $+60$ | $N_2$, $O_2$, $CO_2$, $H_2O$ |

sunburn would be commonplace and severe whenever a person went outside. A description of the filtering process, which includes several chemical reactions, is shown in Figure 4–1. Note that $O_2$ is used in the initial step and produced in the final step, allowing it to start the cycle again. Even though there are relatively few molecules of $O_2$ per cubic foot in the stratosphere, the number is sufficient to maintain the cycle. The production of ozone, $O_3$, also provides additional UV (ultraviolet) filtering in the third step. It is quite remarkable that such a simple cycle plays such a crucial role in protecting the life-supporting region on earth. The opposition to the supersonic transport (SST) was based partially on the belief that the UV filtering process in the stratosphere would be disrupted.

Much of the sunlight that strikes the surface of the earth is reflected back toward space. If it were not for the atmosphere, much of this radiation, basically heat, would be lost to space. The atmosphere traps the reflected radiation, allowing the entire earth to function much like a greenhouse. This effect plus the filtering process mentioned earlier allows conditions suitable for life on earth.

The atmosphere is composed of a mixture of gases. The chemical composition of this mixture varies from day to day and place to place. Changes

$$O_2 + \text{UV light} \longrightarrow O + O$$
$$O_2 + O \longrightarrow O_3$$
$$O_3 + \text{UV light} \longrightarrow O_2 + O$$
$$O_3 + O \longrightarrow O_2 + O_2$$

**Figure 4-1** Filtering of UV radiation in the stratosphere. Note the cycle ultimately converts UV radiation to other forms of energy.

**Table 4-2**
**Chemical Composition of Dry-Air**
**Sample Taken from Near the Earth**

| Compound or Element | Parts per Million[a] (ppm) |
|---|---|
| $N_2$ | 780,900 |
| $O_2$ | 209,400 |
| Ar | 9,300 |
| $CO_2$ | 315 |
| Ne | 18 |
| He | 5.2 |
| $CH_4$ | 1.1–1.2 |
| Kr | 1 |
| NO | 0.5 |
| $H_2$ | 0.5 |
| Xe | 0.08 |
| $NO_2$ | 0.02 |
| $O_3$ | 0.01–0.04 |

[a] Out of one million molecules in a sample of air, 780,900 are $N_2$, 209,400 are $O_2$, and so forth.

depend on meteorological, geographical, and demographical conditions,° but a typical dry-air composition for the troposphere can be determined. This is shown in Table 4-2. (If water were included, it would vary from 10,000 ppm† to 30,000 ppm.) Molecular nitrogen, $N_2$, is by far the most common constituent; in the atmosphere nitrogen is essentially unreactive. Molecular oxygen, the next most common component, plays an essential role in life processes. Notice that $O_3$ is also present in small amounts. Carbon dioxide, $CO_2$, also essential to life, has a concentration of 315 ppm. This seems small when compared with $O_2$ considering the large number of plants that depend on $CO_2$.

Argon, neon, krypton, and xenon appear in this list. They are the inert, or noble, gases and participate in no known chemical reactions in the environment. Methane, $CH_4$, occurs wherever organic materials decay anaerobically.‡ Nitrogen oxides (NO and $NO_2$) are produced by several processes in the environment; these include lightning and volcanic activity.

The relative proportion of the chemicals listed in Table 4-2 has undoubtedly changed with time. The atmosphere of primordial earth is thought to have been largely $CO_2$, with little $O_2$ present. Even over short time periods,

---

° *Demographical conditions* refers to details of population content such as total number of people and their concentration.
† The abbreviation ppm refers to parts per million, a measure of atmospheric concentration.
‡ Without $O_2$ present.

significant changes have occurred. Measurements in Paris indicate a change of 10 percent in the $CO_2$ concentration from 1891 to 1964.[*] This is probably owing to the increased use of fossil fuels, whose major combustion products are $CO_2$ and $H_2O$.

## PROPERTIES OF GASES

More is known about the gaseous state of matter than about solids and liquids. It is possible to explain the behavior of gases in terms of the properties of individual gas molecules.

Gases have played a very important role in the advancement of chemistry. They were among the first pure substances (elements and compounds) isolated. The ability to trap and to measure accurately the volume of gases released during chemical reactions provided seventeenth- and eighteenth-century chemists with their first clues to the composition of matter. Such quantitative measurements, not common in chemistry during that period, led to the conclusion that chemical compounds were unique combinations of elements.

Intuitively, everyone has a feeling for the properties of gases. Aerosol cans carry warnings that at temperatures above 120° they might explode. Everyone has blown up a paper bag and then has smashed it between his hands, with a resounding pop. The rising of dough during the baking of bread is caused by the production and expansion of a gas ($CO_2$) with temperature. To state these observations more concisely, if one heats a gas in a closed container (aerosol can), the pressure will increase. If one decreases the volume of a gas (squeezes a paper bag), the pressure will increase (causing it to pop). If one heats a gas ($CO_2$ in dough), it will expand to a larger volume (causing the dough to rise on cooking). The converse reactions to these processes also occur. If one decreases the temperature of a gas while maintaining the same volume, the pressure decreases. If one lowers the temperature of a gas while maintaining the same pressure, the volume decreases.

These properties of gases help to explain much of the movement of air within the atmosphere. For instance, high and low centers on a weather map refer to localized regions of high pressure and of low pressure. Winds blow horizontally from highs to lows. The motion of air vertically occurs by the rising of warm air masses and the sinking of cooler ones. This occurs because as air warms, it expands, becoming less dense than the air above, and therefore it rises. As air cools, it decreases in volume, becoming more dense than the surrounding air, and it sinks.

For all practical purposes, this is enough information about gases to continue the discussion of the atmosphere. However, to obtain a better understanding of why the atmosphere, and gases in general, behave as they

---

[*] B. Bolin, *Scientific American* 223 (1970), 124.

do let's consider the properties of a gas (pressure, temperature, and volume) in terms of the individual properties of gas molecules.

If careful measurements are made on the behavior of gases with changes in temperature, pressure, and volume, a single, simple algebraic expression is obtained that mathematically expresses gas behavior. This equation is

$$PV = nRT,$$

where $P$ is the pressure, $V$ is the volume, $T$ is the temperature, $n$ is a measure of the number of gas molecules in the sample, and $R$ is a proportionality constant. Volume is measured in liters (one liter is about one quart in volume) and pressure, in atmospheres (one atmosphere equals atmospheric pressure at sea level, or 15 lbs/sq in.). Temperature is measured in degrees absolute. The relation of the absolute, or Kelvin, temperature scale to Fahrenheit and Celsius temperature scales is shown in Figure 4–2. To understand how the number of gas molecules enters into the equation, consider the following example. As air molecules are added to a basketball, the volume the air occupies remains essentially constant but the pressure increases (the ball bounces higher). This equation is called an equation of state because it inter-relates all the macroscopic properties $(P, V, n, T)$ of the gas. *Macroscopic properties* of a gas are those properties of the gas sample as a whole as opposed to the properties of the individual gas molecules, called *microscopic properties*. If the subject under consideration were a quart jar full of beans, we should say that the total volume, the quart, was the macroscopic measurement, and that the volumes of the beans measured individually were the microscopic measurements.

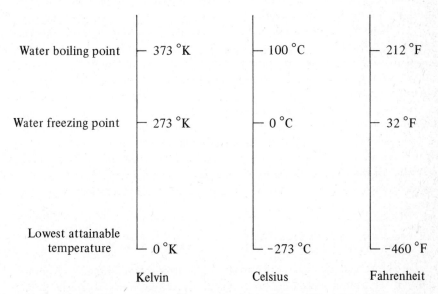

**Figure 4-2** Different temperature scales: $°K = °C + 273$; $°C = \frac{5}{9}(°F - 32)$.

The equation of state is quite remarkable. For a sample of gas, if you are given values for $P$, $V$, and $n$, by using this equation you can calculate the temperature without ever seeing the sample. To put this in perspective, let the sample be a cubic foot of air. This sample contains about $10^{24}$ gas molecules. The simple equation $PV = nRT$ predicts the macroscopic behavior of these $10^{24}$ gas molecules, and the microscopic properties of the individual gas molecules do not even enter into consideration. The complicated behavior of $10^{24}$ molecules is described by one simple equation, a rare occurrence in science.

The equation of state was discovered and verified by a group of experimental scientists. They worked in a laboratory taking measurements, analyzing their results, and drawing conclusions. Another group of scientists, the theoreticians, have also studied the behavior of gases. Theoretical chemists try to explain chemical phenomena in terms of the laws of physics and models of the universe.

For gases, theoreticians proposed a model. They said that gases are composed of widely separated particles (molecules and/or atoms), so that of the total volume occupied by a gas sample, the fraction due to the actual particle volume is small. The particles are supposed to be constantly moving around, colliding with each other and the walls of the container. The speed with which the molecules move depends on the temperature of the gas; the higher the temperature, the faster the movement of the molecules. Next, the theoreticians transformed this model into mathematical equations, and following a long, clever, and intricate derivation, they reached an equation that looked similar to the equation of state for a gas. By a consideration of the microscopic properties of gas molecules, therefore, the macroscopic equation of state $PV = nRT$ was derived.

There now exists a microscopic explanation for the macroscopic behavior of gases. If one heats up a gas in a closed container, the molecules speed up, hit the walls of the container harder and more frequently, increasing the pressure. Compressing the volume of the blown-up paper bag causes the molecules to hit the walls of the bag more often; the pressure increases and the bag pops. Heating the air in a balloon causes the molecules to speed up, thereupon hitting the walls harder and pushing out the walls of the balloon.

Gases are unique in having a simple equation of state. A good equation of state for liquids or solids has not been discovered, experimentally or theoretically. In view of the difficulty of concisely describing the behavior of $10^{24}$ speeding molecules, it is amazing that an equation of state exists even for gases.

## INTRODUCTION TO AIR POLLUTION

This section describes the impact on the atmosphere of man's activities on earth. This impact is described as much as possible from a scientific point of view but other criteria and effects will be included since air pollution also

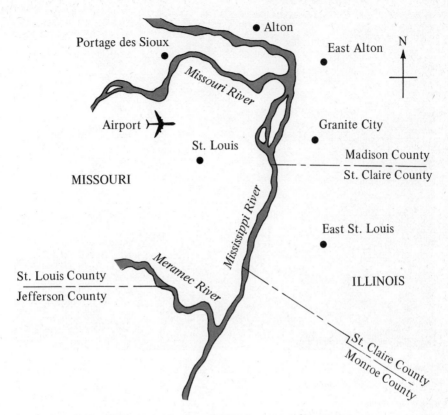

**Figure 4-3** Greater St. Louis metropolitan area.

has a social facet. For a perspective on pollution, a historical section starts the discussion. Throughout this discussion St. Louis, Missouri, will be used as an example. While St. Louis may not be typical of all cities, it does have typical air-pollution problems. Also, the Public Health Service, in the 1960s, made a detailed survey of the impact of air pollution on St. Louis. This PHS survey provides actual figures for an actual city, so the results discussed are real, not hypothetical for some "average" city. In addition, pollution information from other American cities will be discussed and compared with national figures.

## BRIEF HISTORY OF AIR POLLUTION

Today air pollution is front-page news. Statements by political leaders, scientists, environmentalists, and others receive network news coverage daily. Air pollution is not a new problem. It has been with us in various forms and degrees of severity for centuries.

From the beginning of human existence on earth until the fourteenth century, air pollution was a localized phenomenon. Forest fires and volcanoes,

**Figure 4-4**   Even Hawaii has pollution problems. This is a severe case of air pollution in Honolulu. (*ESSA photograph.*)

while producing much pollution, were normally restricted to very small areas. Smoke and soot from heating and cooking were concentrated in the cave or home and produced no widespread discomfort. The population was not concentrated to the degree that pollution was recognized as a problem. If pollution from garbage-dump odors or accumulated soot became too great, one could always move to a cleaner area.

Wood was the primary energy source during the centuries before 1300. It was close by and plentiful, and as long as the population of a community was not too large, it could be distributed with relative ease. But with enlarged populations, the delivery of wood became more difficult; the sources were farther away, and the amount of heat per pound of wood as compared with coal was small. Therefore coal slowly became the primary energy source. This change took time. Coal with its sulfurous odors was thought to be connected with witchcraft; and certain laws in Europe restricted, and even prohibited, its use.

By the year 1300, coal, with its large energy per pound and easy access, was providing the energy needs of a growing European population. It was also providing odors, smoke, and soot. Protests against the use of coal were recorded in the early 1300s and occasional enforcement for misuse was carried out during this period. By 1400, regulations on the use of coal were established in Britain, but little more was done during the ensuing centuries. In 1660,

John Evelyn, a founder of the Royal Society of London (a scientific society), made a survey of air pollution in London. His report outlined all the main pollutants and their sources, and possible remedies. Air-pollution-control devices were proposed during the same century. During the early 1800s, the pollution problem was so prevalent that a parliamentary committee was established to initiate a smoke abatement program. The problem was studied in committee for 130 years, until in 1952 the London killer fog finally moved people to the realization that pollution must be put under control.

London (1952) was not the first air pollution "episode." Examination of vital statistics indicates that similar episodes may have occurred during 1873 in London and 1909 in Glasgow. These large-scale killer episodes indicate that citizens and their governments in the past have not been willing to accept the inconvenience and costs for effective air pollution control.

The history of air pollution in American cities is shorter than that for London, but it is similar. The histories follow a typical pattern. In America, because of the vast coal fields east of the Mississippi River, industry and trade flourished by the end of the nineteenth century. Black smoke in the air was a sign of business activity and of unparalleled prosperity for the citizens. Many were immigrants and the opportunities for work overrode the disadvantages of the smoke and odors.

Initial attempts at smoke abatement were not very successful. In the late 1800s, ordinances prohibiting the emission of black smoke were passed but later declared unconstitutional. In the period between 1900 and 1940, different antismoke ordinances were passed, but enforcement was hampered by financial constraints. Smoke-abatement leagues raised money to initiate a citizens' education campaign, and by the 1940s effective laws were passed. They regulated the quality and size of coal to be used. They also outlawed the use of coal containing highly volatile constituents. The results were quite dramatic and the situation improved rapidly. But an increasing population and more industry required that tighter controls be instituted to maintain acceptable air quality. Low-sulfur coal had to be used and it had to be washed to rid it of dust. Gas release regulation and dust control ordinances were enacted. For a time this was sufficient. By the early sixties, it became obvious that air quality was slipping. The menacing agents now included sulfur dioxide ($SO_2$), fine aerosols, $NO_2$, and hydrocarbons. At this point, the United States Public Health Service surveys were made. Steps are now being taken to try to improve air quality. The future will undoubtedly bring new pollutants with new problems, and in time, new regulations will be enacted to curb them. As long as the population concentration increases and people desire a higher standard of living, this problem will continue.

## AIR POLLUTANTS

An intelligent approach to the study of air pollution requires some idea of what constitutes an air pollutant and what the common pollutants are. Air pollutants are substances released into the atmosphere in sufficient concentration to produce a measurable effect on animals, plants, or materials. The measurable effects may be physical, such as lung damage, or aesthetic, such as odor nuisance. These effects may be due to a new substance released into the air or to increased concentrations of normal constituents of the atmosphere. Alterations in the normal atmospheric composition upset natural cycles in the environment and undoubtedly affect some chemical processes in living organisms. Pollutants fall into two principal categories: primary pollutants and secondary pollutants.

**Primary pollutants** are emitted directly from identifiable sources. Particulates, which consist of fine and coarse solid particles, are emitted from natural and man-made sources. Bacteria, fungi, pollen, and organic compounds are all from living sources. Carbon (soot), metal particles, tars, resins, asbestos fibers, and many other particulates arise from the mechanical devices and processes that are so much a part of modern society. These fine solids may carry an electrical charge, cause water condensation, and form aerosols that limit visibility. They may be inherently toxic, or able to cause mechanical damage to living systems. In man, the mucous membranes cannot filter them out. In addition, they cause soiling and deterioration of buildings, clothes, and other materials. Coarse solids cause similar problems, but settle from the atmosphere quickly and cannot ordinarily reach the lungs of animals. However, particulates are important pollutants, because although they are small in volume, they have tremendous surface area. Instead of being smooth, they have jagged and deeply cut surfaces. The large surface area is important in the production of secondary pollutants.

There is a wide variety of primary chemical pollutants, most of which are molecular gases. Among these substances are sulfur-containing compounds including sulfur dioxide ($SO_2$), sulfur trioxide ($SO_3$), hydrogen sulfide ($H_2S$), and sulfur-containing organic molecules. The oxides of sulfur have a suffocating odor and combine with atmospheric water and oxygen to form sulfurous acid ($H_2SO_3$) and sulfuric acid ($H_2SO_4$):

$$SO_2 + H_2O \longrightarrow H_2SO_3;$$

$$SO_3 + H_2O \longrightarrow H_2SO_4.$$

Hydrogen sulfide ($H_2S$) is a poisonous gas that has played a role in several air pollution episodes. Organosulfur compounds, many of which are emitted in industrial processes such as petroleum refining and paper production, have unpleasant odors. The unpleasant odors of onion and skunk are due to organosulfur compounds. The newspapers report $SO_X$ as a pollutant; this term refers to the combined amounts of $SO_2$ and $SO_3$ and is called total sulfur oxides.

Organic compounds form another group of primary chemical pollutants. Hydrocarbons such as those contained in turpentine, gasoline, and kerosene are important members of this group. Hydrocarbons used as solvents in paints, as cleansing agents, and as fuels play an important role in the formation of secondary pollutants. Organic compounds called aldehydes and ketones are included in the general term *oxidants,* used to designate compounds that are reactive in the atmosphere. Many organic pollutants containing benzene rings, such as pyrene (see p. 71), are thought to be carcinogenic (cancer-producing).

The oxides of carbon and nitrogen are important primary pollutants. Carbon monoxide (CO) results from incomplete combustion of fossil fuels, that is, the burning of the fuels under the condition of a limited oxygen supply. Carbon dioxide ($CO_2$), an important natural constituent of the atmosphere, is produced in large amounts by fossil-fuel combustion.

$$2CH_4 + 3O_2 \text{ (limited)} \longrightarrow 2CO + 4H_2O$$

$$CH_4 + 2O_2 \text{ (excess)} \longrightarrow CO_2 + 2H_2O$$

Nitrogen oxide (NO) and nitrogen dioxide ($NO_2$) are formed under the conditions of high-temperature combustion. The source of nitrogen for the pollutants may be organic nitrogen compounds in fossil fuels or molecular nitrogen in the atmosphere. Nitrogen dioxide, the only important pollutant with a distinct color (yellow brown), reacts with water vapor to produce nitric acid and nitrous acid.

$$2NO_2 + H_2O \longrightarrow HNO_3 + HNO_2$$

There are many other primary chemical pollutants. Hydrogen fluoride (HF) and hydrogen chloride (HCl) as well as metallic fluorides are emitted by metallurgical processes. Lead is produced in the air when leaded gasolines are burned. Usually emitted as a fine particulate, lead can reach the lungs easily. The list could be continued, but the important pollutants, at least important at this time, have been mentioned. In the future, new chemicals from new processes will be eligible for mention on the list. Radioactive gases, for instance, will increase in concentration with the advent of nuclear power production.

**Secondary pollutants** are produced in the atmosphere by the reaction of primary pollutants with each other and with the constituents normally found in the atmosphere. The identity of most secondary pollutants is not known, but a few examples can be discussed. Most primary pollutants are present in the atmosphere in concentrations of 1 to 60 ppm. Any process that will concentrate these pollutants will enhance the number of successful chemical reactions that can occur. Particulates are perfectly designed to do this; they act much like charcoal filters in concentrating impurities by adsorbing them. The large surface area of fine particulates is responsible for this concentrating effect. Concentration of primary pollutants also occurs in water droplets on environmental surfaces.

Most of the reactions in the atmosphere require a little energy to get started. This initial kick is usually supplied by light. One such initial step is a photochemical decomposition of $NO_2$, similar to the UV filtering by $O_2$ in the stratosphere, except that the light involved is visible light. This reaction can be written as follows:

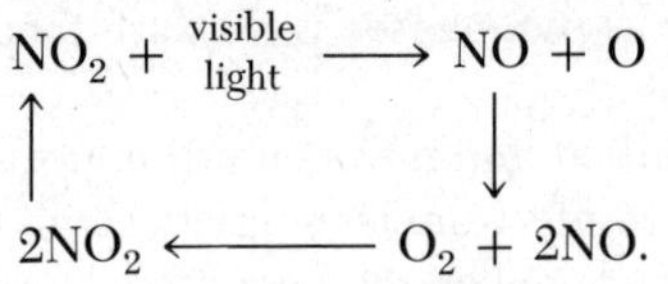

$$NO_2 + \overset{\text{visible}}{\text{light}} \longrightarrow NO + O$$

$$2NO_2 \longleftarrow O_2 + 2NO.$$

Notice that the reaction is one part of a cycle that continuously produces oxygen atoms.

This reaction is extremely important because it is the source of atomic oxygen in the atmosphere. The normal oxygen found in the air is $O_2$. Once oxygen atoms are formed, a large number of reactions can occur. The production of ozone ($O_3$) and sulfur trioxide ($SO_3$) are typical examples:

$$O_2 + O \longrightarrow O_3;$$

$$SO_2 + O \longrightarrow SO_3.$$

Chain reactions involving oxygen atoms, hydrocarbons, and other organic pollutants are also initiated. The net result of these chain reactions is smog, sometimes referred to as photochemical smog because of the role that light plays in initiating its formation. Aldehydes and ketones also play an important role in photochemical atmospheric reactions; they are formed by the reaction of ozone with alkenes.

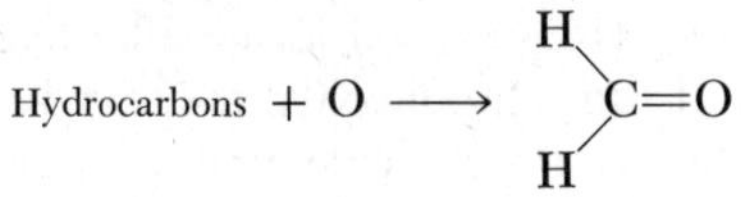

$$\text{Hydrocarbons} + O \longrightarrow \quad \overset{H}{\underset{H}{\diagdown}} C{=}O$$

Formaldehyde

$$or \quad \overset{H}{\underset{H}{\diagdown}}C{=}C\overset{\overset{\displaystyle H}{\diagup}}{\underset{H}{\diagdown}}C{=}O$$

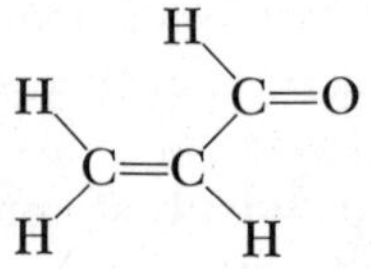

Acrolein

Very reactive peroxides and a very irritating class of compounds called PAN (peroxyacyl nitrate) also result from atmospheric reactions and induce smog formation.

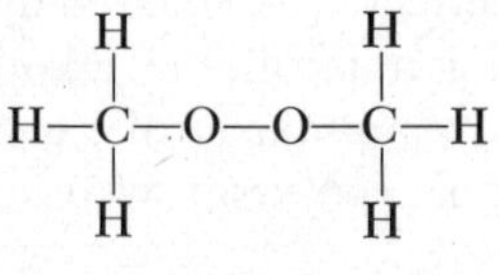

$$H-\overset{\displaystyle H}{\underset{\displaystyle H}{C}}-O-O-\overset{\displaystyle H}{\underset{\displaystyle H}{C}}-H \qquad -\overset{}{\underset{}{C}}-\overset{\displaystyle O}{\overset{\|}{C}}-O-O-NO_2$$

A peroxide 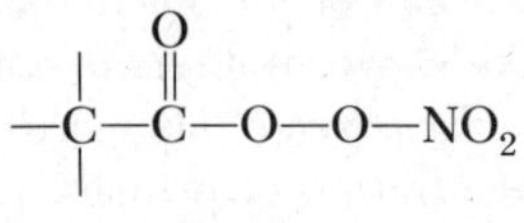                    PAN

Both primary and secondary pollutants cause problems in the environment. Primary pollutants are the easiest to identify and control, because their source can be found and regulated. Secondary pollutants are quite complex in nature and virtually impossible to control once the primary pollutants get into the atmosphere. It is not surprising that vigorous controls on the sources of primary pollutants represent the chief thrust of efforts to improve the quality of the air.

## SOURCES OF AIR POLLUTION

The primary pollutants just discussed were emitted by identifiable sources within the biosphere. Information concerning the amounts and sources of pollutants puts in perspective those pollution sources that must be controlled. It is important to keep in mind that the list of pollutants and their sources and amounts change with time as new chemical processes are developed within the environment. In general terms, air pollution results from the activities of individuals as well as of industry, the utilities, and the government. Different sources are primarily responsible for different pollutants.

The emission inventory determined for St. Louis is a specific case history. The St. Louis metropolitan area covers two states, six counties, over 120 cities and towns, and many overlapping miscellaneous political districts (see Figure 4–4). Of the 3,600 square miles in the area, 95 percent of the population and most industrial activity is located in a 400-square-mile area at the center of metropolitan St. Louis. Similar population profiles are found in other metropolitan areas. Industries in St. Louis include two cement plants, two big chemical companies, four big oil refineries, two big steel plants, eight steel foundries, and several grain-processing plants. Five steam-electric utilities provide the electricity for the area. Transportation (mainly ground vehicles), municipal incineration, solvent evaporation, and many other activities can be identified as pollution sources.

The results of the air-pollutant-emission inventory are summarized in Tables 4–3 and 4–4. Carbon monoxide (CO) is by far the largest pollutant, representing 50 percent (1,390,000 tons) of the pollutants listed. In a consideration of abatement procedures, the following points are of interest.

1. Of the total particulates, 59% are emitted from the combustion of fuel (56% from coal), 25% from industrial processes, and 11% from the open burning of refuse.
2. More than 90% of sulfur oxides come from the burning of coal (87%) and fuel oil (3%).
3. Gasoline and diesel-powered motor vehicles emit 77% of the total CO pollution.
4. Motor vehicles emit 62% of the hydrocarbons released into the environment and open burning of refuse emits 23%.
5. Nitrogen oxides are discharged from the burning of coal (51%), transportation sources (35%), and the burning of natural gas (7%).

**Table 4-3**
**Air Pollutants and Sources for Metropolitan St. Louis, Missouri**
**Percentage of Emissions of Each Type According to Source**

| Source | CO | Hydrocarbon | $NO_x{}^a$ | $SO_x{}^a$ | Particulates |
|---|---|---|---|---|---|
| Transportation | 77% | 62% | 35% | 1% | 5% |
| Combustion of fuels (stationary sources) | 2 | 2 | 62 | 93 | 59 |
| Solid waste disposal | 1 | 23 | <1 | <1 | 11 |
| Industrial processes | 20 | 3 | 3 | 6 | 25 |
| Solvent evaporation | ..... | 10 | ..... | ..... | ..... |
| Total pollutants | | | | | |
| Percentage | 55% | 15% | 6% | 18% | 6% |
| Thousands of tons per year | 1,390 | 374 | 138 | 455 | 147 |

SOURCE: Interstate Air Pollution Study.
$^a NO_x$ = total oxides of nitrogen; $SO_x$ = total oxides of sulfur.

Table 4-4 presents a more detailed summary of emissions from *stationary* combustion sources and refuse burning. Stationary combustion sources are fixed in one place. Of these sources, power production is a principal polluter, contributing 58 percent of $SO_2$ and 22 percent of particulates. Industry produces most of the particulates; this results from a wide range of activities such as metal working, material transfers, milling, grinding, and drying operations. Residential sources produce the most CO, some $SO_2$, and particulates. Open burning is primarily responsible for aldehydes from stationary sources.

The relative importance of pollution sources varies from city to city. Each city has its own mixture of industry, refuse disposal methods, modes

**Table 4-4**
**Detail of Table 4-3. Air Pollutants from Combustion of Fuel and**
**Refuse from Stationary Sources in Metropolitan St. Louis, Missouri**
**Percentage of Each Type According to Source**

| Source | CO | Hydrocarbons | $NO_x$ | $SO_x$ | Particulates |
|---|---|---|---|---|---|
| Industry | 11 | 1 | 26 | 27 | 38 |
| Steam-Electric | 4 | <1 | 62 | 58 | 22 |
| Residential | 71 | 4 | 9 | 12 | 18 |
| Incineration | 1 | 1 | <1 | <1 | 1 |
| Open burning | ... | 93 | <1 | <1 | 15 |
| Other | 13 | 1 | 2 | 3 | 6 |

SOURCE: Interstate Air Pollution Study.

**Table 4-5**
**Variations in Source of Carbon Monoxide by City**
**Percentage of Carbon Monoxide by Source**

| City | Transportation | Stationary Fuel Combustion | Industrial | Solid Waste Disposal |
|------|----------------|----------------------------|------------|----------------------|
| Boston | 90.3 | 1.3 | Negligible | 8.4 |
| Chicago | 90.4 | 5.1 | 2.4 | 2.1 |
| Denver | 93.8 | 0.7 | 4.0 | 1.5 |
| Los Angeles | 98.4 | 0.6 | Negligible | 1.0 |
| New York | 95.5 | 1.1 | 3.0 | 0.4 |
| Philadelphia | 59.4 | 1.4 | 26.8 | 2.4 |
| St. Louis | 77.4 | 1.7 | 19.9 | 1.0 |
| Washington, D.C. | 99.0 | 0.5 | Negligible | 0.5 |

of transportation (cars versus mass transit) and methods and location of electrical and steam-power generation. Table 4–5 lists some large cities and the relative importance of various CO pollution sources. The variation is wide. In Washington, D.C., transportation sources emit 99 percent of the CO; in Philadelphia only 59 percent. Chicago appears to have a substantial CO problem from stationary fuel combustion, and Boston from solid waste disposal.

Table 4–6 presents national air pollution sources and includes several nonindustrial, nonurban factors such as forest fires and agricultural burning, and gasoline marketing (about 4 percent of hydrocarbon emissions). If these

**Table 4-6**
**Air Pollutants and Their Sources in the United States**
**Percentage of Each Type According to Source**

| Source | CO | Hydrocarbons | $NO_X$ | $SO_X$ |
|--------|-----|--------------|--------|--------|
| Transportation | 63% | 52% | 40% | Negligible |
| Combustion of fuels (stationary sources) | 2 | 2 | 48 | 78% |
| Solid waste disposal | 8 | 5 | 4 | $<1$ |
| Industrial processes | 11 | 14 | Negligible | 22 |
| Forest fires and agricultural burning | 7 | 12 | 6 | Negligible |
| Miscellaneous | 9 | 15[a] | 2 | Negligible |
| Total pollutants | | | | |
|    Percentage | 55% | 17% | 12% | 16% |
|    Millions of tons per year | 102 | 32 | 21 | 29 |

[a] Includes 10% from solvent evaporation.

pollution sources were spread out evenly over the country, pollution would be much less noticeable. The concentration of these sources in urban areas magnifies the problem and promotes the formation of significant amounts of photochemical smog.

What is unique about each source and the type of pollutant it emits? Power companies burn coal. Coal contains a certain amount of material that won't burn, which is given off as particulates during combustion. Coal also contains sulfur, which is converted to $SO_2$ during combustion. This explains the high percentage of particulates and $SO_x$ due to stationary sources. Sulfur dioxide from residential sources is due primarily to oil-burning furnaces since sulfur-containing impurities are also found in oil. Nitrogen oxides are produced by both transportation and stationary combustion. Both of these sources have one thing in common, the high temperatures produced during the burning process. The heat produced is enough to rupture the $N_2$ molecule, which is chemically inert under normal environmental conditions. Nitrogen atoms can then react with oxygen to produce NO and $NO_2$. Open burning and municipal incineration, by their very nature, involve the release of particulates and a wide variety of organic compounds.

Transportation, on a national basis typified by the family car, is responsible for 63 percent of the CO, 52 percent of the hydrocarbons, and 40 percent of nitrogen oxides, and use of the automobiles will increase in the future. Diesel and gasoline trucks as well as airplanes are also included in the transportation category as contributors to pollution, with the problems associated with diesel power differing slightly from those with gasoline power. (See Figure 4–5.) Air pollution from the automobile begins before the engine is started. While the car is just sitting there, evaporation of gasoline from the fuel tank and carburetor accounts for over 20 percent of the hydrocarbons emitted by the car. This amounts to about one gallon of gasoline per month. It is unsatisfactory to seal the fuel system, because the pressure of the gasoline vapors in the gas tank would increase considerably as the temperature increases. As a measure in counteracting this loss, fuel vapor collection systems have been developed to collect the vapor and use it as fuel when the car is started.

The power plant of the automobile is the internal combustion engine (ICE). Internal combustion refers to the fact that the combustion process occurs within the engine and that it is not open to the atmosphere. The operation of the ICE is shown in Figure 4–6. Twenty percent of hydrocarbon emissions occur when hydrocarbons escape during the compression and power strokes. Some of the fuel mixture escapes between the cylinder wall and the piston. This fuel and some evaporated oil are referred to as crankcase emissions. Positive crankcase ventilation (PCV) valves are common on cars since the mid-sixties, and are designed to trap crankcase emissions and divert them into the carburetor for combustion. These valves work well as long as they are maintained properly. The remaining 60 percent of hydrocarbon emissions result from unburned fuel discharged into the exhaust system, some of which is partially burned giving rise to the large amount of aldehydes emitted.

The exhaust contains the remainder of the pollutants. Carbon dioxide and water are the normal combustion products and are produced in large quantities. Carbon monoxide results when gasoline is not completely burned. Nitrogen oxides, produced in the heat of the power stroke, are also released. A variety of other pollutants are also emitted. Additives, either in the gas or added extra, contain sulfur and other elements. Leaded gasoline results in lead emissions. Figure 4–7 shows a typical street scene.

Engineers are approaching the problem of automotive air pollution in a variety of ways. They are redesigning engines so as to provide smaller cylinder surface areas, to remove pockets in the cylinder and piston heads, and to allow the use of leaner fuel mixtures. The first two steps would reduce hydrocarbon and $NO_x$ emissions; leaner fuel mixtures mean more oxygen in the cylinder and less partial combustion. Engineers are also conducting experiments with different fuels, such as liquid propane gas. The use of catalytic mufflers, devices designed to complete combustion, could reduce or eliminate CO and hydrocarbon emissions; a similar reduction in nitrogen oxides emissions can also be obtained.

The removal of lead from gasolines is significant for two reasons. Lead emissions poison or foul catalytic mufflers, and moreover, lead itself is a pollutant. To understand why lead is added to fuels, it is necessary to define

**Figure 4-5** Visible or not, airplane emissions will continue. Federal regulations due to be enforced after January 1, 1973, should reduce the visible emissions from aircraft; but not all emissions are visible. (*Reprinted by permission of* Environment. *Photograph by Robert Charles Smith.*)

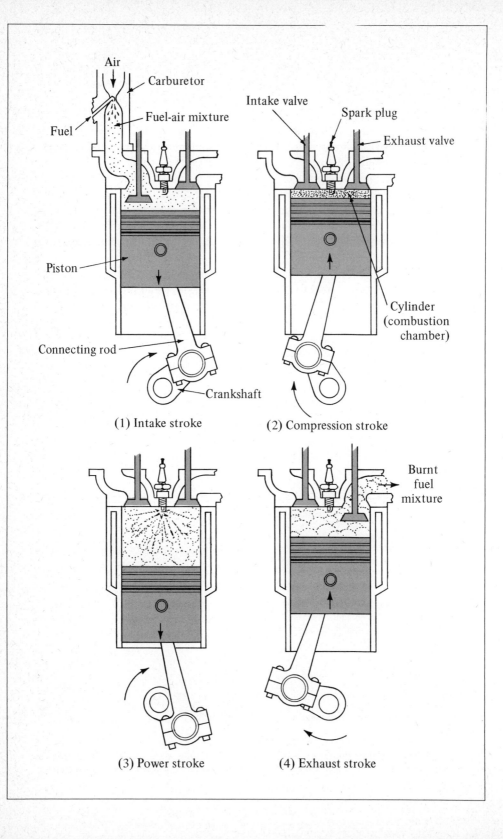

Air

Carburetor

Fuel-air mixture

Fuel

Piston

Connecting rod

Crankshaft

Intake valve

Spark plug

Exhaust valve

Cylinder (combustion chamber)

(1) Intake stroke

(2) Compression stroke

Burnt fuel mixture

(3) Power stroke

(4) Exhaust stroke

**Figure 4-7** Much air pollution results from untuned cars. (*Reprinted, by permission, from* St. Louis Post-Dispatch. *Photograph by Ferguson.*)

octane ratings. The **octane rating** of a fuel is determined by comparing the performance of the fuel to the performance of a pure hydrocarbon, isooctane, in a standard experimental engine. Pure isooctane has an octane rating of 100;

$$
\begin{array}{ccc}
 & \overset{\displaystyle CH_3}{|} & \overset{\displaystyle H}{|} \\
CH_3{-}\overset{|}{\underset{|}{C}}{-}CH_2{-}\overset{|}{\underset{|}{C}}{-}CH_3 \\
 & CH_3 & CH_3
\end{array}
$$

Isooctane, $C_8H_{18}$

most hydrocarbons have an octane rating below 100 (their performance is poorer than isooctane), but aromatic hydrocarbons have octane ratings as high as 120. Gasolines sold in the United States have octane ratings in the range of 90 to 110. The gas used depends on the engine design; for instance, an engine with a compression ratio of 10:1 (a measure of engine design) needs a 98-octane gasoline. Gasolines are a mixture of hydrocarbons and it was found

**Figure 4-6** One cylinder of auto engine during combustion. During the intake stroke, air and gas are drawn into the cylinder. The compression stroke compresses the air and gas, and upon ignition the hot gas expands, driving the piston during the power stroke. The gases are then compressed again and pushed out of the cylinder.

that the addition of three grams of tetraethyl lead, $Pb(CH_2CH_3)_4$, per gallon of gasoline raises the octane rating of the mixture about ten points. It was felt that, in addition, lead helped the engine run more smoothly.

The production of nonleaded gasolines with octane ratings high enough for today's cars requires the use of hydrocarbons with high octane ratings in the gas mixture. Basically this means adding more aromatic hydrocarbons, which cost more to produce. The big oil companies now produce lead-free regular gasolines at a slightly higher cost (5%–10%) than leaded regulars. Unfortunately, nonleaded gasolines have their drawbacks. Recent tests indicate that high-aromatic unleaded gasolines produce 2.5 times as many aromatic emissions as leaded low-aromatic gasolines. These aromatics have proven carcinogenic properties and they facilitate smog production. This is another example of the tradeoffs involved in pollution control: one problem is cured but another is created.

## EFFECTS OF AIR POLLUTION

Many detrimental consequences are attributed to air pollution; but often the relationship between a pollutant and its effects on plants, animals, and property is only qualitative. Furthermore it is difficult to separate the contribution of air pollution in a given instance from the contribution of overcrowding, occupation, personal habits, and weather, say to lung disease or property damage. Often air quality standards are based on esthetic considerations; for instance, people react much more strongly to harmless, strong odors in the air than to more dangerous but odorless pollutants. But concern is also voiced about the damage to health and property caused by chemical agents in the air.

The different effects of an air pollutant are ordinarily associated with different concentrations of the pollutant. Maximum allowed concentrations, MAC for short, are usually set on a citywide, regional, or statewide basis and used to signal dangerous pollutant levels and emergency abatement measures such as limiting auto use and closing industrial plants.

Sulfur oxides aggravate and contribute to the development of respiratory disease. In concentrations of 10–20 pphm (part per hundred million), $SO_2$ aggravates pulmonary disorders. The formation of sulfuric acid ($H_2SO_4$) is a result of atmospheric reactions involving water and sulfur oxides as well as industrial processes. Sulfuric acid, in concentrations found in the air, accelerates metal corrosion by as much as 250 percent. Sulfur oxides also cause damage to statuary, leather, textiles, and paint. In smaller concentrations, $SO_2$ causes changes in reflexes and activities of the brain. Detectable in concentrations as small as 0.3 ppm by taste and smell, $SO_2$ typically has a MAC of 0.3 ppm.

Carbon monoxide in concentrations of 2,000 ppm causes death by interfering with the distribution of oxygen in the body. Hemoglobin, a constituent of blood, normally carries an oxygen molecule, but in the presence of CO, it picks up CO and becomes carboxyhemoglobin. At 30 ppm for 8 hours, about 10 percent of hemoglobin is converted to carboxyhemoglobin; this level affects the ability to think and see clearly. It is the air quality standard for CO in the St. Louis area. In traffic or downtown areas, the CO concentration will be even higher. Carbon monoxide also increases susceptibility to certain diseases. Normal average levels are between 10 and 20 ppm, and the typical MAC is 100 ppm.

Nitrogen dioxide in concentrations of 10 pphm gives rise to the brown tinge in smog and reduces visibility. Nitrogen oxides damage vegetation and perhaps cause damage to the alveoli, the small air sacs in the lung. At concentrations of 13 ppm, eye and nasal irritation are apparent. Pulmonary discomfort and bronchiolitis are symptoms of higher concentrations. The typical MAC for $NO_2$ is 5 ppm; odor detection is 1–3 ppm.

Ozone, characterized by a piercing odor at 0.02–0.05 ppm, causes damage to biological tissues similar to the damage caused by radioactivity; it is also damaging to rubber products and other substances containing carbon-carbon multiple bonds. In concentrations of 0.1–1.0 ppm, ozone irritates the upper respiratory tract and eyes. Ozone also causes chromosomal changes. Ozone has a typical MAC of 0.5 ppm.

Particulates also cause health problems. Visibility is reduced when particulate concentrations are high; air service at metropolitan airports has been interrupted for this reason. People with asthma and sensitivity to aeroallergens (airborne materials causing allergies) suffer from particulates in the air. Damage to vegetation arises when leaves are coated by dust so thick that light cannot reach the leaf surface and induce photosynthesis. Lead particulates from the burning of leaded gasolines are coming under increasing scrutiny as the source of chronic or subchronic levels of lead poisoning in city dwellers. Rush-hour levels of 54 ppb (parts per billion) have been recorded in Los Angeles and the MAC of 1–2 ppb has been exceeded in cities from Richmond, Virginia, to Fairbanks, Alaska. (See Figure 4–8 relative to Denver.)

Hydrogen fluoride (HF), a halogen acid released in some manufacturing, and particularly noted for damage to vegetation and animal life, settles on vegetation, hindering its growth. Animals then eat the vegetation and the fluoride causes fluorosis, a disease characterized by abnormal skeletal growth and deteriorating health.

When two or more pollutants are present in the atmosphere, their combined effects may be quite different from their separate effects. While most effects are additive, several important combinations are not. An antagonistic pair of pollutants has a total effect less than that of the sum of the individual pollutants. For example, HF or HCl mists are damaging to biological materials; in the presence of $NH_3$, their effects are greatly reduced. A

**Figure 4-8** Located at the eastern edge of the Rockies, Denver has pollution problems. (*EPA photograph.*)

synergistic relation exists between two pollutants when their combined effect is greater than the sum of their individual effects; $H_2S$ and $CO$ as a pair are fatal to mice in concentrations that for each alone would produce no toxic effects. Individually, sulfur dioxide and particulates cause respiratory problems; but when they are together, the particulates concentrate the $SO_2$ on their surfaces. Thus when they are inhaled, a high localized concentration of $SO_2$ occurs. Many other synergistic effects probably exist and are undoubtedly of a very complicated nature.

Within the biosphere, that part of the earth in which life can survive, large-scale chemical exchanges are constantly occurring. These exchanges occur as cycles, which carry chemicals necessary to life throughout the biosphere and put them at the disposal of those organisms that need them. Two of these cycles are the carbon cycle and the nitrogen cycle; others exist, such as the oxygen and the water cycles. These are not independent cycles, but act more like coupled cogwheels; the turning of one affects the operations of the others. The interrelations are often subtle and occur on such a grand scale that a thousand years of careful observations might not completely reveal the interconnections. Throughout the development of the biosphere these cycles have evolved into a state of equilibrium, which maintains for the biosphere its life-supporting ability. Upsetting these cycles could conceivably alter those forms of life the earth can support.

The carbon cycle is shown in Figure 4-9; it traces the main routes by which carbon is fixed in living matter, that is, the ways in which carbon is assimilated and combined chemically in plants and animals. This assimilation is absolutely essential since carbon is the chief element in organic materials, providing the backbone for all molecules of biological importance. Proteins, carbohydrates, and genetic materials are examples of such molecules. The carbon cycle could easily be divided into two subcycles: one that occurs between the atmosphere and land life, and the other between the ocean and sea life. The subcycles are tied together at the atmosphere-ocean interface through the exchange of $CO_2$. Most of the earth's carbon is tied up as calcium carbonate sediments buried under the land and seas or as coal and oil. The remainder can be found in living organisms, the atmosphere, and the oceans, and as decaying organic matter.

Carbon is fixed by plants through the process called photosynthesis. Plants use energy from the sun to convert $CO_2$ in the atmosphere to organic compounds in the plant, releasing molecular oxygen, $O_2$. These organic compounds are incorporated into structural materials—cellulose, lignin, and proteins—and into energy sources, such as fats and carbohydrates. The plants may then die or be eaten. Dead plants decay and release $CO_2$ back to the atmosphere. If eaten by animals, the organic compounds synthesized by the plants are used by animals to produce essential chemicals necessary for life

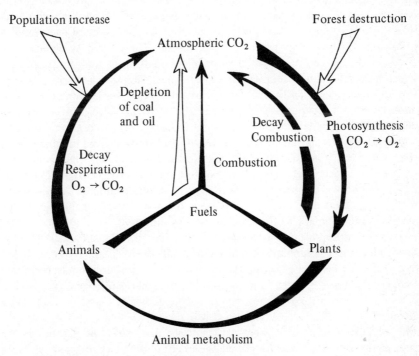

**Figure 4-9** Carbon cycle (———➤); outside perturbations (═══➤).

and energy. The plants might also be used for a variety of other purposes: housing, clothing, paper, and so on. Animal metabolism takes $O_2$ produced by photosynthesis and returns $CO_2$ to the atmosphere, thus completing the cycle. Plant materials and animal waste products are often used as fuels; burning these along with coal and oil releases $CO_2$ into the atmosphere.

Man's activities on earth affect the carbon cycle mainly through the release of $CO_2$ into the atmosphere. The past eighty years have seen a dramatic increase in the rate of consumption of coal and oil and an associated rise in the atmospheric $CO_2$ concentration, a rise of about ten percent. It is estimated by projection from current statistics that the $CO_2$ concentration will reach 375 ppm by the year 2000, compared with the present 315 ppm. The amount of $CO_2$ released in the atmosphere by burning fossil fuels should raise the concentration by 2 ppm annually, but the observed rise is only 0.7 ppm. The remaining 1.3 ppm either is assimilated by an ever growing mass of terrestrial plants and bacteria or is being absorbed by the ocean and deposited as carbonates. It appears that the biosphere can react to increased $CO_2$ concentration but it cannot completely relieve the stress. It is estimated that since the oceans react to stress very slowly, requiring about 1,000 years to establish a new equilibrium, the long-range effects of increased $CO_2$ concentration on the carbon cycle are very difficult to predict. One immediate effect of increased $CO_2$ concentration is its role in blocking the escape of radiant heat from the earth to space. Energy from the sun in the form of light is converted by the earth to heat. Carbon dioxide is primarily responsible for retaining this heat by absorbing it, and an increased amount of $CO_2$ will retain an even greater proportion of the energy received from the sun. The possibility of increased atmospheric temperatures could result in the melting of the polar ice caps, which would flood all occupied coastal lowlands. The atmospheric temperature has decreased slightly during the last fifty years despite increased $CO_2$ concentration. This decrease is believed to be due to an increase in particulate concentration that is preventing sunlight from reaching the earth's surface. An increasing population and the destruction of forests also aid in unbalancing this cycle. The uneven redistribution of dead and decaying plant materials through garbage collection and other waste disposal methods will undoubtedly have an effect on the carbon cycle. Since the reaction time of the biosphere is slow, mankind may be able to correct the ways in which he upsets the carbon cycle or adapt to the changes he has initiated.

The nitrogen cycle is shown in Figure 4–10. Like the carbon cycle, this cycle causes compounds to be made available to living systems. In the case of the nitrogen cycle, the end products are nitrogen compounds, including inorganic nitrates, nitrites, and proteins. Atmospheric nitrogen, $N_2$, is converted to simple nitrogen compounds—nitrates, nitrites, and ammonia—by nitrogen-fixing bacteria (bacteria capable of direct conversion of $N_2$ to nitrates) and by atmospheric processes. Atmospheric processes like lightning and volcanic eruptions account for about 10 percent of the total nitrogen currently fixed each year. Nitrogen-fixing bacteria and other organisms account for about

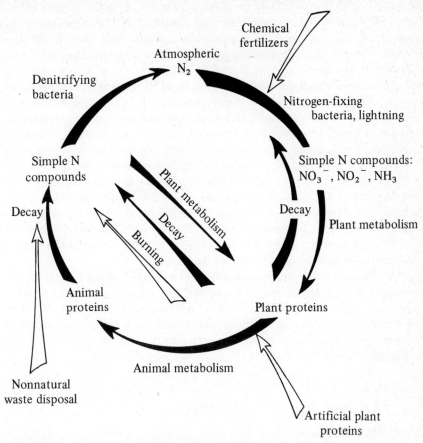

**Figure 4-10** Nitrogen cycle ( ——▶ ); outside perturbations ( ⇒ ).

43 percent of nitrogen fixation; the process by which organisms fix $N_2$ is now being investigated. Nitrogen is also fixed by industrial processes (32 percent) and the widespread use of legume crops (15 percent), which act as hosts for nitrogen-fixing bacteria. Farmers have discovered that planting crops such as beans and peas (legumes) adds nitrogen to the soil.

The simple inorganic compounds put in the soil by the fixation process are incorporated into plants through normal metabolic processes, appearing in amino acids and hence in proteins and genetic material. Plant material is then eaten by animals, or it decays; decay returns simple nitrogen compounds to the soil. Animals need plant protein to survive since animals are incapable of synthesizing all the compounds they need. To continue the cycle, animals give off waste products, are eaten by other animals, and eventually die, creating decay products that return nitrogen to the soil. Burning organic materials also releases nitrogen oxides into the atmosphere. In addition to the fixing of $N_2$, a wide variety of biochemical reactions occur involving nitrogen, which is released into the air.

It is estimated that man, through synthetic $N_2$ fixing processes and legume planting, is fixing about nine million metric tons more of nitrogen than is released in denitrification processes. Chemical fertilizers have been heavily blamed for the eutrophication of inland waters and localized regions of coastal areas. Waste-disposal processes tend to prevent nitrates from returning to the soil by natural decay processes, forcing extra fertilization. Farmers have found that fields rich in nitrates through legume planting are quickly stripped of other nutrients, such as trace metals; this necessitates additional fertilization. Artificial plant-protein synthesis has been suggested and tried as a possible protein source for highly populated areas; if it becomes widespread, this practice would further increase the amount of nitrogen fixed by man. As in the case of the carbon cycle, the long-range effect of the introduction of excess compounds into the environment is hard to predict since nature slowly seeks to reach a new equilibrium with the chemicals man is producing. Care must be taken to recycle nitrogen wastes, so that the reliance on man-made fertilizers will not increase further.

The entire biosphere is a meshing of a number of natural cycles that transport life-needed chemicals to the regions where organisms need them. Perturbing these cycles will cause short- and long-range changes in the types of life that can survive in the new environment. It is up to man to use his resources wisely and to try to minimize the effects of the changes he is causing within the biosphere.

## AN AIR POLLUTION DAY

Within a metropolitan area, the concentrations of air pollutants differ from hour to hour during the day. The exact daily pattern varies from city to city but the factors affecting the pattern are all the same. Topography, meteorology, distribution of industry and people, and other factors influence the buildup and dispersion of air pollution. From the details of the PHS survey conducted in metropolitan St. Louis, we will re-create a typical air pollution day. For this city, two of the most important factors that modify the pollution effects on any given day are the topography and meteorology.

Figure 4–11 gives a general idea of the lay of the land in the St. Louis area. Much of the area is contained within a bowl-shaped region centered on the Mississippi River. To the west, the land rises slowly from 420 feet at river level to about 600 feet on the ridge east of St. Louis International Airport. On the east side of the river, a large lowland area, called the American Bottom, rises to only 440 feet until it reaches a crescent-shaped bluff that rises to 640 feet. There are no large mountains or bodies of water that might affect weather patterns.

A wide variety of climatic conditions affect air pollution levels: temperature, wind, percentage of possible sunshine, precipitation, and cloud cover are a few. Of these, wind and sunshine play an important role every day since

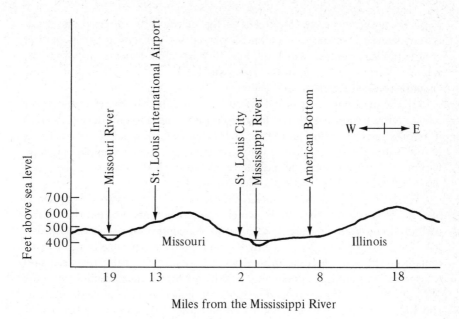

**Figure 4-11** Topography of the Greater St. Louis area.

they are responsible for the principal modes of pollution dispersal. Wind is primarily responsible for horizontal dispersal of pollutants; it is most important in the afternoon. Vertical dispersion is also very important. As air is heated, it expands its volume. As the same amount of air expands into a larger volume, the density of the air decreases, causing the body of air to rise; note that both wind and vertical dispersion depend on the properties of gases. Wind is the movement of a gas (air) from a region of high pressure to a region of low pressure. Vertical dispersion is based on the principle of the expansion of gases on heating. Temperature affects air pollution in a variety of ways. On cold days residential fuel combustion increases, and on hot days demand for electricity increases. Increased power consumption results in increased pollution.

The manner in which temperature changes with altitude is also an important factor in air pollution. If the temperature decreases with increasing altitude, warm air masses, loaded with pollutants, will continue to rise and disperse the pollution vertically. If the temperature increases with altitude, the warm air masses near the surface will not rise but will remain stagnant, and if there is no horizontal dispersion, pollution builds up rapidly. This last effect is called a **thermal inversion,** or just an inversion. Inversions are particularly noticeable in the American Bottom area, where cool air (being more dense than warm air) settles into the low-lying bottoms area. Since the cool air is trapped below the warm area, a natural inversion occurs. As an indication of the frequency of inversions, inversions of 10 °F or greater (when the temperature aloft, at 500 to 1,000 feet, is greater than at ground level by

10 °F or more) occur on 45 percent of the mornings in the St. Louis area. Fifteen-degree inversions occur on 25 percent of the mornings. Stagnation of high-pressure centers, which inhibits the horizontal dispersal of pollutants, is uncommon in St. Louis. Instances in which these centers stagnate for four or more consecutive days are very rare.

The St. Louis area does impose its own effect on the atmosphere in one important way. It acts as a heat reservoir, or heat island; during the day it absorbs solar radiation and at night it releases heat. This has two effects. It pushes the normal nighttime inversion aloft and creates complex air flow within the "bubble" until the bubble breaks through the inversion layer and disperses the pollution vertically. See Figure 4–12.

The typical air pollution day in St. Louis begins with the presence of a temperature inversion aloft, which traps pollutants emitted during the early morning hours. Figure 4–13 presents concentration versus time-of-day data for hydrocarbons, $NO_2$, aldehydes, and $O_3$. The concentrations of these remain constant or increase even when comparatively little industrial and transportation activity is occurring. Results for $SO_2$ would parallel those for $NO_2$. As the morning rush hour begins, the concentration of hydrocarbons and $NO_2$ increases, reflecting increased auto use and continued presence of the inversion layer. Carbon monoxide also increases in concentration. Between 8 A.M. and 10 A.M., the air trapped in the bubble is warmed by the sun, allowing it to break through the inversion layer, and then vertical dispersion reduces the concentration of hydrocarbons and $NO_2$. Around noon the ozone concentration picks up, since $NO_2$ absorbs light, producing NO and an oxygen atom, which attacks $O_2$ and produces $O_3$. The increase in $O_3$ is a reliable measure of photochemical smog production. While the $O_3$ increases, the remaining hydrocarbons are undergoing atmospheric reactions to produce larger molecules, which eventually interfere with visibility. The appearance of aldehydes also indicates the buildup of smog, as hydrocarbons react with atomic oxygen to produce them. The typical composition of photochemical smog is shown in Table 4–7.

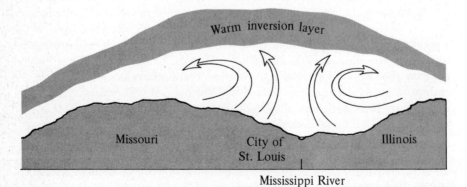

Figure 4-12   Heat island and nighttime inversion.

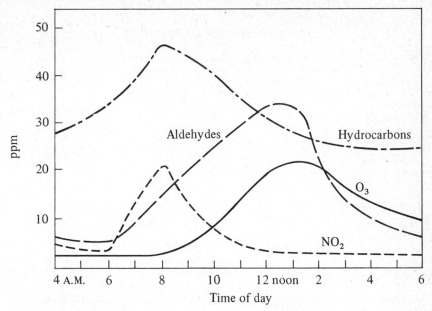

**Figure 4-13** Daily profile of pollution.

Fortunately winds increase after noon and disperse the smog and evening rush hour pollutants to the point at which no bumps in the curves are shown. By late afternoon $O_3$ concentration decreases as the sun is no longer overhead. Evening brings a reduction in activity and pollution concentration. The cycle is now ready to begin again. This pattern is followed in St. Louis each day, subject to variations in precipitation, movement of weather systems and temperature, weekends, and a large number of other factors.

Topography and meteorology are therefore shown to be two factors that control the collection of pollutants. Topography provides the bowl, or reaction vessel, in which primary pollutants collect and react to produce secondary pollutants. Meteorology plays the important role of keeping the lid on the reaction vessel.

Other cities in the United States, particularly those on the coasts, find themselves caught between the ocean and mountains, in a natural container for pollutants. Los Angeles, the prime example, had photochemical smog formation as long ago as the middle 1940s. Los Angeles, characterized by a lack of heavy industry and a large and widely dispersed population, finds itself a victim of large-scale automobile-related pollution.

## PROBLEMS AND SOLUTIONS

The solutions to the air pollution problem are diverse and complicated, and almost always raise problems of their own. Technical problems in developing cleaner industrial processes can and will certainly be solved, with the

### Table 4-7
### Photochemical Smog Composition

| Constituent | Concentration (ppm) |
|---|---|
| $NO_x$ | 2 |
| $SO_2$ | 2 |
| $O_3$ | 5 |
| CO | 400 |
| $CH_4$ | 25 |
| Other alkanes | 2.5 |
| $C_2H_4$ | 5 |
| Other alkenes | 2.5 |
| Benzene | 1 |
| Aldehydes | 6 |
| $NH_3$ | 0.2 |
| $H_2O$ | 20,000 |
| $CO_2$ | 4,000 |

question of cost implementation left to be determined. A better understanding of the chemical details of atmospheric reactions will pinpoint the sources most likely to produce secondary pollutants. Recycling of chemicals now emitted will prove more profitable in the future. Stricter air quality standards and improved detection methods and equipment will aid companies in setting long-range pollution-abatement projects and government agencies in the enforcement of clean air ordinances.

The solutions are more than just technical. Political considerations undoubtedly will play an ever increasing role. In the metropolitan St. Louis area, for instance, there are more than ten political or quasi political units that have to cooperate in the establishment and enforcement of clean air standards. Figure 4-14 points this out clearly; $SO_2$ emissions, though centered in Illinois, affect the entire metropolitan area. The same holds for particulates. Yet there are no areawide standards for regulation of these two pollutants. Recent federal guidelines do arrange for areawide standards, but they do not go into effect until 1975. One of the setbacks to instituting areawide standards was the establishment of an overall environmental agency for the State of Illinois. In the fall of 1969, Missouri Air Conservation Commission members and their counterparts in Illinois had moved close to agreeing on standards. But the establishment of a new agency in Illinois meant delays until the new people involved had become familiar with the proposed agreement and how it would affect the overall policy of the new agency. Without areawide cooperation, real progress cannot be made. For instance, residential incineration and open burning has been banned in Missouri since the fall of 1969; no such rule applies in Illinois, where notorious dump fires have been burning for years.

It is also becoming apparent that citizens must become individually

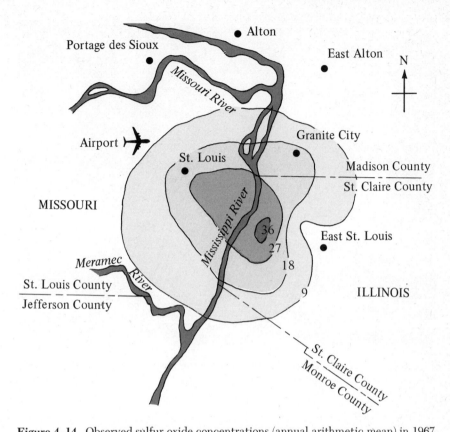

**Figure 4-14**  Observed sulfur oxide concentrations (annual arithmetic mean) in 1967 in East St. Louis–Granite City, Illinois, and St. Louis, Missouri.

responsible for pollution control. Devices designed to reduce air emissions from autos must be maintained. Old cars without these devices must be put out of service as quickly as possible. A continued educational program will be necessary to maintain public interest in the pollution problem.

The solutions to air pollution problems often cause problems of their own. For example, the generation of electricity by nuclear power plants does produce radioactive materials that must be stored; it also causes thermal pollution, the heating of waters used to cool the reactor. The plastic bags used to gather leaves for the garbage man to pick up are often produced from chlorinated hydrocarbon polymers; if incinerated, the bags produce air pollutants containing chlorinated compounds. If not burned, the bag will take years to disintegrate. Proposed electric cars only shift the pollution problem from the streets to the power plants. A little imagination will lengthen this list. Actuality is represented in Figure 4–15.

The future will produce successes in controlling current pollution problems, and at the same time will create and reveal new problems. Carbon dioxide, a natural constituent of the atmosphere, in higher concentrations may

**Figure 4-15** Darkness at noon: smog obscuring the sun. (*Reprinted, by permission, from* St. Louis Post-Dispatch. *Photograph by Spainhower.*)

further unbalance the biospheric equilibrium. New energy sources will produce their own problems. Population concentration will receive more attention as a prime pollution cause. Man will have to set his own course on this planet, but he does need the will developed by pressing dangers and the way developed by advanced understanding of his role in the environment.

## THE ENVIRONMENTAL PROTECTION AGENCY

The Environmental Protection Agency (EPA) was created December 2, 1970, as part of Reorganization Plan 3 to permit a coordinated and effective governmental action to protect the environment by abating and controlling pollution on a systematic basis. The reorganization transferred to EPA a variety of research, monitoring, standard-setting, and enforcement agencies. As a result, EPA is responsible for both coordinating and supporting research and antipollution activities and setting up air- and water-quality standards. As in the case of the Atomic Energy Commission, conflicting activities of the EPA bring it under the watchful eye of worried citizens.

The EPA is divided into five basic units: planning and management, enforcement, air and water programs, categorical programs (pesticides, radiation, solid waste), and research and monitoring. In addition, ten regional offices have been created to coordinate and administer EPA programs.

Since its creation EPA has taken several dramatic stands, including the DDT suspension and a crusade against automobile air pollution. The latter is aimed at reducing automobile air pollution quickly. The agency continually sets optimistic goals, with the hope of ultimately proving that control of automobile air pollution lies in mass transportation.

## POSTSCRIPT: BEWARE OF THE ECOLOGICAL CON MAN

Environmentalists, ecologists, and conservationists represent a powerful force in America today. Their aims of clean air and water and wise use of natural resources are indisputably important ends to be realized. However, vigilance must police the means by which these ends are attained. An alert citizen is constantly bombarded with "facts" from "experts" or movie stars or athletes, and so on. Like any group advocating a point of view, certain environmentalists overstate or distort information and lobby on less than a factual basis. Often a group's or individual's financial and political success depends on a continuing environmental crisis. In many ways, their situation is like a politician's finding issues to publicize himself.

Like any lobbying efforts, from those of the labor unions to those of big business, environmental crusades must be checked as to sources and information. Just because the pollution problem is worthy of solution doesn't mean that all those trying to solve it will use truthful, rational methods.

## PROBLEMS

1. $NO_2$ is one of the major contributors to photochemical smog production. Explain why this is true.
2. What chemical role do microscopic particles play in reactions that occur in the atmosphere?
3. Just what does 10 pphm $SO_2$ in the atmosphere mean? Consider a box of air 1 ft $\times$ 1 ft $\times$ 1 ft at 27 °C. The number of molecules of air in the box is given by

$$n = \frac{PV}{RT}$$

$R$ = gas constant = $4.81 \times 10^{-27} \dfrac{\text{ft}^3\text{-atm}}{\text{molecule-}°\text{K}}$

$T$ = temperature $°\text{K} = °\text{C} + 273$

$P$ = 1 atmosphere (atmospheric pressure at sea level)

$V$ = volume of container

(a) Calculate the number of molecules in the box.

(b) 10 pphm $SO_2$ means that for each 100,000,000 molecules in the air sample, 10 are $SO_2$. How many $SO_2$ molecules are there in the box?

4. Why are you supposed to check the air pressure in your automobile tires when they are cold?

5. Is pollen an air pollutant? Discuss your decision.

6. Describe, in terms of the physical properties of gases, the formation and destruction of thermal inversions.

7. Electricity is advertised as the clean heat. List five pieces of evidence against this notion.

8. Make a list of the primary air pollutants you have directly caused in the past week. Consider all activities in terms of air pollution potential.

9. Consider the cost of air travel to the environment. Include the cost of mining, the production of new materials, and the power consumption of the industry.

10. Each time you breathe you take in a fixed volume of air, let's call it one lungful. If air is uniform and contains 20% oxygen no matter where you are, why is it "more difficult to breathe" in mountainous areas. Be specific. (HINT: The key point will also explain why atmospheric pressure decreases as you go to higher altitude.)

11. The equation that expresses gas behavior is $PV = nRT$ where $R$ is a constant and $n$ = number of molecules. Explain what happens to $P$, $V$, $n$, and $T$ as the gas or gases involved (increase, decrease, remains the same) when the following situations occur.

(a) A balloon filled with helium gas is put in the refrigerator.

(b) A tire blows out.

(c) An aerosol can containing Freon gas is thrown into a lighted incinerator.

(d) A mixture of propane gas ($C_3H_8$) and oxygen gas in a closed reaction vessel is allowed to react to form $CO_2$ gas and $H_2O$ gas as follows:

$$\underset{\text{gas}}{C_3H_8} + \underset{\text{gas}}{5O_2} \longrightarrow \underset{\text{gas}}{4H_2O} + \underset{\text{gas}}{3CO_2} + \text{Energy.}$$

# BIBLIOGRAPHY

1. *Chemistry and the Atmosphere.* Special report. *Chemical and Engineering News,* 1969.

2. Coalition for the Environment. *Save Our Spaceship—Earth.* St. Louis, Mo.: Universal Printing Company.

3. J. T. Middleton. *American Scientist* 59 (1971): 188.

4. *Air Pollution Primer.* National Tuberculosis and Respiratory Disease Association. New York: 1969.

5. *Scientific Statesmanship in Air Pollution Control.* PHS Publication No. 1239. Superintendent of Documents, Washington, D.C., 20402.

6. E. D. Pearce. "Volunteer Organization and Air Pollution Control." Paper presented at HEW National Conference on Air Pollution, December 1966.

7. *Scientific American,* September 1970.

8. Arthur C. Stern, ed. *Air Pollution,* vols. 1, 2, and 3. New York: Academic Press, 1968.

9. *Interstate Air Pollution Study,* United States Public Health Service. Washington, D.C.: U.S. Government Printing Office, 1966.

10. *Proceedings of the Second Air Pollution Conference.* University of Missouri–Columbia. Columbia, Mo.: November 1969.

11. K. E. Weaver. *National Geographic* 135 (1969): 593.

# Liquids, Solutions, and Problems

# 5

*America! America!*
*God shed His grace on thee,*
*And crown thy good with brotherhood*
*From sea to shining sea!*

Unfortunately, the shine on today's sea may be the glare of sunlight reflected from a floating beer can. Our streams, lakes, and rivers have been converted to sewage reservoirs, which are no longer safe for swimming or fishing. In the land of the most technologically advanced population on this planet runs the Cuyahoga River (Ohio), so overrun with volatile industrial discharges that it caught fire recently and burned two railroad trestles. How did we allow a river to become a fire hazard? How do we allow it to continue? What will happen in the future?

Among all liquids, the most important liquid on earth is water. Water, like all the other resources on this planet, is present in finite amounts. Fortunately, water is plentiful, although inexpensive, clean water in the form of streams and lakes is rapidly being depleted. As our need for such water increases, so will its price. Who uses the water, what is put into it, what must be taken out of it, and who will pay for the cleaning? These intricate questions will be discussed under the general heading of water pollution, which is subdivided to carefully focus on several important facets of the problem: detergents—over 5 billion pounds per year down the drain but not gone forever; industrial and municipal pollution—no longer separate problems; oil and water—two liquids that do not mix well; mercury pollution of water—a hazard concentrated through the food chain.

A knowledge of the properties of liquids and liquid solutions, and of the chemistry of water and water purification, will provide the reader with a better understanding of those many processes and activities involved in our quest for clean water.

## THE LIQUID STATE

There are three phases of matter—solid, vapor, and liquid. In its existence, the liquid phase must maintain a relationship with the other two phases, sometimes interchanging with them. Liquids have the following properties:

1. Practically total incompressibility, because there is very little space between molecules. Liquids are therefore used in hydraulic pressure systems.

**Down by the old paper-mill stream.** A polluted stream containing paper-mill waste. (*TVA photograph.*)

2. Retention of volume. One quart of milk poured into a vessel of any size is still a quart of milk (whereas a gas tends to fill the volume of its container).
3. No characteristic shape. Liquids assume the shape of their container. There is some order to the structure of liquids but not as much as that of solids.
4. Slow diffusion of molecules. Molecules in the liquid phase do not mix so quickly as those in the gas phase, owing to the greater crowding of the molecules of the liquid.
5. Evaporation. Those molecules with enough kinetic energy to overcome the attractive forces of neighboring molecules escape into the gas, or vapor,° phase; the liquid **vaporizes.**

In Figure 5–1, the fraction of molecules vs. kinetic energy is plotted for a given temperature. As shown, a small fraction of molecules in the liquid phase have very low or very high energy; most molecules have an energy close to the average. The minimum amount of energy needed to escape from solution is shown in the darkened area. The small fraction of molecules that have energies in this range may escape. If the highly energetic molecules leave, those left behind have a lower average kinetic energy. Since kinetic energy is directly related to absolute temperature, the temperature of the liquid left behind drops, thus the "cooling" effect when liquids evaporate. The resulting change in the plot is shown by the dotted lines.

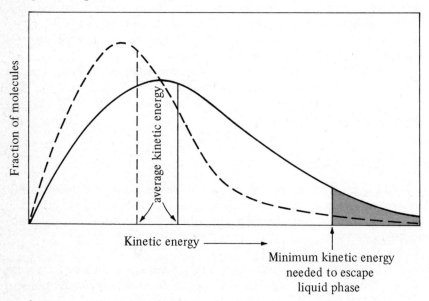

**Figure 5-1**  Kinetic energy distribution of molecules. Solid lines represent the distribution at a given temperature; dotted lines represent the distribution at a lower temperature.

---

° The terms gas and vapor will be used interchangeably.

A similar effect has been noted for economically depressed areas of the United States. One of the biggest factors in the continued decline of impoverished areas is that those members of the community with enough skill, drive, and ambition to leave do so, the number of natural leaders and progressive citizens among those remaining thereby becoming reduced.

The accompanying diagram depicts a liquid in a beaker enclosed by

a bell jar. Some molecules will escape from the liquid phase into the vapor phase, but the vapor will be confined to the enclosed area. The molecules that escape will eventually return to the liquid phase. These two processes continue simultaneously. When the number of molecules leaving the liquid equals the number of molecules returning to the liquid phase a **dynamic equilibrium** exists, that is, a condition in which two changes exactly oppose each other, with no net change occurring.

Molecules of a liquid in the vapor phase exert a pressure, in the manner of all gases. At equilibrium, this pressure is characteristic of the liquid and is called the equilibrium **vapor pressure** (a physical property of a pure liquid at a given temperature). The magnitude of the pressure depends on two factors: (1) Nature of the liquid—attractive forces between molecules differ for different compounds. If there is a large mutual attraction of molecules in the liquid phase, there will be a lesser tendency to escape. (2) Temperature—if the temperature of the solution is increased, the average kinetic energy of the molecules is increased and more molecules have sufficient energy to escape to the vapor phase. These relationships are described in Figure 5–2.

You will note that as the temperature increases, the equilibrium vapor pressures for both water and chloroform increase but not in linear fashion. You will also note that at any given temperature, the vapor pressure of chloroform is higher than the vapor pressure of water. This is due to the fact that in the liquid phase, molecules of water have a greater mutual attraction than molecules of chloroform.

The temperature at which the vapor pressure of a liquid is equal to the prevailing atmospheric pressure is called the **boiling point.** Since atmospheric pressure varies from place to place, a standard had to be adopted.

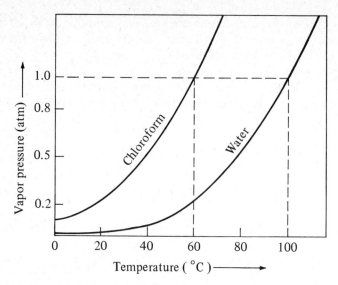

**Figure 5-2** Vapor pressure vs. temperature diagram for liquids.

Therefore, the boiling point of a liquid is described as the temperature at which the vapor pressure of the liquid exactly equals one atmosphere of pressure. For example, the vapor pressure of pure water is exactly one atmosphere (atm) at 100 °C.

It is possible to separate mixtures of liquid compounds by taking advantage of their different vapor pressures. The compound with the highest vapor pressure and lowest boiling point will be the most readily converted to vapor, which can be condensed and collected (a process known as distillation). By careful heating and continuous redistillation of volatile fractions, one can achieve extremely good separation. This process of fractional distillation is used effectively by the petroleum industry. Compounds from petroleum boiling in different temperature ranges are collected and used appropriately; thus, there is a gasoline fraction (hydrocarbons boiling in the range about 98° to 150°), a kerosene fraction (ca. 174°–215°), and so on.

### Humidity

The air in our living and working space contains water vapor but equilibrium between the liquid source and atmospheric water vapor is seldom reached. The ratio of the *actual* pressure of water vapor to what it would be at equilibrium at a given temperature is called the **relative humidity:**

$$\text{relative humidity at } T \ (\%) = \frac{\text{observed pressure of water at } T}{\text{equil. v. p. of water at } T} \times 100.$$

Relative humidity has a bearing on our physical comfort, the deterioration of materials, and the growth of molds and bacteria. When the temperature of air at a given atmospheric pressure is changed, the observed pressure, or actual concentration, of the water in the air remains unchanged. The equilibrium vapor pressure depends on the temperature and changes accordingly. The equation for relative humidity shows that an increase in the temperature will result in lower relative humidity ("drier" air) since the equilibrium vapor pressure will increase while the observed pressure of the water vapor remains unchanged. In the same fashion, a decrease in the temperature results in a higher relative humidity. When 100 percent relative humidity is reached, equilibrium is established and condensation (known as dew) begins; the temperature at which this occurs is called the dew point.

## EQUILIBRIUM

We have just been considering a system in equilibrium, liquid going to vapor and vapor to liquid. When a system is at equilibrium, any changes that occur exactly oppose each other resulting in no net change. *If a stress is added to a system at equilibrium, the system readjusts if possible, to reduce the stress.* This is known as the principle of Le Chatelier, and can be exemplified by the following equilibrium equation:

$$\text{liquid} + \text{heat} \rightleftharpoons \text{vapor}.$$

This system is composed of liquid water and water vapor. At equilibrium, the number of molecules of liquid that become vapor is exactly equal to the number of vapor molecules that become liquid. The equation has arrows in both directions ($\rightleftharpoons$), showing that both forward and reverse reactions are occurring. If a stress is applied (for example, additional heat), the system adjusts to reduce the stress (that is, to use up the heat). According to the equation, heat can be used up by having some liquid convert to vapor; the system would still be at equilibrium but there would be more water vapor and less liquid water than before. Alternatively, if heat is suddenly removed (if the system is cooled), there is a stress on the system. To relieve that stress, the system adjusts to produce heat. According to the equation, heat will be produced when water vapor is converted to liquid water. Thus, some vapor condenses to liquid, removing the stress.[*]

The phrase "upsetting the equilibrium" is not restricted to chemistry; the law can apply in a social context. If there is a sudden shortage of food (for example, the Irish potato famine), the population responds to relieve the stress by reducing itself (say by perishing, or migrating to another country that has an abundance of food, or maintaining a lower birthrate).

---

[*] Reference will be made to the principle of Le Chatelier throughout the chapter.

## CHANGES OF STATE

Let us imagine that we are starting with solid water (ice) and adding heat energy to it at a constant rate. We plot a graph using temperature as one axis and elapsed time, or energy added, for the other axis, Figure 5–3. At the beginning $(t_0)$, the ice happens to be at $-80$ °C. Continual addition of heat $(t_0 \rightarrow t_1)$ causes a rise in the temperature of the solid as the molecules in the highly ordered crystalline structure vibrate at a faster rate (the kinetic energy of the molecules is increasing). Finally, the vibrations become so vigorous that any added heat loosens the binding forces between neighboring water molecules (at $t_1$). From time $t_1 \rightarrow t_2$, the added heat energy is used by the molecules for a change of phase, to convert from solid to liquid. Energy is needed for this change because molecules in the liquid phase have more potential energy than molecules in the solid phase. Therefore, the potential energy of the system (solid + liquid water) is increasing. There is no increase in the average kinetic energy of the molecules, so there is no change in temperature (it remains constant). During this period $(t_1 \rightarrow t_2)$, solid and liquid water coexist in equilibrium according to the following equation:

$$\text{solid} + \text{heat} \rightleftharpoons \text{liquid}.$$

As more heat is added, the system is under stress. This stress is relieved by using the heat to have additional solid molecules convert to liquid molecules. The temperature at which solid and liquid molecules coexist in equilibrium is referred to as the **melting point** of the compound (which is 0°C for water at 1.0 atm pressure). The amount of heat needed to convert one gram of a solid to the liquid phase is called the **heat of fusion,** and for water this is that amount of heat added between $t_1$ and $t_2$ (80 calories/gram or 36,320 calories/pound).

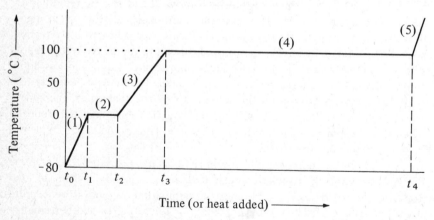

**Figure 5-3** Heating curve for water (at 1.0 atm pressure): (1) solid phase only; (2) solid and liquid phases in equilibrium; (3) liquid phase only; (4) liquid and vapor phases in equilibrium; (5) vapor phase only.

From $t_2 \rightarrow t_3$, the added heat increases the kinetic energy of the liquid, so the temperature increases (the potential energy of the liquid is not changing). From $t_3 \rightarrow t_4$, the added heat is used to overcome the attractive forces of the liquid molecules enabling them to escape to the vapor phase. When a molecule goes from liquid to vapor, its potential energy increases—that is, the vapor has more energy available for work than the liquid. For example, the steam engine is driven by the potential energy lost by the water vapor in condensing to liquid water. From $t_3 \rightarrow t_4$, liquid and vapor coexist in equilibrium; the average kinetic energy does not change, so the temperature remains constant. This temperature is referred to as the normal boiling point of the liquid ($100°C$ for water at $1.0$ atm pressure). The temperature remains constant until all of the liquid has been converted to vapor (at $t_4$), then added heat energy increases the temperature of the vapor. The amount of heat needed to convert one gram of a liquid to the vapor phase is called the **heat of vaporization.** It is a measure of the attractive forces between liquid molecules, and for water would be the amount of heat added from $t_3$ to $t_4$ ($540$ calories per gram, or $245,000$ calories per pound). Water has a high heat of vaporization, absorbing energy when going from liquid to vapor and releasing energy when changing from vapor to liquid. Oceans cover $70$ percent of the earth's surface. Of the solar energy absorbed at the sea surface, about $50$ percent is used for the evaporation of seawater, and the energy is thereby stored in the water vapor. When the sun sets, the water vapor condenses to liquid, giving off the energy initially absorbed. With water as an energy buffer, the narrow ranges of temperature necessary for support of life can be maintained. Desert areas are very cold at night because there is not much water vapor present to release energy upon condensation; and they are hot during the day because there is not much water to absorb heat by evaporation.

The curve we have been discussing (Figure 5–3) is a heating curve. For liquids to "boil," there are two conditions that must prevail; the molecules must:

1. Have enough average kinetic energy so that added energy will be used only for a change of phase.
2. Be close to the liquid-vapor boundary to escape the solution.

Occasionally, the first factor will have been met (the liquid is at the boiling point) but the molecules that could leave are trapped within the liquid and cannot escape. In this special case, added heat is converted to kinetic energy, the temperature rises, and a nonequilibrium condition known as superheating results. The sudden ability of the molecules to escape can cause a violent eruption (for example, a stone containing trapped air dropped into the liquid would cause bubbling). Adding heat energy to a boiling liquid will not cause the vapor pressure to increase past the prevailing atmospheric pressure unless vapor is prevented from escaping. Heating water already at

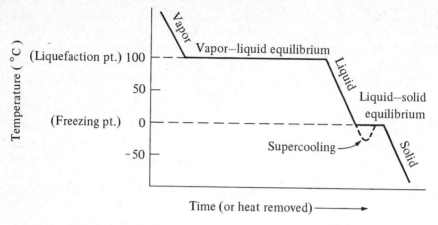

**Figure 5-4**   Cooling curve for water (at 1.0 atm pressure).

100 °C in a sealed container will cause the vapor pressure of the water to exceed atmospheric pressure and the temperature of the water to exceed 100 °C. This is the principle of the pressure cooker.

If we start with a vapor and continually remove heat energy, we obtain a cooling curve, as shown in Figure 5-4. Now the vapor-liquid equilibrium point is called the **liquefaction point** (numerically the same as the boiling point) and the liquid-solid equilibrium point is the **freezing point** (melting point).

Sometimes the liquid does not crystallize at the freezing point. The kinetic energy of the liquid is low, the molecules are not very mobile, and they have not lined up in a characteristic crystal arrangement. If heat is continually removed, the kinetic energy continues to decrease, so that the temperature decreases below the freezing point and we have a special non-equilibrium condition known as supercooling. Suddenly, a few molecules do line up correctly and a small crystal forms. The potential energy that is given off in going from liquid to solid is converted to kinetic energy. As crystallization continues, the kinetic energy increases until the temperature reaches that value associated with the melting point. Sometimes, a permanent supercooled state exists, for example, common glass—there is no crystallization but there is more order to the molecules than normally found in the liquid phase.

### Phase Diagrams

The relationship among solid, liquid, and vapor states of a given substance can be described on a graph called a **phase diagram.** The phase diagram for water (Figure 5–5) is drawn out of proportion to dramatize particular aspects of the curve. On this graph, various points represent the states of water at different temperature and pressure. Each one of the three designated regions (solid, liquid, vapor) corresponds to a one-phase system.

The boundary lines represent equilibria between the phases involved and all three regions intersect at one point only (called the **triple point**), where the three phases are in equilibrium.

By looking at the phase diagram, we can obtain a great deal of information. For example, we see that at 1.0 atm, starting from the left and going horizontally to the right (toward higher temperature), water exists only as a solid below 0°. At 0°, we reach the solid-liquid boundary, or equilibrium line, which is the melting point. Between 0° and 100° at 1.0 atm pressure, water is a liquid. At 100°, we reach the liquid-vapor boundary or the boiling point. Above 100°, water exists only in the vapor phase.

At 0.5 atm, by going from left to right we reach the solid-liquid boundary at a temperature above 0° (actually only 0.005° but grossly exaggerated in this diagram). The liquid-vapor boundary is reached at a temperature of 82°, which is therefore the boiling point of water at 0.5 atm. At 0.003 atm, starting at low temperature where water exists only as a solid and going to slightly higher temperature, we reach a solid-vapor boundary, where solid and vapor coexist in equilibrium. At a higher temperature, water exists only in the vapor phase. Thus, it is possible to go from solid to vapor without passing through the liquid phase if the pressure is kept at 0.003 atm. This is the principle behind "freeze-drying," where water is removed from a material (for example, a coffee solution), to produce dehydrated material, at a temperature so low that other chemical reactions do not occur that would change the "natural flavor."

We can also observe, for example, that at 50° the liquid-vapor boundary is reached at approximately 0.1 atm. Therefore, at that pressure, water boils at 50°; and above that pressure, water at 50° exists as a liquid; below that

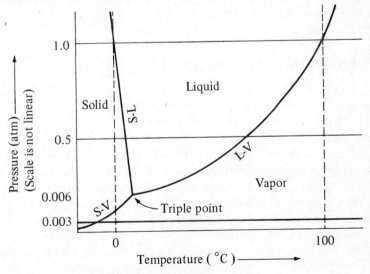

**Figure 5-5** Phase diagram of water (not drawn to scale).

pressure, as a vapor. Much more information can be obtained from the phase diagram. Note that the solid-liquid boundary line tilts to the left (for most compounds the tilt is to the right). Thus, an increase in the pressure on solid water (ice) at constant temperature results in a change of phase to liquid water. Skating on ice entails applying tremendous pressure (the weight of the body) to a small surface area (the blade), converting the ice under the blade to the liquid water on which the blade slides. Would it be more difficult to skate on ice at $-50°$ as compared with $0°$?

Can solid, liquid, and gas coexist in equilibrium? Yes, under only one set of conditions, identified by the triple point. For water, the triple-point temperature is $0.01°C$ and the pressure is $0.006$ atm. (Remember that Figure 5-5 is not drawn to scale.)

Ice occupies a larger volume than liquid water (the expansion in volume as water freezes causes cracks in water pipes, automobile radiators, and so on). Consider the equilibrium described below:

$$\text{solid} + \text{heat} \rightleftharpoons \text{liquid.}$$

If a stress is applied, such as additional heat, more of the solid converts to liquid to use up the added heat, with consequent reduction of stress. Increased pressure introduces a stress on the system that can be relieved by reducing the volume—converting from solid to liquid water. However, liquid water has a higher potential energy than solid water. This necessary energy is obtained at the expense of kinetic energy, so at high pressure the average kinetic energy of solid and liquid water in equilibrium is lower, or the melting point is lower. The phase diagram clearly shows that the melting point (on the solid-liquid equilibrium line) decreases as the pressure increases.

Consider what would happen to life as we know it if, because it were more dense than water, ice sank to the bottom of lakes and oceans.

## DESALINATION: HEAT IT OR COOL IT?

Nondrinkable seawater accounts for nearly 98 percent of the water on earth. An unlimited supply of usable water would be available if seawater could be desalted at a reasonable cost. Two possible physical methods to this end are distillation and freezing.

The principle of distillation is to use heat to convert liquid water to water vapor causing dissolved salts to be left behind. The vapor is then condensed to give pure water. Nature is the number one distiller, accounting for approximately $10^{21}$ gallons per year. The principal cost of distillation is the energy required for heating the liquid. Solar energy seems to be an inexpensive source, but it is difficult for man to harness this energy, which, moreover, has limited use (especially in cloudy climates). Other possibilities for fuels include both nuclear energy and garbage. An exciting possibility for the use of undesirable coastal areas is a combined nuclear power plant and desalting plant. This dual-purpose facility would generate high-pressure steam for producing electric power and low-pressure steam for evaporating seawater. A community could be developed and provided with the necessary quantities of electrical energy and water.

Freezing is an attractive alternative to distillation because it takes less energy to freeze water than to boil or evaporate it (see Figure 5–3). When impure water freezes, it tends to form ice crystals of pure water, which "squeeze out" impurities. Unfortunately, all of the technological problems involved with this technique have not been successfully overcome.

The cost of usable water is usually quoted in cents per 1,000 gallons. Municipalities charge about 40¢; water for irrigation costs 1¢–35¢ depending on the quality. The best present desalting methods cost 65¢–$1 but it is hoped that the cost can be lowered to 20¢ per 1,000 gallons in this decade as better technological methods are developed.

## SOLUTIONS

A **solution** can be defined as a mixture of two or more substances dispersed as molecules. Wine, gasoline, and water containing dissolved salt are all examples of solutions, but not tomato juice or clay, which represent mixtures of suspended and colloidal particles in a liquid. This discussion will be confined to liquid solutions, and for simplicity, examples will usually concern two component solutions. When one component is a gas or a solid and the other a liquid, the former is referred to as the **solute** and the latter the **solvent.** If two or more liquids compose the solution, the more abundant component is called the solvent. Two familiar terms are continually used but poorly defined: **dilute,** indicating a small amount of solute; and **concentrated,** indicating a large amount of solute. The difficulty in definition is due to the ambiguity in the terms *small* and *large.* A solution containing one part in ten thousand of a physiologically active substance such as mercury may be considered concentrated, whereas a solution of one part alcohol in one hundred parts of water may be considered dilute.

Whether or not two substances will form a solution depends on their physical properties. For example, in many molecules, a separation of charge within the molecule leads to a polarity of individual bonds and establishment of sites of positive and negative charge. It is possible for these charged sites to attract one another so that the negative site of one molecule attracts the positive site of another molecule, as shown in the accompanying figure. Water

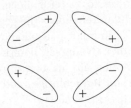

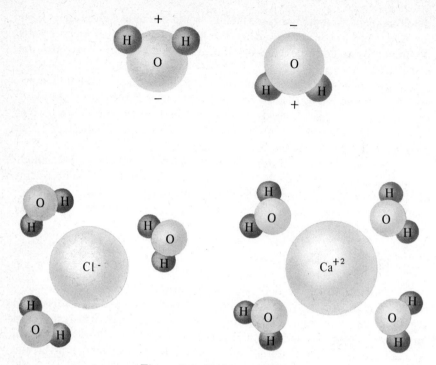

**Figure 5-6**  Hydration of ions.

is a polar molecule that has strong attractive forces in the liquid phase. These attractive forces and hydrogen bonding (see p. 66) account for some of the special properties of water, for example, its unpredictably high boiling point (100 °C compared with −88 °C and −62 °C for the heavier and less polar $C_2H_6$ and $H_2S$ molecules, respectively).

Polar molecules can attract ions. The negative sites attract positive ions; the positive sites attract negative ions. Thus, a polar molecule like water can attract both positive calcium ions and negative chloride ions (see Figure 5-6). When a molecule or ion is surrounded by solvent molecules it is said to be **solvated**; when the solvent is water, it is **hydrated.**

Sugar molecules are also attracted by water molecules and can be pulled away from other sugar molecules to an environment in which they are surrounded by water molecules. Thus, we say that sugar dissolves in water. Oil does not dissolve in water. The atoms of hydrocarbons are more attracted to one another than they are to water molecules. Thus, oil is said to be insoluble in water. It is useful to remember the generality—like dissolves like. Polar liquids tend to dissolve in other polar liquids, since the molecules have an attraction for one another. Nonpolar liquids tend to dissolve in other nonpolar liquids. Thus, nonpolar oils are best dissolved in nonpolar cleaning solutions (for example, carbon tetrachloride) and not in a polar liquid such as water.

### Ground Water

Both dissolved and undissolved matter can be contaminants of water. Included in the latter category are suspended particles (which can normally be filtered out or soon settle out, such as iron powder and coal dust) and colloidal particles (which are so small that they do not settle out and cannot be easily removed by filtration).

Water evaporates from the land and the seas, condenses into clouds, and returns to the earth as rain, sleet, or snow. This fresh water falls into the sea, is frozen into the polar ice caps, or runs off the land into the streams, rivers, and lakes and eventually into the sea. The fresh water dissolves mineral salts as it runs through the land and carries them to the sea. Seawater contains an average of 35,000 ppm of dissolved salts, approximately three-fourths of which is sodium chloride. When the water evaporates from the sea, the salts are left behind.

### Properties of Solutions

Almost all of the water on earth is found as the major component of aqueous solutions. For a complete understanding of water, therefore, we must understand the properties of solutions as related to the properties of pure liquids.

One vessel containing pure water and one identical vessel containing an aqueous solution (for instance, a salt solution such as brine or sodium phosphate in water) are placed in a closed system, as shown in Figure 5-7. Initially, the levels of liquid in both vessels are equal. As time passes, we can observe that the level of the pure water drops and that of the solution increases. This is because *the escaping tendency of liquid molecules in pure liquid is greater than that of liquid molecules in a solution,* while molecules returning to the liquid phase will have an equal tendency to return to both

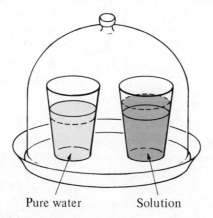

Pure water          Solution

**Figure 5-7**   The difference in the escaping tendency of water molecules from the two beakers results in a transfer of water from the left beaker to the right beaker.

beakers. The lower rate of evaporation of molecules from a solution can be rationalized by considering that molecules close to the surface have the best chance of escaping; the solution has fewer water molecules at the surface since molecules of the other component are there as well. The vapor pressure of the solution depends on the relative number of solvent molecules in the liquid phase, so that the vapor pressure of water above the vessel containing pure water is greater than that above the vessel containing the solution. We defined boiling point as the temperature at which the vapor pressure is equal to atmospheric pressure. For a solution (as compared with the pure liquid) to get the vapor pressure to atmospheric pressure, we must add more heat; this causes an **elevation of the boiling point** of the solvent.

The ability of the liquid solvent to convert to the solid phase is decreased by the presence of solute. To obtain pure solid solvent from a solution it is necessary to lower the temperature of the solution below the freezing point of the pure solvent. This is referred to as the **lowering of the freezing point,** or the freezing point depression, of a solution. The magnitude of the elevation of the boiling point and of the lowering of the freezing point is directly proportional to the concentration of solute particles, independent of what they are. A water-soluble antifreeze (for example, ethylene glycol) in the radiator of an automobile not only prevents the engine-cooling solution from freezing at low temperatures by lowering the freezing point, but should also prevent "boiling over" at high temperatures by raising the boiling point of the component with the lowest boiling point, in this case water.

### Osmotic Pressure

A membrane that allows one kind of particle but not another to pass through is called semipermeable. The walls of living cells are permeable to water and other small molecules but not to larger molecules. If an aqueous solution and the pure solvent (water) are separated by a **semipermeable membrane** (one through which water molecules may pass but solute particles may not), more water molecules will pass through the membrane into the solution than from the solution into the pure solvent. This phenomenon is called osmosis, and the force causing the pure water to migrate through the membrane into the solution is called osmotic pressure. Osmotic pressure results from the higher vapor pressure of pure solvents compared with their solutions at the same temperature. Figure 5–7 portrays a simple example of osmosis with air acting as a semipermeable membrane.

The principle of osmosis has been applied in reverse for the important task of obtaining pure water from aqueous solutions (desalination). Using a semipermeable membrane, the system is arranged so that a high external pressure is placed on the solution but not on the pure solvent. The osmotic pressure is more than overcome and more water molecules pass through the membrane from the solution into the solvent than from solvent to solution. This procedure is expensive at present but a potentially attractive method for the treatment of heavily salted water. (See Figure 5–8.)

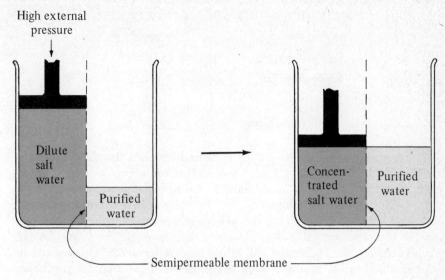

**Figure 5-8**  Desalination by reverse osmosis.

## ACIDS AND BASES

Acids and bases are related as indicated by the following reversible equation

$$\underset{\text{Acid}}{HA} \rightleftharpoons \underset{\text{Proton}}{H^+} + \underset{\text{Base}}{A^-},$$

which shows the dynamic equilibrium between the acid and its proton and anion (negatively charged ion) in solution. An acid is any species (molecule or ion) that can provide a proton; a base is any species that can accept a proton. Carboxylic acids (for example, acetic acid) and mineral acids (hydrochloric acid) dissociate in water to produce protons:

$$\underset{\text{Acetic acid}}{H_3C-\overset{\displaystyle O}{\overset{\|}{C}}-O-H} \xrightarrow{H_2O} H^+ + \underset{\text{Acetate ion}}{CH_3-\overset{\displaystyle O}{\overset{\|}{C}}-O^-}; \qquad HCl \longrightarrow H^+ + Cl^-.$$

Notice that the acetate ion is composed of a group of atoms rather than a single atom (like $Cl^-$), but the **group** often behaves like a single entity.

Acids and bases react to **neutralize** one another. The common alkali bases contain hydroxide ions, which accept protons from acids to give neutral water:

$$\underset{\text{Sodium hydroxide}}{NaOH} \xrightarrow{H_2O} Na^+ + \underset{\text{Hydroxide ion}}{OH^-}; \qquad OH^- + H^+ \longrightarrow H_2O.$$

$$NaHCO_3 \xrightarrow{\ H_2O\ } Na^+ + HCO_3^-$$

Sodium bicarbonate                          Bicarbonate ion

$$HCO_3^- + HCl \longrightarrow H_2O + CO_2 + Cl^-$$

Bicarbonate as base

$$HCO_3^- \underset{\ H_2O\ }{\rightleftharpoons} H^+ + CO_3^{-2}$$

Bicarbonate as acid                          Carbonate ion

The bicarbonate ion is a base since it neutralizes acids. It can also behave as an acid by donating a proton. A species that can accept or lose a proton can behave like a base or an acid and is said to be **amphoteric.**

The acidity of a solution is given by the proton concentration, listed as a **pH value.** For simplicity, the pH scale is related to ppm (by weight) of $H^+$ in water (see Table 5–1, especially the footnote). The lower the $H^+$ concentration, the higher the pH value. Pure water at 25 °C has a pH of 7 and is said to be "neutral." Most life processes depend on the regulation of the acid concentration (pH) and cannot tolerate large fluctuations; the pH of mammalian blood, for instance, is constantly maintained at approximately 7.4.

The dissociation of acids is a reversible reaction:

$$HA \rightleftharpoons H^+ + A^-.$$

At equilibrium, if the dissociation is extensive (giving $H^+$), the acid is called a strong acid; if dissociation is limited (mainly undissociated HA at equilibrium), the acid is called a weak acid. A 1% aqueous solution of HCl (1 gram of HCl to 99 grams of water) is almost 100% dissociated, but a 1% acetic acid solution is only approximately 1% dissociated.

The reversible equation for the dissociation of acetic acid is written

$$CH_3COOH \rightleftharpoons CH_3COO^- + H^+.$$

At equilibrium at 25 °C, in a 1% aqueous solution, most of the acetic acid is not dissociated, that is, there is much more $CH_3COOH$ than $H^+$. If we start with 1,000 molecules of acetic acid at equilibrium, there will be 990 molecules of acetic acid and 10 protons and 10 acetate ions. If more acetate ion is added from an external source, a stress is applied to the system. The equilibrium will shift to relieve the stress, in other words, to use up the acetate ion. This is done by having the acetate ion combine with the available protons to give even more acetic acid, thus lowering the $H^+$ concentration or increasing the pH. Now, if we suddenly add a large number of protons to the equilibrium, another stress is produced. To relieve the stress, the protons must be used up. This can be done by having acetate ions react with the protons to give more acetic acid. Thus, the added acetate ion acts as a "buffer" protecting the system from a sudden change in proton concentration. Aspirin, or acetylsalicylic acid, can cause upset stomach because of its acidity. Some preparations of aspirin come mixed with a buffer so that there is no sudden change in the pH of the stomach. Commercial stomach antacids are buffers as well, that is, they contain the base of a weak acid such as bicarbonate.

$$H_2CO_3 \rightleftharpoons H^+ + HCO_3^-$$

Carbonic acid            Bicarbonate
(a weak acid)

    The reversible equation for the dissociation of pure water is $H_2O \rightleftharpoons H^+ + OH^-$. There are an equal number of $H^+$ ions and $OH^-$ ions in pure water. At equilibrium at 25 °C, the pH is 7, so the concentration of $H^+$ ions (as well as hydroxide ions) is approximately 0.1 part per billion (see Table 5–1). If the $H^+$ ion concentration of an aqueous solution is greater than that of pure water, the solution is said to be acidic (the pH will be less than 7). If the $H^+$ ion concentration of an aqueous solution is less than that of pure water, the solution is said to be alkaline, or basic (the pH will be greater than 7). Thus, mammalian blood is slightly alkaline (pH = 7.4).

    If sodium carbonate is dissolved in water, the $H^+$ ion concentration becomes less than that of pure water, so the sodium carbonate solution is said to be basic. The proton concentration in water is reduced because some of the carbonate ion is combining with water to produce bicarbonate ions and hydroxide ions.

$$2Na^+ + CO_3^{-2} + H_2O \rightleftharpoons HCO_3^- + 2Na^+ + OH^-$$

          Carbonate               Bicarbonate
            ion                   ion

## Table 5–1
## pH or $H^+$ Concentration

| pH | Parts per million of $H^+$ in water (by weight)[a] |
|---|---|
| 1 | 100 |
| 2 | 10 |
| 3 | 1 |
| 4 | 0.1 |
| 5 | 0.01 = 10 parts per billion |
| 6 | 0.001 = 1 ppb |
| 7 | 0.0001 = 0.1 ppb |
| 8 | 0.00001 = 0.01 ppb |
| 9 | 0.000001 = 0.001 ppb |

[a] In the chapter on gases, concentration was given in ppm by species so the term "1 ppm of $SO_2$ in air" means that one molecule in a million is $SO_2$. The concentration of aqueous solutions is usually given in ppm by weight, so the term "1 ppm of aqueous hydrogen ion" means that there is one gram of hydrogen ion per million grams of solution (999,999 grams of which is water). To convert from a weight relationship to a species relationship, one must consider the relative mass of the two species. The mass of water is 18 times greater than that of a hydrogen ion, so 1 ppm by weight is the same as 18 ppm by species.

$$\frac{1 \text{ gram } H^+}{999,999 \text{ grams } H_2O} \times \frac{18}{1} = \frac{18 \; H^+ \text{ ions}}{999,999 \text{ molecules } H_2O} = 18 \text{ ppm.}$$

Relative mass
of $H_2O$ to $H^+$

## THE CHEMISTRY OF WATER

When a fat is hydrolyzed with aqueous sodium hydroxide, the sodium salt of the acid formed is called a soap. Common soaps are sodium and potassium salts of palmitic, stearic, and oleic acids (see accompanying figure).

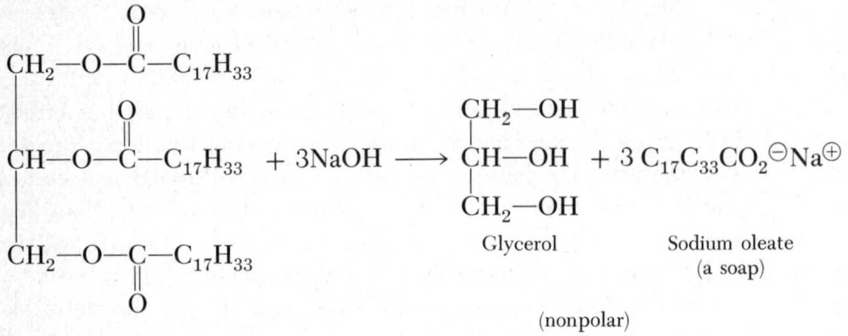

Glycerol   Sodium oleate (a soap)

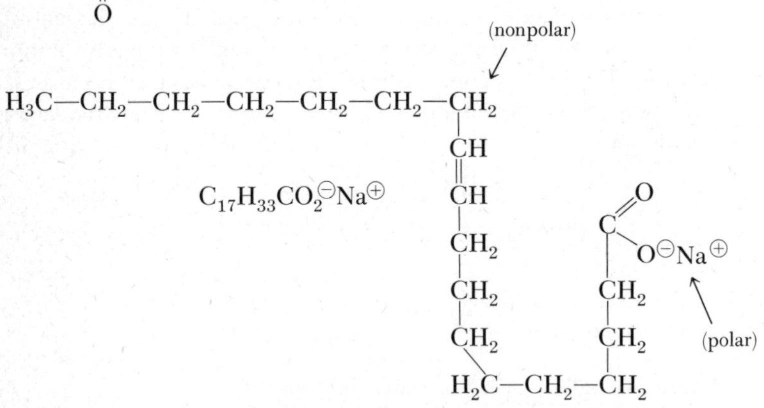

The soap molecule is then composed of a large nonpolar hydrocarbon portion, which is repelled by water, and a carboxylate salt end, which is water-soluble. When soap is added to water, several molecules cluster together forming spherical droplets with the nonpolar hydrocarbon portions pointing in towards the center; the negatively charged polar ends pointing out. These microscopic droplets are called **micelles.** In the cleansing process, oil or grease in the water dissolves in the nonpolar hydrocarbon portion of the micelle. The micelles can be washed away carrying the water-insoluble grease with them.

Unfortunately, calcium and magnesium salts of carboxylic acids tend to be insoluble in water. Thus, in areas where the water is rich in calcium and magnesium ("hard" water), soaps become insoluble and washing is difficult; for example, water-soluble sodium oleate is converted to insoluble calcium oleate. One method of alleviating this problem is to remove the hardness from the water, that is, remove the calcium and magnesium ions, or "soften" the water. Before we can learn this operation in detail, some knowledge of the composition of natural water and the chemistry of its constituents will be useful.

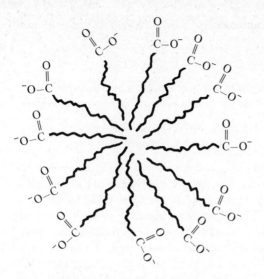

*Soap micelle.* The hydrocarbon portions of the soap molecules are pointed toward the center of the sphere.

### Chemical Constituents of Water

The dissolved impurities in water consist mainly of the salts of three metals—calcium, magnesium, and sodium. There are also trace amounts of iron and manganese salts, and also large amounts of potassium salts. Since the potassium salts behave so like sodium salts (they are both in Group IA in the periodic chart), they will be omitted for simplicity. Calcium and magnesium salts behave in a similar fashion (both in Group IIA) and often only the chemistry of one of them will be mentioned to save duplication. The metal salts dissolve in water to form positive metal ions (cations) and negative ions (anions). There are mainly three anions present in natural water—bicarbonate, sulfate, and chloride. There are then nine possible combinations of cation and anion (3 × 3). These combinations compose almost 100 percent of the dissolved salts in most natural waters. If water containing these six ions is evaporated, the first salt to precipitate is calcium carbonate, which is the least soluble in water; it is followed by magnesium carbonate. Where did the carbonate come from? What happened to the calcium bicarbonate?

| $Ca^{+2}$ | $Mg^{+2}$ | $Na^+$ | $HCO_3^-$ | $SO_4^{-2}$ | $Cl^-$ |
|-----------|-----------|--------|-----------|-------------|--------|
| Calcium ion | Magnesium ion | Sodium ion | Bicarbonate ion | Sulfate ion | Chloride ion |

Calcium carbonate (limestone) is insoluble in water, yet in nature it is transferred from one site to another by running water (limestone caves, stalactites, stalagmites). This is because in the presence of slightly carbonated water, the calcium carbonate is converted to the water-soluble bicarbonate:

$$CaCO_3 + H_2O + CO_2 \longrightarrow Ca(HCO_3)_2.$$

Insoluble                           Soluble

When the water is heated or evaporates, calcium carbonate is reprecipitated by the reverse reaction:

$$Ca(HCO_3)_2 \longrightarrow CaCO_3\downarrow + H_2O + CO_2\uparrow.$$

(*Note:* The arrow pointing down refers to a precipitate, that pointing up refers to a volatile component.) A similar reversible reaction occurs with $MgCO_3$:

$$MgCO_3 + H_2O + CO_2 \rightleftharpoons Mg(HCO_3)_2.$$

Insoluble                                                     Soluble

Thus by boiling of the water, the Mg and Ca ions of the bicarbonates can be removed as insoluble carbonates. The presence of Ca and Mg ions in water is referred to as **hardness.** The hardness caused by the Ca and Mg bicarbonates is called "temporary hardness." The hardness caused by other Ca and Mg salts (for example, sulfates and chlorides) is called "permanent hardness." Bicarbonates are mildly alkaline and neutralize acids with the liberation of $CO_2$. Thus baking soda ($NaHCO_3$) and an acid in baking powder neutralize each other producing $CO_2$, which is the raising agent in baked goods. Noncarbonate salts of Mg and Ca are "hard" but neutral. Sodium bicarbonate is not hard but alkaline. NaCl and $Na_2SO_4$ are not hard and not alkaline. It is important to realize that when salts dissolve to form ions, the original source of the ions is not important. $CaSO_4 + NaCl$ gives a mixture of $Ca^{+2}$, $Na^+$, $SO_4^{-2}$, and $Cl^-$ ions but so will $CaCl_2 + Na_2SO_4$. The quality of water is determined by three factors—hardness, alkalinity, and total weight of dissolved solids:

$$HCO_3^- + H^+ \longrightarrow H_2CO_3 \longrightarrow H_2O + CO_2\uparrow.$$

Carbonic
acid

## Water Softening: Chemical Methods

The oxides of Ca, Mg, and Na are called lime, magnesia, and soda, respectively. Calcium from any source can be precipitated as $CaCO_3$ and Mg from any source can be precipitated as $Mg(OH)_2$. The lime-soda process does just that. Calcium hydroxide can be formed by the reaction of lime with water (lime-slaking):

$$CaO + H_2O \longrightarrow Ca(OH)_2.$$

Lime

The $Ca(OH)_2$ is used to remove carbonate hardness, as follows:

$$Ca(HCO_3)_2 + Ca(OH)_2 \longrightarrow 2CaCO_3\downarrow + 2H_2O.$$

$$Mg(HCO_3)_2 + 2Ca(OH)_2 \longrightarrow Mg(OH)_2\downarrow + 2CaCO_3\downarrow + 2H_2O.$$

It also removes any $CO_2$ in the water:

$$Ca(OH)_2 + CO_2 \longrightarrow CaCO_3\downarrow + H_2O.$$

To remove noncarbonate hardness, $Ca(OH)_2$ and $Na_2CO_3$ (soda ash) are used as follows:

$$CaSO_4 + Na_2CO_3 \longrightarrow CaCO_3\downarrow + Na_2SO_4.$$

$$CaCl_2 + Na_2CO_3 \longrightarrow CaCO_3\downarrow + 2NaCl.$$

$$MgSO_4 + Ca(OH)_2 + Na_2CO_3 \longrightarrow Mg(OH)_2\downarrow + CaCO_3\downarrow + Na_2SO_4.$$

Notice that the water is softened (Ca and Mg removed) but it still contains dissolved salts ($Na_2SO_4$ and NaCl). The important difference between softening carbonate and noncarbonate hardness is the greater amount of dissolved salts remaining in the latter process.

The precipitated $CaCO_3$ is recycled by converting it to lime:

$$CaCO_3 \xrightarrow{\text{heat}} CaO + CO_2 \uparrow.$$
$$\text{Lime}$$

### Cation Exchangers

Picture a porous material containing $Na^+$ ions held firmly enough to resist washing out with water but loosely enough to be replaced by the $Ca^{+2}$ or $Mg^{+2}$ ions or some other positive ion in the water flowing through it. The cation exchanger ("Zeolite") now becomes loaded with $Ca^{+2}$ and $Mg^{+2}$ ions instead of $Na^+$ ions. The water is now loaded with sodium ions instead of $Ca^{+2}$ and $Mg^{+2}$ ions so it is softened. Exchange is very efficient. Eventually all of the $Na^+$ ions in the exchanger are exhausted. To regenerate the exchanger, a concentrated solution of NaCl is passed through the material. The reverse reaction occurs due to the high concentration of $Na^+$ ions. $Na^+$ ions now replace the $Ca^{+2}$ and $Mg^{+2}$ ions in the exchanger, which are washed out in the flush water and thrown away. The exchanger is now ready for use again. The softened water will contain sodium bicarbonate, sulfate, and chloride. Now picture another cation exchanger that does not contain $Na^+$ ions but instead contains $H^+$ ions, or protons. Not only will the $Ca^{+2}$ and $Mg^{+2}$ ions exchange but so will $Na^+$ ions. Moreover, the bicarbonate ions will react with the protons to give $CO_2 + H_2O$. By using a "hydrogen Zeolite," the softened water will contain no metal cations, and therefore no dissolved solids. However, the sulfate and chloride ions will react with the protons to give sulfuric and hydrochloric acids, so the softened water will have a lower pH. To correct this problem both cation exchangers and anion exchangers are used. The anion exchanger will have hydroxide ions attached to the porous material. The softened water from the cation exchanger containing HCl and $H_2SO_4$ is passed through and the acid is neutralized by the hydroxide ions in an exchange process. (See Figure 5–9.)

It is therefore possible to chemically remove *all* dissolved solids by

1. Converting the salts to acids by hydrogen exchange.
2. Converting the acids to water by anion exchange.

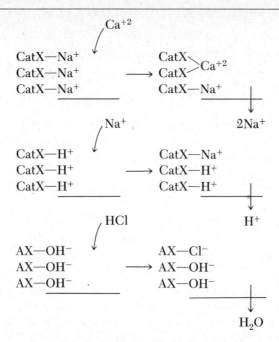

**Figure 5-9**  Exchange columns (schematic).

The result is deionized water, which is needed for cooling water in nuclear power plants, among other uses, because dissolved materials can become radioactive. There are many different highly specialized exchangers that are commercially available. These exchangers *do not* remove un-ionized impurities such as neutral organic matter (sugar, ethylene glycol, DDT).

### Chelating Agents

A chelating agent (another type of water softener) can surround a metallic ion and tie it up in a water-soluble form or "complex" so that it can no longer behave as a free metallic ion. The effect of the agent to a metal has been compared to the effect of a cage to a tiger. When in the cage, the tiger is still available but rendered harmless. The chelating agent prevents $Ca^{+2}$ and $Mg^{+2}$ ions from reacting with soap. These agents are employed in soaps and detergents. They are usually sodium phosphates, for example, sodium tripolyphosphate. Unfortunately, phosphates are excellent nutrients for tiny plants (algae), which thrive in the streams containing wash-water effluent. Compared with other water-softening processes mentioned earlier, the use of phosphate chelating agents is more expensive but more convenient when only a few gallons of water need to be softened at one time.

# WATER PURIFICATION

### What Else Is in the Water?

Other contaminants of water include silica (sand, $SiO_2$) and alumina ($Al_2O_3$), which are constituents of common clay, and also a dangerously increasing amount of nitrate ion ($NO_3^-$) from nitrate fertilizer runoff. The U.S. Pharmacopoeia (USP) defines distilled water as containing up to 10 ppm of dissolved solids (if you evaporate the water from 100,000 gallons of a 1-ppm salt solution, you will obtain one pound of salt). The U.S. Public Health Service says that good water for drinking should not contain more than 500 ppm of total dissolved solids. (See Table 5-2 for comparison of different water sources.) Maximum levels for specific impurities are 0.2 ppm for iron, 0.05 ppm for manganese, and 0.01 ppm for lead and arsenic. There are also contaminants from industrial and municipal wastes.

Particles of suspended matter that cause turbidity of water can settle out to become sediments. If the particles are very fine, they are held in suspension (colloidal) and will not settle out, so coagulation is necessary. Aluminum hydroxide is used to bind the fine particles into clumps of material, called **flocs,** that can be filtered. It is generated by adding aluminum sulfate to naturally alkaline water:

$$Al_2(SO_4)_3 + 3Ca(HCO_3)_2 \longrightarrow 2Al(OH)_3 + 6CO_2 + 3CaSO_4.$$

Settled matter is called sludge. Flocs form faster and are larger if they are formed in the presence of previously precipitated flocs of the same kind. The recycling of previously formed sludge is known as the activated sludge method of removing suspended matter from water.

### Filters

Most community water purification plants use a layer of sand on layers of gravel for both gravity and pressure filters. Sometimes anthracite coal is used as a filter material. Granular activated charcoal is used as a purifier,

**Table 5-2**
**Comparison of Different Water Sources**

| Water Source | Ppm Dissolved Solids |
| --- | --- |
| New York City (Catskill supply) | 41 |
| Chicago (Lake Michigan supply) | 165 |
| Jacksonville, Fla. (from wells) | 410 |
| Seawater | More than 25,000 |

SOURCE: Adapted from A. S. Behrman, "Water is Everybody's Business," Doubleday, 1968.

absorbing impurities that cause objectionable tastes and odors. Diatomaceous earth is so fine a filter that it removes cysts containing the amoebae that cause amoebic dysentery.

Clear water may be loaded with disease-causing organisms that can cause cholera, typhoid fever, dysentery, and hepatitis. Thus water for consumption must be disinfected. Chlorine is the standard disinfectant and usually used in the last step of the purification process. It is effective against most bacteria and other microorganisms, inexpensive, easily detectable in excess, and has a history of safe use. Chlorine is a gas at room temperature and atmospheric pressure and is poisonous (it was one of the "poison gases" used in World War I). It can be used in liquid form or as an aqueous solution of hypochlorite, for example, sodium hypochlorite, which is household bleach (approximately 5% aqueous solution) prepared by treating chlorine with sodium hydroxide:

$$2NaOH + Cl_2 \longrightarrow \underset{\substack{\text{Sodium} \\ \text{hypochlorite}}}{NaOCl} + H_2O + NaCl.$$

The addition of $Cl_2$ to neutral water produces an acidic solution.

$$Cl_2 + H_2O \rightleftharpoons HCl + \underset{\text{Hypochlorous acid}}{HOCl}$$

Chlorine and the hypochlorites are **bactericides.** It is believed that they kill bacteria by stopping the activity of certain molecules essential to their life processes. Other halogens (Group VIIA) are also effective in this way. Iodine has been used by the U.S. Army to disinfect canteen water but its relatively higher cost will limit its use to small water-treatment plants.

### Chemistry in Your Swimming Pool

Although chlorine is used as a bactericide in pools because it is cheap and also an algicide, it can irritate. The use of chlorine lowers the pH of the water (see equation above and note the formation of HCl) so alkali must be added to maintain the pool water at pH 7.2 to 7.8, or slightly alkaline. Sodium hypochlorite solutions are alkaline and raise the pH of the pool water so hydrochloric acid must be added to maintain the proper acidity.

Iodine has some advantages for use in pools but it is only slightly soluble in water (340 ppm at 25 °C). Solid iodine will not disperse quickly in water. It can be effectively dispersed by starting with water-soluble NaI and adding chlorine in small amounts, which causes the following reaction:

$$2NaI + Cl_2 \longrightarrow 2NaCl + I_2.$$

To kill algae in pools, $CuSO_4$ (which gives a blue aqueous $Cu^{+2}$ ion) has been used, often as a mixture with citric acid. The citric acid is a chelating agent, which keeps the copper ion in solution.

$$CuSO_4 \longrightarrow Cu^{+2} + SO_4^{-2}$$

Water of good quality should be clear, safe to drink, soft, and free from unpleasant taste and odor due to microscopic plants and animal life, dissolved $H_2S$, and industrial wastes. Activated charcoal from wood, coal, or other carbon-containing material is often used as a purifier. It can remove much of the organic matter in water and even excess chlorine (used for disinfecting), as shown:

$$2Cl_2 + 2H_2O + C \longrightarrow 4HCl + CO_2.$$

The amount of HCl formed is negligible and neutralized by the natural alkalinity of most community water supplies. To remove dissolved $H_2S$, the water is aerated:

$$2H_2S + 3O_2 \longrightarrow 2H_2O + SO_2,$$

$$H_2O + SO_2 \longrightarrow H_2SO_3,$$
Sulfurous acid

$$H_2SO_3 + Ca(HCO_3)_2 \longrightarrow CaSO_3 + 2H_2O + 2CO_2.$$
Calcium<br>sulfite

Iron is removed from water by aeration as well. The iron oxides and hydroxides formed account for unsightly "red water."

$$4Fe(HCO_3)_2 + O_2 + 2H_2O \longrightarrow 4Fe(OH)_3 + 8CO_2,$$

$$2Fe(OH)_3 \longrightarrow Fe_2O_3 + 3H_2O.$$

### Sewage

Organic matter in sewage is decomposed in water by aerobic bacteria and other organisms in a process using the oxygen dissolved in the water. If the dissolved oxygen is being used faster than it is being replaced, the animal life in the water that depends on the oxygen must compete for the remaining supply. Bacteria, sludge worms, carp, and other unpopular forms compete more favorably than trout and bass. The organic matter therefore affects the quality of the water and determines its inhabitants. Bacterial decomposition of organic matter (containing carbon, hydrogen, oxygen, nitrogen, and sulfur) can occur under conditions of aerobiosis, that is, life in an environment containing oxygen or air:

$$\text{Organic matter containing C,H,O,N,S} + O_2 \longrightarrow CO_2 + H_2O + \underset{\text{Nitrate}}{NO_3^-} + \underset{\text{Sulfate}}{SO_4^{-2}};$$

and it can also occur under conditions of anaerobiosis, which is life in an environment without oxygen or air:

$$\begin{array}{c}\text{Organic matter} \\ \text{containing} \\ \text{C,H,O,N,S}\end{array} + H_2O \longrightarrow CO_2 + \underset{\text{Methane}}{CH_4} + NH_4^+ + \underset{\substack{\text{Hydrogen} \\ \text{sulfide}}}{H_2S} \quad .$$

The anaerobic decomposition of sugars is fermentation, that of proteins is putrefaction.

A measure of pollution of water by organic nutrients is the rate at which the nutrient matter in the water can consume oxygen by bacterial decomposition, or the **biochemical oxygen demand (BOD).** The rate of oxidation is expressed in terms of the half-life of the nutrient, or the time required for half the nutrient to decompose. The greater the quantity of degradable organic matter, the higher the BOD. All lakes have a life cycle; they age and eventually die. The whole process is called eutrophication; it normally takes thousands of years.

Algae and other aquatic plants consume $CO_2$ and release $O_2$ during photosynthesis. To survive they need inorganic nutrients: nitrogen, potassium, phosphorus, and sulfur. When the inorganic nutrients are abundant, these plants grow rapidly. When the plants die, they become food for oxygen-depleting bacteria. If the lake can no longer support many important life forms, it is eutrophic. When conditions are such that many plants are dying, the natural slow evolution of a lake to a swamp and finally to a meadow is drastically hastened (many of our lakes are turning into swamps in one generation). The increasing abundance of plant nutrients in recent years can be attributed to phosphates in detergents and nitrates in fertilizers that have been washed or dumped into the waterways. Such conditions have caused some very troublesome problems for which answers will have to be found. They will not be simple answers and the projected costs may be discouraging. The Environmental Protection Agency (EPA) estimates that by 1975 it will cost Americans $8.1 billion per year to bring just water pollution to safe and acceptable levels. Pollution control of all types will cost the average American $140, or a nationwide $49 billion per year, by 1975.

In general, there are two potential means of control:

1. Monitor what is being dumped into the water.
2. Treat all sewage before returning it to the environment.

### Treatment of Sewage

The commonly used primary treatment is a mechanical process (for example, large settling tanks and crude filters) that screens solid objects or allows them to settle out of sewage (see Figure 5–10). It removes an average of 30%–40% of the BOD in organic wastes and none of the dissolved or nonsettleable matter (for example, fine sediment, bacteria, dissolved phosphates, oils, lead).

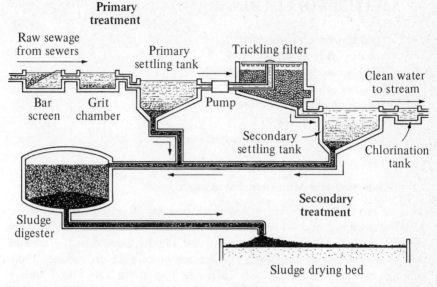

**Figure 5-10** Treatment of sewage, showing primary and secondary treatment. (*From* The Living Waters, *U.S. Public Health Service Publication No. 382.*)

A highly efficacious secondary treatment involves passing the effluent through sand and gravel filters. Aerobic bacteria consume organic matter in the wastewater from the primary treatment. It can remove up to 90% BOD but the average for most water treatment plants is about 60 percent. Dissolved phosphates and nitrates can be partially removed (ca. 60 percent). Suspended matter in the effluent is precipitated via the activated-sludge process. The water is then drawn off to be treated with chlorine before it is returned to the sender. A more advanced tertiary treatment is recommended for heavily industrial areas. By means of special filters, this treatment can remove most pollutants, including some poisons (like phenols, cyanides, and arsenates); but it cannot remove oils, pesticides, radioisotopes, nor any bacteria. This is one of the reasons why the last-mentioned pollutants must be kept out of the water. The type of treatment used should fit the type of pollution; it often does not.

A recent study by the EPA showed that in the U.S., 966 communities serving 7.4 million people dump raw sewage into waterways. Another 2,214 communities, with a population of 54.8 million provide only primary treatment, removing no more than 40 percent of the impurities in the water. Most of the toxic pollutants come from 240,000 water-using industrial plants. Of their total discharge of more than 14 trillion gallons a year, only 4.3 trillion gallons receive any purification at all. More than half the total comes from four industry groups: paper, petroleum and coal, organic chemicals, and primary metals.

## WATER POLLUTION

*"Water, water, everywhere,*
*Nor any drop to drink.*
*The very deep did rot: O Christ!*
*That ever this should be!*
*Yea, slimy things did crawl with legs*
*Upon the slimy sea. . . ."*
Coleridge, *Rime of the Ancient Mariner*

### Industrial and Municipal Pollution

Every day, the average American uses 60–70 gallons of water for drinking, washing, and other household purposes. By contrast, industries in the U.S. use 940 gallons per person per day, slightly more than the amount used by agriculture for irrigation. This is a seventeenfold increase since 1900, although the population has only increased two and a half times. Ninety percent of the water used by industry is for cooling. To manufacture a ton of steel requires 135 tons of water; to make one ton of paper requires 920 tons of water. Water is as essential to industry as material and labor, and the availability of water, more often than the other categories, determines the location of the production facilities. Industry has always looked for large quantities of cost-free high-quality water, often preferring relocation to paying for it.

Industrial needs regarding amount and quality of water vary. The electric power industry is by far the greatest user of water by a factor of ten over each of the two next largest users, steel plants and the chemical industry. The water is mainly used for cooling and does not have to be of high quality. By contrast, the water used in food-packing plants and the synthetic fiber industry must be of higher quality. The water intake of a plant may be reduced if water is used several times over, especially in its use as a coolant. However, some water is constantly lost due to evaporation, which means that remaining reused water is of poorer quality, and contains an ever increasing percentage of dissolved minerals. Reused water gets hotter with each cycle, resulting in an extremely warm effluent when it is finally discharged into the local waterway. The warmed water may upset the aquatic life in the river.

Water is used by industry to dispose of waste products of processing. These wastes include organic wastes, mainly from food-packing plants and pulp mills, soluble inorganic wastes consisting of chemicals (acids and cyanides, for instance), and insoluble particles such as mineral tailings, which may make the water turbid or may settle at the bottom, smothering water-purifying organisms. The indiscriminate use of waterways as dumping grounds has resulted in intolerable levels of pollution, upsetting the delicate balance among plants, insects, and fish and posing problems regarding quality of water for human consumption.

**Biochemical oxygen demand** (BOD) is the standard unit for measuring the polluting capacity of organic wastes. In the process of metabolizing organic materials in water microorganisms use up the oxygen dissolved there that is

**Figure 5-11**  Waste effluent from an industrial company. (*Reprinted, by permission, from* St. Louis Post-Dispatch. *Photograph by Ferguson.*)

necessary to sustain life. Marine biologists say that most fish cannot live long in water that has less than five parts per million (ppm) of dissolved oxygen. The purest drinking water contains eight ppm. Industry is now releasing three times as much organic wastes as domestic sources are, much of it untreated. Some wastes are treated through municipal waterways at a cost to the public. This same public has been apathetic about treating their own sewage. Outdated, overworked sewage plants are not being modernized, with the result that tapwater from downstream is of dubious quality. The different types of treatment, the pollutants they remove, and those they don't remove all compose a situation that is generally not well understood. It is difficult to believe that otherwise the public would tolerate the present situation.

In the summer of 1964, government scientists found that the bottom of Lake Erie was almost completely devoid of oxygen and unable to support most forms of aquatic life. By the summer of 1965 this was confirmed and an effort was begun to save this huge body of "fresh" water. Meanwhile, into nearby Lake Superior, the second largest body of fresh water in the world, the Reserve Mining Company near Silver Bay was discharging an average of 67,000 tons per day of waste iron-ore grindings, slowly but surely building up an ever thicker sludge, according to the Nader Task Force Report on Water Pollution. Warnings by scientists and conservationists to remember Lake Erie have only recently prompted serious efforts to investigate the growing pollution caused by this agent and others.

Why does it take so long for remedial processes to begin? One of the main problems in correcting the pollution of waterways is the reluctance of public officials. A vigorous campaign to correct the problem means:

1. Asking the local industry to clean up or shut down.
2. Asking citizens to pay for new sewage treatment facilities.

The economy of some towns is dependent on a job-producing industry. If the industry decides to move rather than clean up, a serious local recession may ensue. Obviously, some compromise must be reached where the industry can meet its obligations without bankrupting costs. But who will decide? If public officials ask local citizens to update waste disposal facilities they will probably forfeit reelection. Voters may agree to finance new schools or park improvements but few will relish the idea of paying for garbage. If the knowledgeable public officials do not act, who will? State agencies and officials are reluctant to incur the wrath of politically supporting industries and many government agencies and officials find it a thankless or unrewarding task.

Former Interior Secretary Stewart Udall, in testifying before the Senate Subcommittee on Air and Water Pollution, pinpointed the difficulty, "I think that some have been quite correct in saying that water pollution is one part water and one part politics."

While some problems are local and subject to correction by regional or state authorities, others are not, and require a coordinated, or federal, effort.

The Federal Water Quality Administration recently issued a list of the ten most polluted rivers. It included the Houston Ship Channel and the Ohio River; the former is 12 miles long but the latter is nearly 1,000 miles in length!

Ten years ago, San Diego Bay was polluted with phytoplankton, the salt-water equivalent of the algae that have caused eutrophication of fresh-water lakes. The plankton thrived on sewage (60 million gallons a day) dumped into the bay. Dissolved oxygen was badly depleted. All except trash fish left the bay. Beaches were quarantined. Through state and local efforts and some federal assistance, a study was made of sewage disposal needs on an area basis. Four years after the voters approved a $42 million bond issue, the new system was in operation. Within one year, the phytoplankton disappeared and most ocean fish had returned. The blue waters of San Diego Bay are a testimonial to the fact that water pollution is a solvable problem.

Until a few years ago, fighting pollution ranked very low on the list of corporate priorities. That appears to be slowly changing. Self-interest is obviously one primary reason (pollution is damaging to business). Another reason is public pressure. However, many companies ease the pressure through deceptive reasoning and "good guys" advertising. For example, a good share of a company's advertised pollution control expenses may have been used to clean up the incoming water it uses rather than the outgoing effluent (most companies refuse to delineate these expenditures). Research into and development of new nonpolluting technology should be given high priority. Yet, in 1970, the major electric utilities collectively spent seven times as much on advertising and promotion—despite their monopoly—as they spent on research and development.° On the topic of pollution control, *Business Week* (April 11, 1970) quoted an industrialist, "We are living in a fool's paradise if we think industry will do anything until it is forced to." For the most generous industries, pollution control spending of all types amounts to less than 0.7 percent of revenues (less than that spent on advertising the "good guy" image in 1969).

Many industries honestly point to their competitors. If pollution control is not standard for the industry, then surely the cost of production for the "clean" company will be higher, necessitating a higher-priced, less competitive product. However, cleanup or modernization often leads to greater efficiency and new profitable processes.

In 1863, public outcry led England's Parliament to enact an anti-pollution bill to diminish the industrial discharge of gaseous hydrochloric acid into the air. Although industry insisted that the standards were impossible to meet, new processes exceeding these standards were quickly developed. The roles played by the public, government, and industry in the matter of pollution control do not appear to have altered with time or location. Public pressure is needed to get things started. One of the conclusions of the 1971

---

° *St. Louis Post-Dispatch*, Dec. 13, 1971.

Nader Task Force Report on Water Pollution is that "the major problem in pollution control is the vast economic and political power of large polluters, . . . [who] have more influence over government than do those they 'pollute'."

Since utilities are principal polluters in many cities, the development of cleaner nuclear plants is clearly of high priority. Most such plants now are so-called boiling-water reactors, whose discharge of warm water into nearby streams or lakes is a threat to fish and waterway vegetation. One technique used by some plants is to pour the water first into a catch basin, where it cools before flowing into the stream. Another method employs cooling towers for the water, although this boosts the cost of a generating plant by between $5 million and $10 million.

Is it possible to use the hot water constructively? Apparently so. The process of warming cold winter waters with heated-water discharges from power plants has reportedly increased the oyster population in Long Island Sound (New York), and in Japan, has allowed the Japanese to extend their shrimp-growing season. Hot water from a power plant in Oregon's Willamette Valley was piped to orchards and vegetable farms to increase agricultural yields. Indeed, we may lean more to the control of our environment and to use of our pollutants than we have in the past. Water rich in nutrient pollutants can be used for breeding fish in fish ponds. Since ancient times, water has been filtered through sand to remove some pollutants. Recent attempts involving living-soil treatment have produced excellent results. The living soil not only filters the water efficiently but becomes enriched. In these cases, it is important to know what pollutants are present. Because of fear of patent disclosures and the revealing of secret processes with concomitant jeopardy of their investment, many industries do not disclose what is in their effluents. Many of the compounds that are finding their way into our waters are extremely toxic, carcinogenic, mutagenic (causing genetic damage), or teratogenic (causing birth defects). These include pesticides, detergents, acids and minerals (from mine runoff), radioactive compounds and disease-bearing organisms (bacteria and virus). With incentive and funds to conduct research in these and related areas, it may soon be possible to convert pollutants to ecologically harmless materials via economically benefiting processes.

### Detergents

Six billion pounds of soap and synthetic detergents were sold in 1969, earning $1.7 billion for their manufacturers. Soap, made from natural fats and oils, is biodegradable (capable of being broken down biologically into harmless end products such as $CO_2$ and water). Synthetic detergents, which composed 84 percent of the market in 1969, contain compounds that are not so easily degradable to harmless products.

A detergent is composed of a surfactant (the surface active agent, which is useful in wetting both the cloth and the soil) and a detergent builder. Builders by themselves possess little or no detergent effectiveness, but when used in combination with surfactants, produce marked improvements in performance by "softening" the water and furnishing and maintaining the necessary alkalinity for cleaning.

$$H_3C-\overset{\overset{\displaystyle H}{|}}{\underset{\underset{\displaystyle CH_3}{|}}{C}}-\overset{\overset{\displaystyle H}{|}}{\underset{\underset{\displaystyle H}{|}}{C}}-\overset{\overset{\displaystyle H}{|}}{\underset{\underset{\displaystyle CH_3}{|}}{C}}-\overset{\overset{\displaystyle H}{|}}{\underset{\underset{\displaystyle H}{|}}{C}}-\overset{\overset{\displaystyle H}{|}}{\underset{\underset{\displaystyle CH_3}{|}}{C}}-\overset{\overset{\displaystyle H}{|}}{\underset{\underset{\displaystyle H}{|}}{C}}-\overset{\overset{\displaystyle H}{|}}{\underset{\underset{\displaystyle CH_3}{|}}{C}}-\bigcirc-SO_3^{\ominus}Na^{\oplus}$$

ABS

$$H_3C-\overset{H}{\underset{H}{C}}-\overset{H}{\underset{H}{C}}-\overset{H}{\underset{H}{C}}-\overset{H}{\underset{H}{C}}-\overset{H}{\underset{H}{C}}-\overset{H}{\underset{H}{C}}-\overset{H}{\underset{H}{C}}-\overset{H}{\underset{H}{C}}-\overset{H}{\underset{H}{C}}-\overset{H}{\underset{CH_3}{C}}-\bigcirc-SO_3^{\ominus}Na^{\oplus}$$

LAS

Ten years ago, nonbiodegradable sudsing surfactants such as alkyl-benzylsulfonates (ABS) were used resulting in foam-covered waterways and foamy tap water. A convenient change to sudsless linear alkyl sulfonates (LAS), compounds in the same class as ABS, apparently alleviated the problem. Effective advertising that had been used to convince buyers that sudsing detergents were good readily convinced the same buyers that sudsless detergents were good. However, populous Suffolk County on Long Island, New York, recently passed a law banning LAS due to "its proven contamination to ground water in geographical regions with sandy soil."[*] Other coastal regions may follow suit, and once again, specific answers to problems prove not to be universal answers.

A number of inorganic compounds have been used as builders, among them carbonates, bicarbonates, borates, and silicates. None has proven as effective as the phosphates, the most commonly used of which is sodium tripolyphosphate ($Na_5P_3O_{10}$). Phosphates are used by detergent producers to the extent of two billion pounds per year.

The largest contributors of phosphate to the waterways in the United States are household detergents. Their contribution to total phosphate in municipal sewage is estimated to be 70 percent. The balance originates from human waste and agricultural and industrial effluents. Phosphorus, available in limited supply in nature, is one of the fertilizing elements that stimulates the growth of algae and aquatic weeds in freshwater lakes and rivers. These algal growths, called blooms, and the subsequent death and decay of large amounts of odorous plant life, age a body of water. Dissolved oxygen is exhausted, eutrophication sets in, slime and scum appear, and the water is unable to support fish. Phosphorus may be the limiting nutrient for algal growth—the element in shortest supply relative to the nutritional requirements of algae. Research at the Woods Hole Oceanographic Institution suggests that

---

[*] *Chemical & Engineering News*, November 15, 1970, p. 15.

while phosphates appear to be the critical nutrient in many freshwater lakes, it is nitrogen that appears to be critical in coastal waters and estuaries that receive roughly half of the country's sewage. Thus, nitrogen-containing replacements for phosphates could increase the problem. In any event, the removal of phosphates from sewage should result in a decrease in the rate of eutrophication of our waterways. Unfortunately, the removal of phosphates from municipal sewage is costly, necessitating advanced treatment and equipment that would not normally be available for many years (present techniques remove 80%–90% at most).

Phosphates occur in different commercial products to varying degrees. Some phosphate detergents have 30 ppm of arsenic and may be the main sources of arsenic in river waters. A glance at the periodic chart will suggest why this may occur. Arsenic is just below phosphorus in Group V. Thus, the elements have similar properties and form similar compounds, which are difficult to separate.

It is estimated that proper use of present phosphate-containing detergents could decrease phosphate needs by 50 percent. A 1962 U.S. geological survey showed that only 27 percent of the population of the 100 largest U.S. cities is served by hard water. In soft-water areas, much less phosphate is needed. Thus, by selling detergents regionally, packaged according to the already known hardness of the water, the problem could be reduced. To date, detergent producers refuse not only to diversify their packages in this fashion but also even to state the phosphate content on the package.

An obvious solution to the phosphate problem would be the substitution of another compound or formulation. Phosphates are cheap and effective. Any substitute would have to be competitive in both categories. Borates and highly alkaline silicates have been substituted for phosphates but they produce other more dangerous hazards, many of which are unknown to the consumer, who is deluged with misleading advertising and half-truths, for example, "non-polluting, phosphate-free."

One must be cautious. Excessive emotion about one facet of the problem can lead to neglect of others. Phosphate toxicity to humans is known to be low. Possible substitutes would have to be tested extensively, and even then, the sudden effect of large amounts of a new material in our waters may produce unexpected consequences. A possible substitute for phosphates, called nitrilotriacetic acid (NTA), gave excellent results in initial tests, prompting huge investments by several industrial companies.

$$N \begin{cases} CH_2CO_2H \\ -CH_2CO_2H \\ CH_2CO_2H \end{cases}$$

NTA

Nitrilotriacetic acid can "tie up" metal ions by behaving as a chelating agent. Thus, it can pick up and transport different metals by reacting with them to form water-soluble chelates. However, further tests showed that these metals could be liberated in other areas when the chelate is decomposed, or the chelate could cause synergistic effects. Moreover, NTA appears to increase the ability of mercury to cross the placenta and produce congenital abnormalities in rats and mice. This disclosure resulted in a temporary ban on the extensive use of NTA pending further studies and caused substantial financial losses to the chemical companies involved. Possible replacements for present detergents include builderless systems that depend on increased use of surfactants, and phosphate replacements such as polycarboxylates, citric acid derivatives, carbonates, and silicates. It has also been suggested that the best solution might be advanced sewage-treatment systems on a national scale that reduce all pollutants. Overzealous citizens and public officials in some areas have passed laws limiting the use of phosphate detergents without considering substitutes. The best intentions of mice and men. . . .

What is wrong with soap? It is an excellent cleansing agent, biodegradable and relatively nontoxic. However, in "hard" water, soap doesn't foam until enough has been added to precipitate the calcium and magnesium ions present (among others). The precipitate is gummy and accounts for "bathtub ring," which is tolerable in the bathtub but not in the washing machine, where it leaves a film on clothes and can clog the machine. Prior addition of a "softening agent," a compound that precipitates the "hard" ions to form harmless granules, allows the use of soap in hard water. An inexpensive agent is baking soda, or sodium bicarbonate. Will this procedure gain popularity? Synthetic detergents were developed for use in hard water and are easier to use. Soaps are obtained from natural fats and oil, the limited supply of which is in heavy demand by food and feed industries. Over two billion pounds of tallow fat from cattle and sheep would be required to meet the market demand. Current U.S. production is 5 billion pounds per year but the U.S. ships 1.8 billion pounds of surplus abroad to be used as food and feeds 1 billion pounds to animals. At times when the world is worried about malnutrition, it may be difficult to interrupt the food supply of humans and animals to solve the detergent problem. However, tallow is only one of many natural fats that may be applicable to soap production. Also, it may be possible to produce the fats synthetically. The problem of reduced edible tallow could then be met by using fish from cleansed waters as a substitute food. A food-grade fish protein concentrate that fulfills animal protein requirements is already commercially available.

### Mercury Pollution

Mercury, of the winged sandals and hat, is the Roman messenger to the gods; it is also the name of the only metallic element that is liquid at room temperature. There have been several disasters involving mercury, which prompted an immediate investigation on the toxicity of mercury compounds.

Public pressure has resulted in a reduction, but not elimination, of the hazards. The nature of the problem provides a good example to study so that it may serve as a guideline and warning for other metallic pollutants.

Mercury is not a new environmental pollutant. Emerging in the industrial revolution, its use increased considerably with the development of chlor-alkali plants, the introduction of organic mercury compounds (as agricultural fungicides), and the enlarged consumption of coal and oil (which contain mercury) as fuels. Its toxicity was not unknown. A dramatic example was the nineteenth century use of mercury in processing felt hats; workers breathing the fumes suffered brain damage, hence the phrase "mad as a hatter."

How does mercury get into the environment? Direct mercury release includes the waste effluent from manufactured products and that mercury released at some time after its incorporation into a product. Mercury is used in more than eighty different industrial processes, the most outstanding of which is the chlor-alkali industry. In this industry, chlorine gas and alkali are produced via an electrolytic process in which mercury is used and discharged. Other users include:

> Cellulose industries, for preserving the wet pulp from bacterial and fungal biodeterioration.
> Plastic industries, for catalytic reactions.
> Electric industries, for the production of relays, switches, batteries, rectifiers, lamps, etc.
> Pharmaceutical industries, for the production of diuretics, antiseptics, cathartics, some contraceptives, and drugs for the treatment of congestive heart failure.
> Paint industries, mainly for the production of anticorrosive paints.
> Metallurgy, for metal refinement by amalgamation.
> Industrial plants treating nuclear wastes for electrolytic purification.
> Industries producing industrial and control instruments such as thermometers, barometers, mercury pumps, etc.

The 1969 consumption in the United States was approximately 3,000 tons; world production is approximately 9,000 tons per year. Practically all mercury for commercial use is obtained by the ancient method of air or lime oxidation from the ore cinnabar, $HgS$:

$$HgS + O_2 \longrightarrow Hg + SO_2;$$

$$4HgS + 4CaO \longrightarrow 4Hg + 3CaS + CaSO_4.$$

The Oak Ridge National Laboratory reported that in the U.S. in 1968, 26% of the total mercury consumption was for "dissipative" purposes, in which mercury is dispersed to the environment; 23% was released as wastes from chlor-alkali plants; and 23% was of unknown ultimate disposition (for instance, thermometers). Only 18% of the mercury produced was reclaimed for further use.

The chlor-alkali industry, a heavy polluter, has been encouraged to reduce its losses (600 tons per year). Apart from their effluent, some mercury contaminates the alkali produced (at least 1 ppm). This alkali is used by the paper industry (among many other industries), which can result in mercury incorporation in paper products. Incineration of the paper results in the release of mercury into the atmosphere, whence it can settle into normally undisturbed remote areas. A similar indirect release of mercury to the atmosphere arises from the burning of fossil fuels; coal may contain 0.5 ppm and petroleum 2 ppm of mercury.

Mercury compounds have been shown to be effective killers of fungi. Seeds of cereal grains, particularly susceptible to fungus attack, were treated with alkyl-mercury compounds (for example, Panogen, which is methyl-mercurydicyandiamide). Treated seeds were eaten by birds and fed to animals, whose concentration of mercury compounded the problem. In recent studies, U.S. and Canadian wheat samples were analyzed and reported as having 0.02 to 0.08 ppm mercury; the World Health Organization's recommended safety limits are 0.05 ppm. Samples of Asian rice, from fields heavily treated with mercury fungicides, in analysis showed 0.2 ppm mercury.

To prevent the spread of mold in lawns, a variety of fertilizers, herbicides, and fungicides are used that contain phenylmercury acetate (PMA). Similar mercury-containing compounds are used by commercial laundries and manufacturers of paper and paints to prevent mold and mildew. The use of these organomercurials has now been sharply reduced in the U.S.

Of primary concern is water pollution and the contamination of fish. The amount of natural mercury in streams and lakes averages 0.03 parts per billion (ppb); oceans average 0.03 to 5.0 ppb. Approximately 50 percent of the mercury released into water is caused by nature, with 5,000 tons per year of mercury transferred from the continents to the oceans by the rivers following continental weathering.

*It is extremely important to differentiate between different kinds of compounds of mercury:* inorganic mercury salts and organomercury compounds (for example, methylmercury). The organomercury compounds present the greatest danger to man. Less than 2 percent of ingested inorganic mercury

$$\langle \text{phenyl} \rangle - Hg - O - \overset{\overset{\displaystyle O}{\|}}{C} - CH_3 \qquad\qquad H_3C - Hg - CH_3$$

PMA                    Methylmercury

is absorbed into the bloodstream but over 90 percent of monomethyl and dimethyl mercury (collectively referred to as methylmercury) is absorbed, only slowly excreted, and extremely toxic. Thus, many mercury compounds, such as calomel, or mercurous chloride, which has been much used in medicine as a purgative, are not extremely toxic, nor are the mercury amalgam fillings in your teeth. Unfortunately, both inorganic mercury salts and elemental forms of mercury can be converted aerobically to toxic methylmercury by microorganisms (bacteria, specifically) in the water. This means that the mercury and mercury salts in the mud at the bottom of the sea can be continually converted to water-soluble, toxic methylmercury. Industries are no longer discharging mercury into the waterways but the concern now is how to recover the vast amounts near the industrial plants that will be slowly and continuously converted to methylmercury.

Mercury in water is concentrated in fish after it enters through the gills during respiration or after the ingestion of phytoplankton containing absorbed mercury, a main source of food for fish. Over 90 percent of the mercury in fish is in the form of methylmercury compounds which is apparently so tightly bound to the protein that it is not disrupted by freezing, boiling, or frying. As the food chain continues, birds or game fish eat the small fry and the concentration of mercury increases (pike in contaminated waters will be carrying methylmercury in their bodies at a concentration 3,000 times that of the surrounding water). The greatest predator, man, then enters the chain by eating the game fish (tuna, swordfish) containing excessive concentrations of mercury.

Effects of mercury poisoning can be reversible, irreversible, or fatal, and include headaches, numbness, loss of motor control (hands, speech) and sensory operations (auditory and visual).

As a temporary guideline based in part on the general assumption that the average American fish intake is less than two meals per week, the U.S. Food and Drug Administration set a maximum allowable level of 0.5 ppm mercury contamination for fish. A vast majority of fish sampled show concentrations well below this level with only occasionally higher levels. However, 95 percent of samples of swordfish recently tested by the FDA exceeded 0.5 ppm, which almost resulted in the closing of that industry. High mercury concentrations in fish from Lake Erie (up to 7ppm) resulted in bans that have caused an estimated $1 billion loss in commercial and sport fishing in that area. The cost from pollution will undoubtedly exceed the cost of pollution control. Careful monitoring of fish and fish products is essential to prevent accidental massive doses.

Of several methods of analysis available, one of the most popular is neutron activation analysis (NAA), which consists of bombarding the sample with neutrons, and converting the mercury to a radioactive isotope that can then be easily measured. This method does not distinguish between organomercurials and inorganic mercury, an important difference.

### Biological Effects

Current studies on humans show a wide range of sensitivity to mercury, with reports of mercury poisoning symptoms in instances of low body concentration and lack of effects in people with high concentration. The biological effect of mercury strongly depends on its concentration, chemical composition of the compound, the temperature and acidity of the medium, and the organism itself.

Organomercury compounds accumulate in the nervous system, provoking severe functional disorders. They cause damage to yeast cells; the surface membrane loses its semipermeability resulting in the loss of potassium ions. In fish, these compounds affect the skin and gills, and the central nervous system. Genetically, it is believed that organomercury compounds cause chromosome damage in plant cells and fruit flies. No evidence of chromosome damage due to mercury has been found in humans.

*Future Problems*

It is obvious that control over mercury emissions must be enforced, and that research must be encouraged toward removing or safely deactivating the excess amount of mercury and its compounds already in living systems and bodies of water. But also bear in mind that mercury is only one of the so-called heavy metals. A glance at the periodic table will show that mercury (at. no. 80) is followed by thallium (at. no. 81), which has highly toxic inorganic salts and which is followed by lead (at. no. 82), which. . . .

## Cadmium: A Forewarning

Above mercury on the periodic chart is cadmium and above that is zinc. Cadmium is corrosion-resistant; it has no distinct ore but is a contaminant of zinc ore (ca. 0.5%). It is used for plating boats and chassis, and for pigments (CdS), plastic stabilizers, and batteries. Most of its uses are dissipative. The United States consumes about 13 million pounds per year. In remelting scrap iron, about 2 million pounds of cadmium are driven off into the environment, part of the 6 million pounds per year that are dumped in the U.S. Cadmium and its compounds are toxic, persistant, and cumulative. If inhaled, it is supertoxic, causing direct lung damage. Ingested or absorbed material leads to emphysema and kidney failure. It is possible to get chronic poisoning from long-term low-level exposure. It is estimated that 100–200 micrograms ($100 \times 10^{-6}$ grams) per day leads to hypertension, a disorder that already affects 23 million Americans. The U.S. Public Health Service claims that the maximum amount in potable water should be 0.01 ppm. This is a recommendation; there are no enforcement procedures. Cadmium is found in fossil fuels (1–2 ppm), and has recently been discovered in tobacco. The banning of cadmium for dissipative uses (there is 18 percent cadmium in silver solder, for example) could lead to increased cadmium pollution. This peculiar turnabout is brought on by the economy of cadmium. Present in zinc in trace amounts and having a lower boiling point, it would be allowed to go "up the stack" and into the environment during the purification of zinc if not for the fact that its present commercial value is about $3 per pound, compared with 22¢ per pound for zinc. At these prices it is worth collecting. If there were no incentive for collecting the cadmium (that is, no market), more would be allowed to dissipate into the atmosphere.

The crux of the problem is that the burden of pollution must be transferred from the pollutee to the polluter. Instead of society's proving after the fact that the product, process, or by-products are harmful, the potential polluter should be forced to prove that the material is harmless before allowing it to be sold or dumped. We are a long way from that philosophy. But the mercury matter has prompted some early investigations (pre-disaster) into the dangers of other metals, as can be seen in Table 5–3.

### Oil Pollution

Massive oil spills such as those from the grounding of the *Torrey Canyon* (1967) and the Santa Barbara Channel blowout (1969) received international headlines and provided the impetus for careful investigation of potential oil pollution measures—both preventive and corrective. Once the oil is on the water, it becomes a difficult and expensive operation to remove it. Lighter than water and nonmiscible, the oil may spread in a film of $\frac{1}{100}$ inch thickness over a 25-square-mile area in 8 hours, a difficult expansion to contain. Skimming and vacuum-line equipment must be able to perform in high seas, against strong winds and currents. Chemical dispersants are often toxic and add an additional pollutant to the surroundings. Cleanup costs average $1–$3 per gallon, which may be negligible when compared with cost from property damage. The Santa Barbara cleanup costs ran to $4.5 million, but legal actions against Union Oil approximate $2.5 billion.

The problem is not so much the big spills as the cumulative effect of continual accidental and deliberate discharges. Oil tankers are unique in that they normally carry cargo only one way. After unloading their cargo, they fill the almost empty tanks with seawater for ballast for the return voyage. Upon approaching home port, to prepare for reloading, these tanks are emptied of the seawater and the oil residues that had remained in the tanks. This accepted dumping is supposed to be done miles away from port. Added to this are the continual spills during loading or unloading of the oil.

Some international regulations will have to be adopted. As a preventive measure, vessels should be designed to fit their particular cargo. Safety features such as double hulls should be considered. Requirements for personnel who supervise cargo transfers should be updated, together with harbor facilities. Unfortunately, safety is often sacrificed for profit. Ships are often registered in countries with poor inspection laws and carry crews of dubious experience.

Inland waters are not immune. An American-Canadian commission investigating potential oil pollution in Lake Erie was appalled at the vast amounts of oil in the Lake Erie basin. For example, during 1969 more than 1,000 barrels per day of oils and greases were discharged into the Detroit River, the principal tributary of Lake Erie. At any one time during the

**Table 5-3**
**Trace Metals That May Pose Health Hazards in the Environment**

| Element | Sources | Health Effects |
|---|---|---|
| **Nickel** | Diesel oil, residual oil, coal, tobacco smoke, chemicals and catalysts, steel and nonferrous alloys | Lung cancer (as carbonyl) |
| **Beryllium** | Coal, industry (new uses proposed in nuclear power industry, as rocket fuel) | Acute and chronic system poison, cancer |
| **Boron** | Coal, cleaning agents, medicinals, glass making, other industrial | Nontoxic except as boran |
| **Germanium** | Coal | Little innate toxicity |
| **Arsenic** | Coal, petroleum, detergents, pesticides, mine tailings | Hazard disputed, may cause cancer |
| **Selenium** | Coal, sulfur | May cause dental caries, carcinogenic in rats, essential to mammals in low doses |
| **Yttrium** | Coal, petroleum | Carcinogenic in mice over long-term exposure |
| **Mercury** | Coal, electrical batteries, other industrial | Nerve damage and death |
| **Vanadium** | Petroleum (Venezuela, Iran), chemicals and catalysts, steel and nonferrous alloys | Probably no hazard at current levels |
| **Cadmium** | Coal, zinc mining, water mains and pipes, tobacco smoke | Cardiovascular disease and hypertension in humans suspected, interferes with zinc and copper metabolism |
| **Antimony** | Industry | Shortened life span in rats |
| **Lead** | Auto exhaust (from gasoline), paints (prior to about 1948) | Brain damage, convulsions, behavioral disorders, death |

NOTE: Bismuth, tin, and zirconium are also present as pollutants from industry, coal, and petroleum, respectively. C&EN has no data on their possible health effects. Titanium, aluminum, barium, strontium, and iron are air pollutants that occur naturally in dust from soils. Other metals known to be in the environment—chromium, manganese, cobalt, copper, zinc, and molybdenum—are essential to human health and probably pose no danger at current levels.

SOURCES: Battelle Memorial Institute, Dartmouth Medical School. Reprinted from *Chemical and Engineering News*, July 19, 1971, p. 30. Copyright 1971 by the American Chemical Society. Reprinted by permission of the copyright owner.

navigation seasons, freighters and tankers carry an estimated 120,000 barrels of oil as either fuel or cargo on Lake Erie and some leakage will occur. The polluting oil is not always on top of the water. It is estimated that in Cleveland Harbor 17,600 tons of oil and grease were included in the 660,000 tons of dry solids removed during the 1966–67 dredging operations. The oil had adsorbed onto suspended materials and deposited as sediment.

Natural oil leakages add to the problem. Active oil seepage in the Santa Barbara channel, a geologically unstable area, has been estimated to amount to 50 barrels a day.

The destructive effects of oil pollution have been concisely described by a governmental commission investigation of the Torrey Canyon oil spill°:

> Oil slicks on the water seem to have an irresistible attraction for water birds. Once a bird alights on the oil mass its feathers become matted and oil-soaked. The almost inevitable result is death by drowning through loss of buoyancy, by toxicosis from ingested oil, or from exposure caused by loss of body heat insulation; unable to fly, the birds may slowly starve to death or be eaten by predators. Successful treatment of rescued birds is extremely difficult.
>
> When surface-feeding fishes swim into the floating oil, the bodies and gills become coated. If death does not occur from such contact, their flesh absorbs the taste and odor-producing fractions of the oil, rendering them unfit for human consumption for a long time afterward.
>
> As an oil mass moves landward, toxic oil fractions can bring death to both larval and adult forms of invertebrate marine life that inhabit the shallow, near-shore areas. Marine life, valuable to man as a food resource, may be totally destroyed or, at the least, is likely to acquire disagreeable tastes and odors from the oil. Beds of seaweeds, valuable as a food or industrial material, can be totally destroyed by the oil.
>
> Finally, as the oil hits the shorefaces and collects in harbor or port areas, it blankets everything in its path. The usefulness of beaches for recreation suddenly ends. Navigational and fire hazards are created in harbors, ports, and marinas. Shorefront properties are despoiled, and the air reeks with the fumes.
>
> In less heavily affected areas the odor may be less, but grime and stain abound. Snow-white cruisers and picturesque sailboats show a dark smear at the waterline; small children playing at the beach come home with oily feet; swimmers are coated with oil patches which cling to their skins and mat their hair. Removal requires thorough scrubbing with detergent and kerosene.
>
> When chemical compounds are used in the shallow littoral areas to emulsify, precipitate, or sink oily materials, or otherwise "cleanse" the surface of the water, the effect on the aquatic community may be more deadly than the floating oil itself.

---

° U.S. Secretary of the Interior and U.S. Secretary of Transportation, *Oil Pollution: A Report on Pollution of the Nation's Waters by Oil and Other Hazardous Substances,* February 1968.

*Oil that has been removed from the sea and from the shore areas
presents yet another problem. If it is thoughtlessly disposed of near drain-
ages, the next rain could return the pollutant into water supplies and fish-
ery grounds. If it is buried where it could affect groundwater, water supplies could be contaminated. Beyond these aspects lie potential costs whose
magnitude cannot yet be readily estimated.*

## POSTSCRIPT

*No man is an island, entire of itself;
every man is a piece of the continent,
a part of the main. If a clod be
washed away by the sea, Europe is the
less, as well as if a promontory were,
as well as if a manor of thy friends or
of thine own were; any man's death
diminishes me, because I am involved in
Mankind, and therefore never send to know
for whom the bell tolls; it tolls for thee.*

**John Donne,** *Devotions* **(XVII)**

Pollution problems are symptoms of environmental decay. To deal with
the symptoms, we must understand the underlying causes. One major cause
is the failure of our economic system to account for the cost of pollution;
to include it as part of the cost of goods sold, just as raw materials, labor,
and equipment are included. If it is possible to dump wastes without charge,
then there is no economic incentive for control. Therefore it is not reasonable
to depend on voluntary action. The fairest approach appears to be firm
standards uniformly enforced. For some industries, international agreement
would be required. But agreements are needed at all levels. It does a city
little good to pass a strict water pollution law if another city upstream runs
its raw sewage into the river. People cannot seal themselves up as individuals
or nations or species—they depend on each other to a greater extent than
is immediately obvious.

Possible solutions, by their very nature, will involve not only science
and technology, but law, sociology, politics, and economics. Agreement in
these areas is never easily reached. Strong pressure on the part of a knowledgeable public and an agreeable goal for the common good will be needed
to overcome individual, independent dissatisfaction and political apathy. In
the long run, the cost of environmental rehabilitation will be borne by the
citizen, but he is bearing the cost already through lower community health,
lower water quality, and the destruction of resources such as fisheries and
wildlife areas. Moreover, the need for new, cleaner processes and the com-

petitive advantage of reduced pollution costs may result in greater efficiency and lower prices of products.

But the key to it all appears to be cooperation. The noted microbiologist René Dubos claims, "We must adopt a new social ethic . . . based on harmony with nature as well as man, instead of the drive for mastery." However, he also notes that "modern man can adjust to environmental pollution, intense crowding, deficient or excessive diet, as well as to monotonous and ugly surroundings." Thus, it has been stated that "the problem may not be one of the extinction of *Homo sapiens* but his mutation into some human equivalent of the carp now lurking in Lake Erie's fetid depths." °

---

° *Newsweek*, January 26, 1970.

## PROBLEMS

1. Define or explain the following terms and give an example of each:
   (a) solution                      (b) melting point
   (c) primary treatment of sewage    (d) dynamic equilibrium
2. Why would a large barrel of water placed in a barn help prevent the animals within from freezing?
3. The phase diagram for carbon dioxide is shown below. What is the triple point temperature and pressure? What pressure is needed to have solid $CO_2$? What temperature is needed to have liquid $CO_2$? At what $T$ will solid and vapor coexist in equilibrium at 1.0 atm pressure (the sublimation point)?

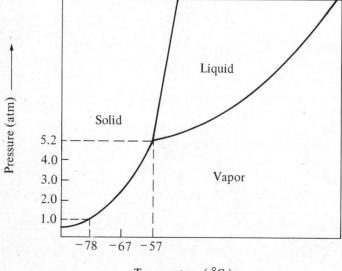

4. Which liquid, water or chloroform, would you predict would require more heat energy to go from liquid to vapor? (See Figure 5–2.)

5. The diagram depicts vapor pressure vs. temperature curves for pure water and salt water. Which one represents pure water? Explain.

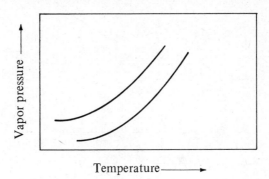

6. From the heating curve in the text (Figure 5–3), what do you judge happens to the potential and kinetic energy of the system from

$$t_1 \longrightarrow t_2 \text{ and } t_2 \longrightarrow t_3?$$

7. According to the reversible equation

$$\text{solid} + \text{heat} \rightleftharpoons \text{vapor},$$

at equilibrium the addition of heat energy to the system will convert solid to vapor. If heat is removed from the system, will there be (a) more solid and less vapor? (b) more vapor and less solid? (c) a change in temperature? Will the potential energy of the system (a) increase? (b) decrease? (c) not change?

8. How many pounds of ice at 0 °C could be melted with the energy obtained from the condensation of one pound of water vapor at 100 °C?

9. What is the purpose of adding salt to ice-covered surfaces?

10. Which will freeze faster, a freshwater pond or a saltwater pond?

11. Is skin a semipermeable membrane?

12. Define "hardness" in water.

13. Vinegar is an aqueous solution of acetic acid. Do you expect the pH of vinegar to be greater than 7.0 or less than 7.0?

14. Explain why evaporation has a cooling effect.

15. According to the phase diagram (below) for the compound Umselite:
    (a) What happens to its melting point as pressure is increased (increases, decreases, remains the same, cannot determine)?
    (b) What is the normal boiling point?
    (c) What is the triple point temperature and pressure?
    (d) At 1.2 atm pressure, liquid Umselite will exist at what temperature(s)?
    (e) At 30°, solid Umselite will exist at what minimum pressure?

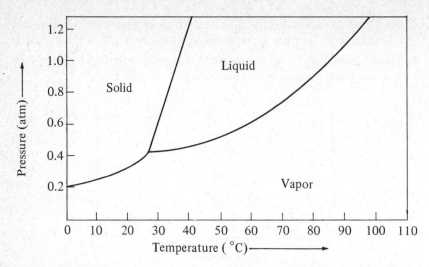

16. The heating curve for the compound Z is shown herewith. Starting with solid Z at 25°, heat is added at a continuous rate resulting in the formation of liquid Z, and finally gaseous Z. The melting point of Z is 40°; the boiling point is 120°

   (a) More heat is required to raise 1 gram of liquid Z by 1° than is required to raise 1 gram of solid Z by 1°. True or false?

   (b) More heat is required to convert liquid Z to vapor than is required to convert solid Z to liquid. True or false?

   (c) If the experiment is repeated starting with twice as much mass of solid Z at 25° as was used originally, what changes would occur in the heating curve?

   (d) Explain what is happening to the potential and kinetic energy of the system during the time that has elapsed for *each* straight line, for example, (1), (2), (3), (4), (5).

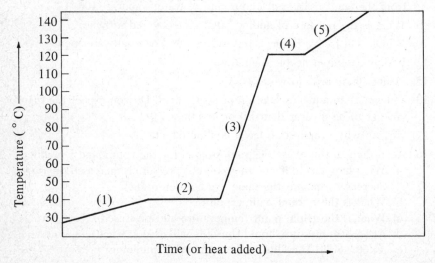

17. List the destructive effects of oil pollution.

18. At room temperature, the equilibrium vapor pressure of water is 0.03 atm. If the observed pressure of water in your house is 0.01 atm, what is the relative humidity in the house?

19. Assume that pressure is constantly maintained at 1 atm. Are the statements (a)–(e) "true" or "false?"

$$\text{liquid} + \text{heat} \rightleftharpoons \text{vapor}$$

According to the equation above, at equilibrium:

   (a) The number of molecules going from liquid to vapor is equal to the number of molecules going from vapor to liquid.

   Heat is suddenly and briefly added to the system but the system adjusts to return to equilibrium. As a result of this change:

   (b) There is now more vapor and less liquid.
   (c) There must still be liquid present.
   (d) The kinetic energy of the system has not changed, nor has the temperature.
   (e) The potential energy of the system has increased.

20. Draw a possible phase diagram for a substance that has the following properties:
   (a) Changes of pressure have no effect on the melting point, which is 25°; the triple point temperature and pressure is 25 °C and 0.6 atm.
   (b) The boiling point is 85°.
   (c) The solid phase will not exist at any temperature if the pressure is below 0.1 atm.

## BIBLIOGRAPHY

1. D. R. Zwick, ed. "Water Wasteland." Nader Task Force Report on Water Pollution, vols. 1 and 2. Washington, D.C.: Center for Study of Responsive Law, 1971.

2. *Water and Industry.* United States Department of the Interior, Washington, D.C.: U.S. Government Printing Office, 1972. 0-457-795 available for $0.20 from Supt. of Documents, Stock #2401-2095.

3. J. C. DeHaven, L. A. Gore, J. Hirshleifer. *A Brief Survey of the Technology and Economics of Water Supply.* Santa Monica, Cal.: Rand Corp., 1954.

4. "Cleaning Our Environment." Summary of report by the American Chemical Society Committee on Chemistry and Public Affairs. *Chemical and Engineering News,* September 8, 1969, pp. 58–69.

5. S. Keckes and J. K. Miettinen. "Review on Mercury as Marine Pollutant," for F.A.O. Technical Conference on Marine Pollution, Rome, December 1970.

6. L. Dunlap. "Mercury: Anatomy of a Pollution Problem." *Chemical and Engineering News,* July 5, 1971, pp. 22–34.

7. D. H. Klein and E. D. Goldberg. "Mercury in the Marine Environment." *Environmental Science and Technology* 4 (September 1970): 765–67.

8. W. Grant. "Mercury in Man." *Environment* 13, no. 4 (1971): 2–15.

9. T. Aaronson. "Mercury in the Environment." *Environment* 13, no. 4 (1971): 16–23.

10. S. Novick. "A New Pollution Problem." *Environment* 11, no. 4 (1969): 2–9.

11. G. Löfroth and M. E. Duffy. "Birds Give Warning." *Environment* 11, no. 4 (1969): 10–17.

12. J. McCaull. "The Black Tide." *Environment* 11, no. 9 (1969): 2.

13. U.S. Secretary of the Interior and U.S. Secretary of Transportation. *Oil Pollution: A Report on Pollution of the Nation's Waters by Oil and Other Hazardous Substances.* February 1968, pp. 3–4.

14. J. McCaull, "Building A Shorter Life (Cadmium)." *Environment* 13, no. 7 (1971): 2.

15. *Special Report on Potential Oil Pollution.* Third Interim Report on Pollution of Lake Erie, Lake Ontario and the International Section of the St. Lawrence River by the International Joint Commission, April 1970.

16. "Can the U.S. Win the War Against Pollution?" *U.S. News and World Report,* March 20, 1972, pp. 84–89.

17. A. S. Behrman. "Water Is Everybody's Business." Garden City, N. Y.: Doubleday, 1968.

Information was gleaned from various journals and news media including *Chemical and Engineering News, Environmental Science and Technology* and the January 26, 1970, issue of *Newsweek.*

A compendium of data on water pollution can be found in the following paperback texts:

18. A. Turk, J. Turk, and J. T. Wittes. *Ecology, Pollution, Environment.* Philadelphia: W. B. Saunders, 1972.

19. H. S. Stoker and S. L. Seager. *Environmental Chemistry: Air and Water Pollution.* Glenview, Ill.: Scott, Foresman, 1972.

An excellent reference for all aspects of environmental pollution is:

20. Thomas R. Detwyler, ed. *Man's Impact on Environment.* New York: McGraw-Hill, 1971.

An excellent 30-minute film, "The Gifts," can be rented free from the Environmental Protection Agency's regional public information offices.

# Biochemistry

# 6

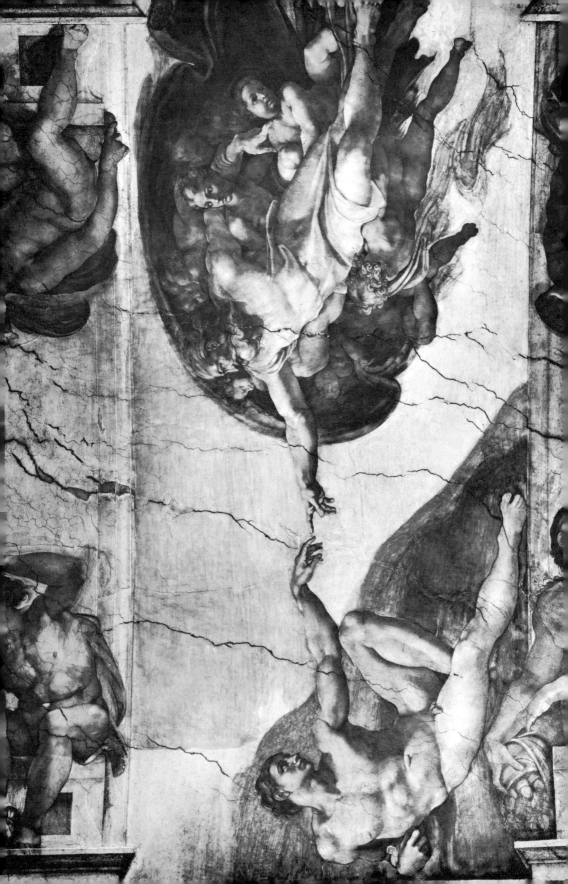

*The problem is: How does energy drive life?*

Albert Szent-Györgyi (Nobel Laureate)

The study of life and the living process is truly a fascinating subject. What differentiates **animate** from inanimate matter? In the past hundred years, an enormous amount of information has been obtained concerning the mechanism of life, yet the surface has barely been scratched. We have learned how to treat a multitude of afflictions that have relentlessly plagued mankind but our understanding of basic processes that often cause the problem is usually embarrassingly meager. All of us have taken aspirin for a headache or for pain, yet the details of how aspirin works still remain a secret. Only recently has the state of the art been advanced sufficiently so that details concerning the mechanism of drug action can be rigorously studied.

What controls the growth and reproduction of cells? How is genetic information transferred from generation to generation? What is the aging process? What does it involve? These are a few of a host of questions that biochemists are now asking. Although detailed answers are not known, enough information is available in some cases to permit us to sketch a rough picture of what is involved. This picture will undoubtedly be changed with time as more information is gathered but it represents man's first attempts to collate, understand, and rationally explain some of the processes occurring within himself and his environment.

A few hundred years ago, the prevailing attitude in science was one that attempted to differentiate animate from inanimate matter (the terms *organic,* meaning "derived from living organisms," and *inorganic,* or "from lifeless or nonliving systems," originated in this manner). It was believed that some vital force was present in organic matter that differentiated this substance from other material. Although it was known that both organic and inorganic matter could be induced to undergo chemical reactions, interconversion of inorganic to organic matter was thought impossible.

The first to disprove this concept was Friedrich Wöhler, who in 1838 removed the veil of mysticism surrounding organic compounds by synthesizing urea, an organic substance present in urine, from an inorganic salt, ammonium cyanate.

$$NH_4OCN \longrightarrow NH_2-\overset{\overset{\textstyle O}{\|}}{C}-NH_2$$

Urea

In a statement to Jöns Berzelius, a founder of modern chemistry, Wöhler wrote, "I can prepare urea without requiring a kidney, or an animal, either man or dog." Wöhler's observation was the first of a series of observations

**"The Creation of Man" by Michelangelo.** (*Courtesy of Alinari–Art Reference Bureau.*)

that ultimately resulted in the acceptance of the concept that the reactions and composition of animate and inanimate matter were regulated by the same laws. Thus the question of division of matter was resolved and the birth of modern biochemistry ensued.

Let us begin the discussion of biochemistry by focusing on the basic composition of living systems. The bulk of organic matter present in such systems can be classified under three main categories: **proteins, carbohydrates,** and **lipids.** Nucleic acids, a fourth category, compose only a small fraction of the total organic matter present in cells, but they are very important in directing the operations of the cell, and are involved in the transmission of genetic information. The chemical composition of nucleic acids is mostly C, H, N, O, and P. Protein, the main constituent of animal cells, is composed primarily of the elements of C, H, N, O, S, and often P. Carbohydrates are composed of C, H, and O, and the composition of lipids is similar to that of protein.

The manner in which the basic elements of matter are combined in nature to form a cell is shown diagrammatically in Figure 6–1. The basic components, carbon dioxide, water, and nitrogen, enter a complex series of reactions to give amino acids, fatty acids, sugars, and nitrogen bases (intermediates in protein, lipid, carbohydrate, ribonucleicacid, and deoxyribonucleic-acid synthesis, respectively). The combinations of these basic building blocks ultimately lead to the molecules of a cell, which in turn lead to the cell itself. The molecular component of a cell, that for the bacterium *E. coli*, is shown in Table 6–1.

**Table 6-1**
**Molecular Components of the Bacterium *E. coli***

|  | Percentage of Total Weight | Approximate Number of Each Kind |
| --- | --- | --- |
| Water | 70 | |
| Proteins | 15 | 3,000 |
| Nucleic acids | | |
| DNA | 1 | 1 |
| RNA | 6 | 1,000 |
| Carbohydrates | 3 | 50 |
| Lipids | 2 | 40 |
| Building-block molecules and intermediates | 2 | 500 |
| Inorganic ions[a] | 1 | 12 |

[a] The inorganic ions consist mainly of $Na^+$, $K^+$, $Mg^{+2}$, and $Cl^-$, with trace ions of such metals as Mn, Fe, Co, Cu, and Zn.
SOURCE: Adapted from Albert L. Lehninger, *Biochemistry* (New York: Worth Publishing Co., 1971).

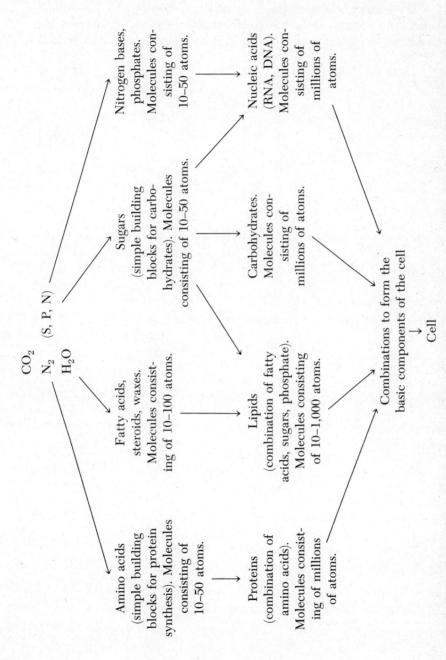

**Figure 6-1** Rough representation of ways in which the basic elements C, H, O, N, S, and P are combined to form cells. (*Data from Albert L. Lehninger, Biochemistry (New York: Worth, 1971).*)

## PROTEINS

The term **protein** originates from the Greek *proteios*, meaning "of the first rank." All protein can be broken into smaller units, called **amino acids.** Each protein consists of a combination of many amino acids linked to form larger molecules.

There are twenty amino acids found universally in protein. The structures of all twenty amino acids are shown in Table 6–2. Note that all of the amino acids except proline can be represented by the following general structure:

$$\boxed{R}-\overset{\overset{\displaystyle H}{|}}{\underset{\underset{\displaystyle NH_2}{|}}{C}}-CO_2H$$

Amino acid

Proteins are synthesized by the body by combining the amino (—$NH_2$) portion with the carboxyl portion (—$CO_2H$) of a different amino acid to form a **peptide bond.** Note that amino acids such as glycine and alanine can be

$$NH_2-\overset{\overset{\displaystyle H}{|}}{\underset{\underset{\displaystyle H}{|}}{C}}-\overset{\overset{\displaystyle O}{\|}}{C}-OH \;+\; NH_2-\overset{\overset{\displaystyle H}{|}}{\underset{\underset{\displaystyle CH_3}{|}}{C}}-\overset{\overset{\displaystyle O}{\|}}{C}-OH$$

Gly                     Ala

NOTE THE
DIFFERENCES
IN STRUCTURE
(Shaded Area)

$$NH_2-\overset{\overset{\displaystyle H}{|}}{\underset{\underset{\displaystyle CH_3}{|}}{C}}-\overset{\overset{\displaystyle O}{\|}}{C}-N-\overset{\overset{\displaystyle H}{|}}{\underset{\underset{\displaystyle H}{|}}{C}}-\overset{\overset{\displaystyle O}{\|}}{C}-OH$$

Ala–Gly

$$NH_2-\overset{\overset{\displaystyle H}{|}}{\underset{\underset{\displaystyle H}{|}}{C}}-\overset{\overset{\displaystyle O}{\|}}{C}-N-\overset{\overset{\displaystyle H}{|}}{\underset{\underset{\displaystyle CH_3}{|}}{C}}-\overset{\overset{\displaystyle O}{\|}}{C}-OH$$

Gly–Ala

$$+\; NH_2-\overset{\overset{\displaystyle H}{|}}{\underset{\underset{\underset{\displaystyle OH}{|}}{\underset{\displaystyle CH_2}{|}}}{C}}-\overset{\overset{\displaystyle O}{\|}}{C}-OH$$

Ser

$$NH_2-\overset{\overset{\displaystyle H}{|}}{\underset{\underset{\displaystyle H}{|}}{C}}-\overset{\overset{\displaystyle O}{\|}}{C}-N-\overset{\overset{\displaystyle H}{|}}{\underset{\underset{\displaystyle CH_3}{|}}{C}}-\overset{\overset{\displaystyle O}{\|}}{C}-N-\overset{\overset{\displaystyle H}{|}}{\underset{\underset{\underset{\displaystyle OH}{|}}{\underset{\displaystyle CH_2}{|}}}{C}}-\overset{\overset{\displaystyle O}{\|}}{C}-OH \;+\; H_2O$$

Gly–Ala–Ser

combined to form Gly-Ala or Ala-Gly. The differences between these two molecules is that in Gly-Ala the intact carboxyl group ($-\overset{\overset{\textstyle O}{\|}}{C}-OH$) belongs to the alanine unit while in Ala-Gly it belongs to the glycine unit. The convention that will be used in this book, unless otherwise noted, is to infer that in an amino acid chain, the amino acid on the far left contains the intact amino group ($-NH_2$) and that the amino acid on the far right contains the intact carboxyl group. Thus, Gly-Ala-Ser can be shown to differ from Ser-Ala-Gly.

The combination of amino acids in the body proceeds until hundreds of such units are condensed in chains to form protein. Each protein regardless of its source or function is formed by the combination of these same twenty basic units. The specific function of a protein is governed by the **sequence** in which these twenty basic building blocks are combined. The sequence of amino acids in a small protein, that of bovine insulin, is shown in Figure 6-2. Interestingly, the structures shown by samples of insulin isolated from other animals differ only slightly in composition; and in some cases, such as whale and pig insulin, the structures are identical. The similarity in

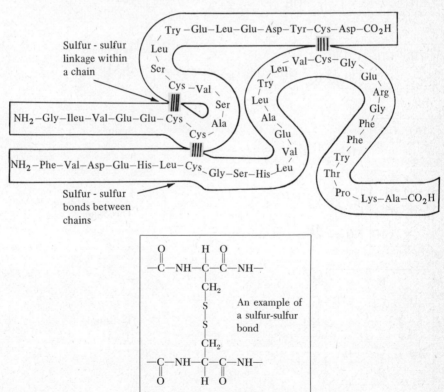

**Figure 6-2** Amino acid sequence in bovine insulin. Shaded area represents an S—S linkage between two cysteine amino acids.

**Table 6-2**
**The Commonly Occurring Amino Acids**

| Structure | Amino Acid[a] |
|---|---|
| 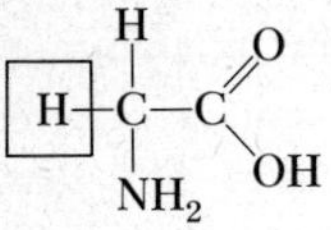 H—C—C(=O)OH with H and NH$_2$ (boxed H) | Glycine |
| 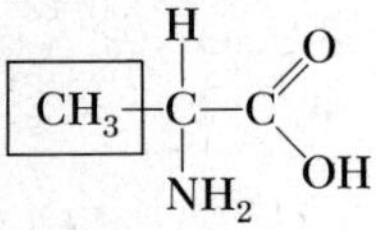 CH$_3$—C—C(=O)OH with H and NH$_2$ (boxed CH$_3$) | Alanine |
| 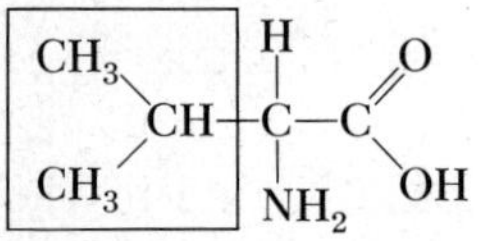 (CH$_3$)$_2$CH—C—C(=O)OH with H and NH$_2$ | Valine (EAA)[b] |
| 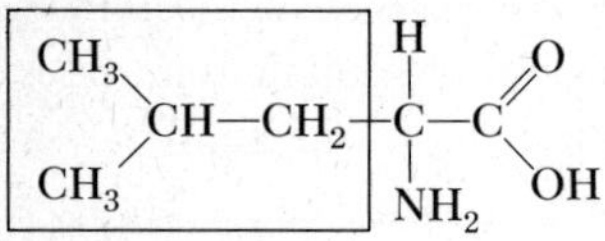 (CH$_3$)$_2$CH—CH$_2$—C—C(=O)OH with H and NH$_2$ | Leucine (EAA) |
| 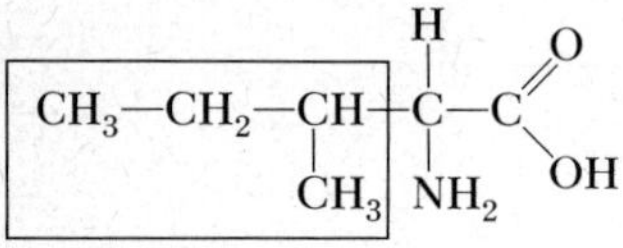 CH$_3$—CH$_2$—CH(CH$_3$)—C—C(=O)OH with H and NH$_2$ | Isoleucine (EAA) |
| 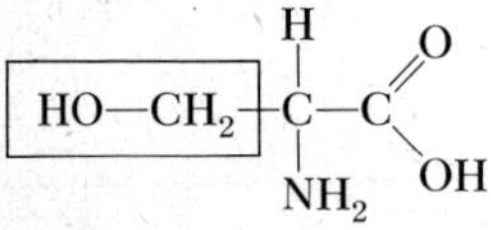 HO—CH$_2$—C—C(=O)OH with H and NH$_2$ | Serine |
| 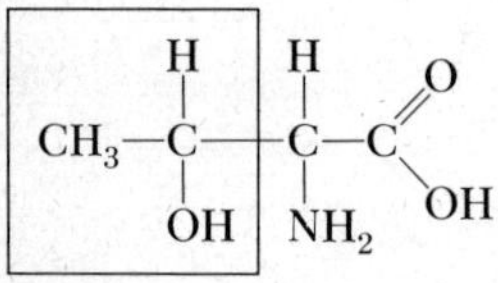 CH$_3$—C(H)(OH)—C—C(=O)OH with H and NH$_2$ | Threonine |
| 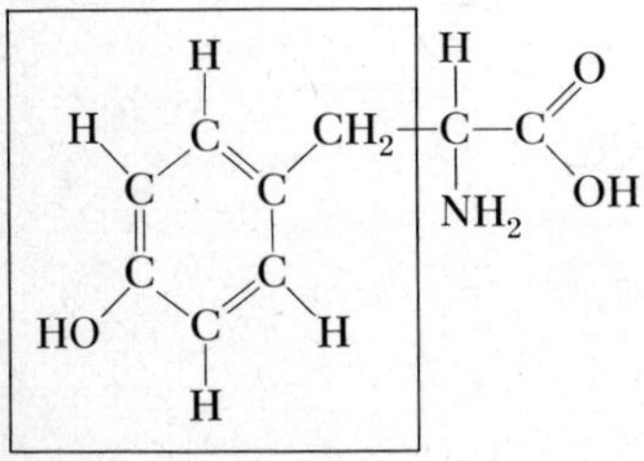 HO—C$_6$H$_4$—CH$_2$—C—C(=O)OH with H and NH$_2$ | Tyrosine (EAA) |

---

[a] The portion underlined is the generally accepted abbreviation for the amino acid.
[b] EAA = essential amino acids, amino acids not synthesized by man.

**Table 6-2 (*Continued*)**

| Structure | Amino Acid[a] |
|---|---|

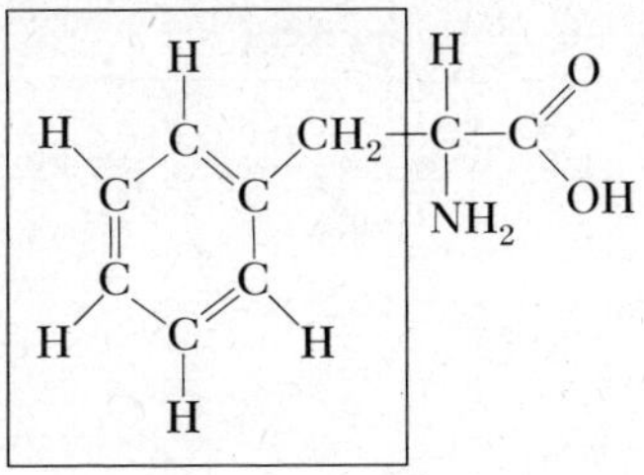

Phenylalanine
(EAA)

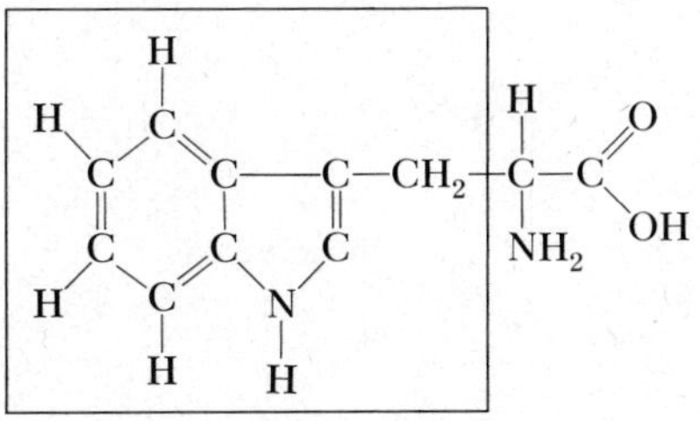

Tryptophan
(EAA)

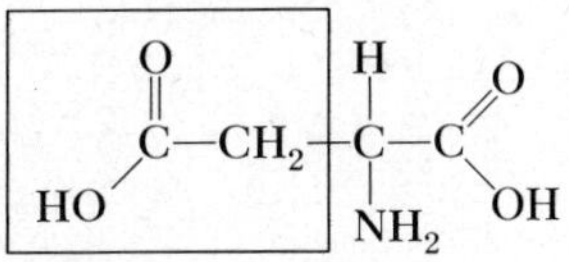

Aspartic Acid

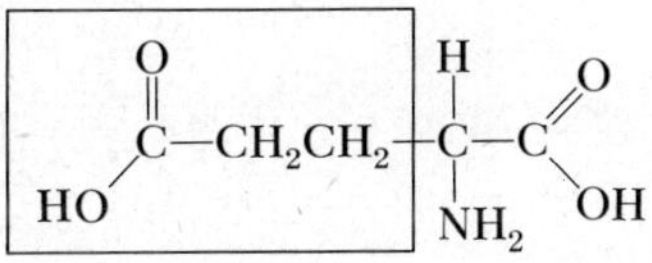

Glutamic Acid

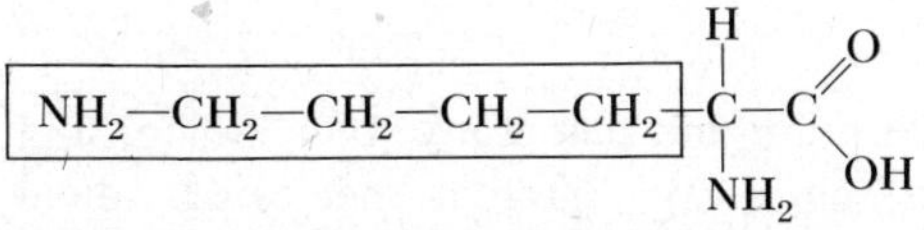

Lysine
(EAA)

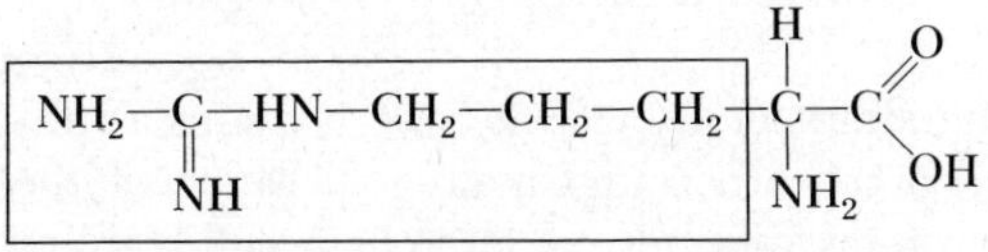

Arginine
(EAA)

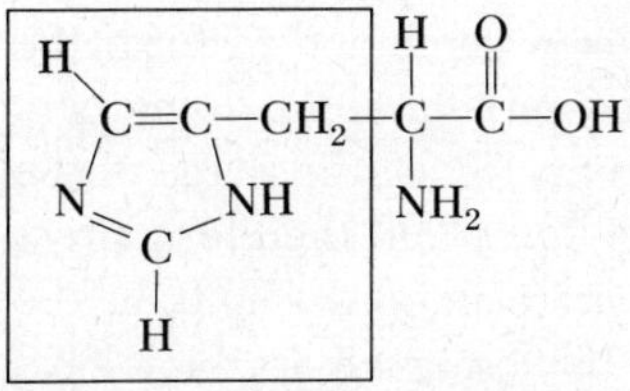

Histidine
(EAA)

**Table 6-2 (Continued)**

| Structure | Amino Acid[a] |
|---|---|
| 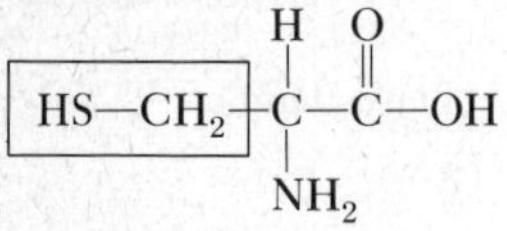 HS—CH$_2$—$\overset{\text{H}}{\underset{\text{NH}_2}{\text{C}}}$—$\overset{\text{O}}{\text{C}}$—OH | Cysteine |
| 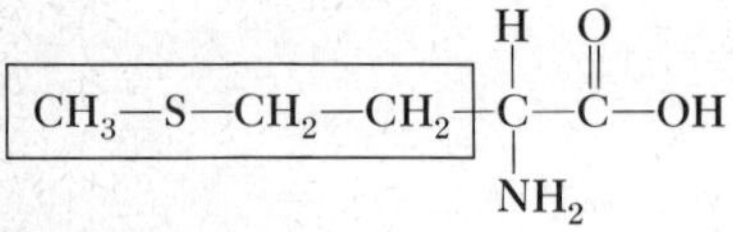 CH$_3$—S—CH$_2$—CH$_2$—$\overset{\text{H}}{\underset{\text{NH}_2}{\text{C}}}$—$\overset{\text{O}}{\text{C}}$—OH | Methionine (EAA) |
| 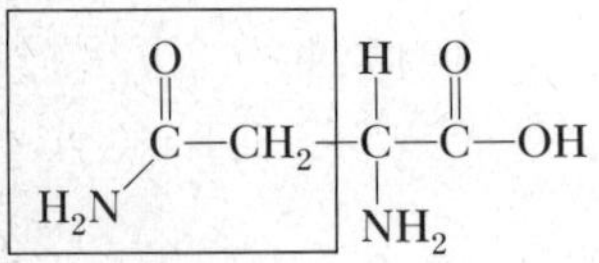 $\overset{\text{O}}{\underset{\text{H}_2\text{N}}{\text{C}}}$—CH$_2$—$\overset{\text{H}}{\underset{\text{NH}_2}{\text{C}}}$—$\overset{\text{O}}{\text{C}}$—OH | Asparagine (Asp-NH$_2$) |
| 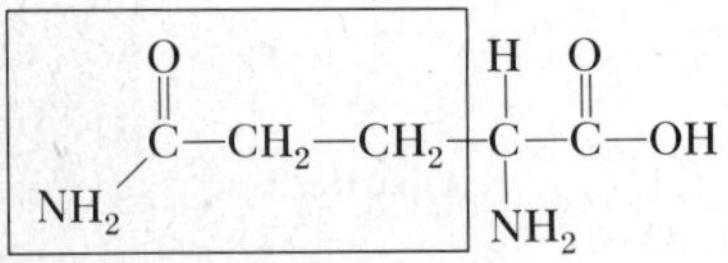 $\overset{\text{O}}{\underset{\text{NH}_2}{\text{C}}}$—CH$_2$—CH$_2$—$\overset{\text{H}}{\underset{\text{NH}_2}{\text{C}}}$—$\overset{\text{O}}{\text{C}}$—OH | Glutamine (Glu-NH$_2$) |
| 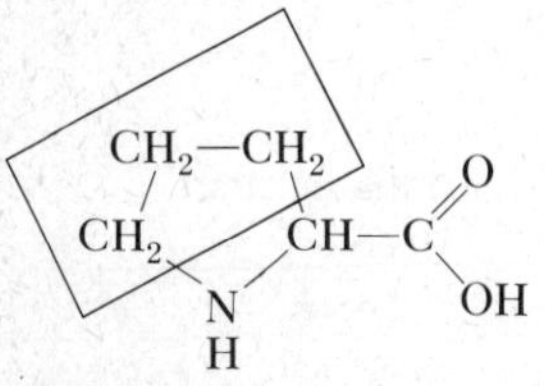 $\begin{array}{c} \text{CH}_2\text{—CH}_2 \\ \text{CH}_2 \qquad \text{CH—C} \\ \text{N} \qquad \text{OH} \\ \text{H} \end{array}$ | Proline |

molecular architecture as found in insulin (and numerous other biologically important molecules) isolated from completely different species of animal life, furnishes the most compelling evidence in support of the theory of evolution.

Note in Table 6–2 that the amino acid cysteine (cys) contains a sulf-hydryl group, sulfur singly bonded to carbon and hydrogen. Much of the three-dimensional structure of proteins depends on interchain sulfur-sulfur bonds formed from the sulfhydryl groups present in cysteine. These bonds are usually formed after the amino acids have been combined to form the protein, and they help to hold the shape of the protein rigid. Interchain sulfur-sulfur bonds are essential for biological activity. Since the shape of the molecule determines this activity, any loss of rigidity would therefore affect activity. Bovine insulin, for example, would separate into two units in the absence of interchain sulfur-sulfur bonds, thereby losing all its biological activity. Note that bovine insulin also contains an intrachain sulfur-sulfur bond. This linkage serves a similar function.

Another type of interchain bonding that is very important in maintaining the shape and activity of the protein molecule is the hydrogen bond. In hydrogen bonding, a hydrogen atom, usually attached to a nitrogen in one peptide chain, is attracted to and shared by an oxygen or nitrogen atom in an adjacent peptide chain. This can occur at many peptide bonds in the chain (dotted lines in Figure 6–3); the great number of these weak bonds make adjacent peptide chains cling to each other. Hydrogen bonds in nucleic acids are also very important, as we shall see later.

The relative abundance of the twenty amino acids and the sequences in which they are combined differ depending on their source. The amino acid composition of some familiar proteins is shown in Table 6–3. The protein in milk, for example, differs in composition from the protein found in gelatin and wool. Differences in the composition of these sources of protein arise primarily because of the different function each protein performs. Consequently, complex protein from one organism cannot be used by another organism unless the protein is broken down into smaller units. This process occurs during digestion and involves a cleaving of the protein, usually from water-insoluble to smaller water-soluble units. Cleavage occurs at the peptide bond to yield free amino acids and smaller proteins. The reaction is called **hydrolysis** and is exactly the reverse of the reaction used to form the protein from the amino acids. The digestive process that occurs can be compared to the restructuring of a building. The building initially dismantled into smaller reusable units, such as bricks, can be reassembled to some new structure using the same materials. The digestive process can be viewed as the dismantling of complex molecules. In most living systems, enzymes aid in the digestive process. The enzyme pepsin, for example, found in gastric juice of the stomach, is capable of splitting a wide variety of proteins into smaller fragments or

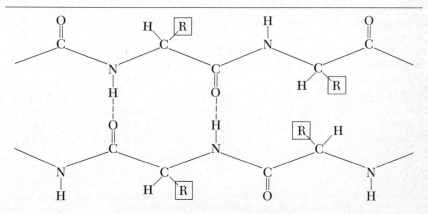

**Figure 6-3** Hydrogen bonding between adjacent protein chains.

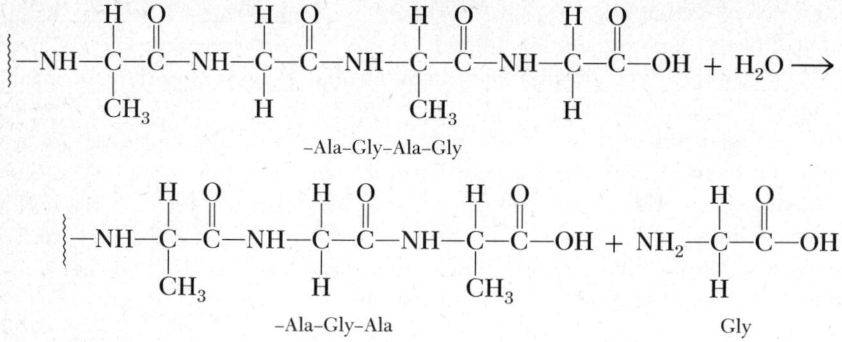

$$\{-NH-\underset{\underset{CH_3}{|}}{\overset{\overset{H}{|}}{C}}-\overset{\overset{O}{\|}}{C}-NH-\underset{\underset{H}{|}}{\overset{\overset{H}{|}}{C}}-\overset{\overset{O}{\|}}{C}-NH-\underset{\underset{CH_3}{|}}{\overset{\overset{H}{|}}{C}}-\overset{\overset{O}{\|}}{C}-NH-\underset{\underset{H}{|}}{\overset{\overset{H}{|}}{C}}-\overset{\overset{O}{\|}}{C}-OH + H_2O \longrightarrow$$

–Ala–Gly–Ala–Gly

$$\{-NH-\underset{\underset{CH_3}{|}}{\overset{\overset{H}{|}}{C}}-\overset{\overset{O}{\|}}{C}-NH-\underset{\underset{H}{|}}{\overset{\overset{H}{|}}{C}}-\overset{\overset{O}{\|}}{C}-NH-\underset{\underset{CH_3}{|}}{\overset{\overset{H}{|}}{C}}-\overset{\overset{O}{\|}}{C}-OH + NH_2-\underset{\underset{H}{|}}{\overset{\overset{H}{|}}{C}}-\overset{\overset{O}{\|}}{C}-OH$$

–Ala–Gly–Ala　　　　　　　　　　　　　　　　　　　Gly

amino acids depending on the contact time and the **acidity** of the stomach.

The broad nonspecificity of pepsin as a **proteolytic** (protein-splitting) enzyme also causes problems. Most animal cells contain considerable amounts of protein. Pepsin itself is a protein. Those cells producing pepsin would ultimately be destroyed by their own creation if during the synthesis of the

**Table 6-3**
**Approximate Percentage Composition of Some Naturally Occurring Proteins**

| Amino Acids | Gelatin | Keratin, Wool | Albumin, Human Serum | Casein (Protein from Milk) |
|---|---|---|---|---|
| Alanine | 9.3 | 4.1 | 6.2 | 3.2 |
| Arginine | 8.5 | 10.4 | 9. | 4.1 |
| Aspartic Acid[a] | 6.7 | 7.2 | 0.7 | 7.1 |
| Cysteine | 0 | 11.9 | 5.6 | 0.3 |
| Glutamic Acid[a] | 11.2 | 14.1 | 17.0 | 22.4 |
| Glycine | 26.9 | 6.5 | 1.6 | 2.0 |
| Histidine | 0.7 | 1.1 | 3.5 | 3.1 |
| Isoleucine | 1.8 | ··· | 1.7 | 6.1 |
| Leucine | 3.4 | 11.3 | 11. | 9.2 |
| Lysine | 4.6 | 2.7 | 12.3 | 8.2 |
| Methionine | 0.9 | 0.7 | 1.3 | 2.8 |
| Phenylalanine | 2.5 | 3.7 | 7.8 | 5.0 |
| Proline | 14.8 | 9.5 | 5.1 | 10.6 |
| Serine | 3.2 | 10 | 3.3 | 6.3 |
| Threonine | 2.2 | 6.4 | 4.6 | 4.9 |
| Tryptophan | 0 | 1.8 | 0.2 | 1.2 |
| Tyrosine | 1.0 | 4.7 | 4.7 | 6.3 |
| Valine | 3.3 | 4.6 | 7.7 | 7.2 |

[a] Glutamine and Asparagine are reported as combinations with Glutamic and Aspartic acids, respectively, the two amino acids to which they are easily converted.
SOURCE: Adapted from West and Todd, *Textbook of Biochemistry* (New York: Macmillan Co., 1961).

enzyme, the **active site** (that portion of the enzyme that is intimately involved in cleaving the peptide bond) of the enzyme were not obstructed. The enzyme is activated when it is mixed with the acidic gastric juices in the stomach. From then on until it is deactivated (when the contents of the stomach are mixed with the alkaline bile acids of the small intestine), pepsin essentially devours all protein in sight including parts of the stomach wall and itself when two molecules of the enzyme meet.

Following the digestive process, the water-soluble amino acids and small proteins are transported through the intestine walls and reassembled in cells to form a protein with a new sequence of amino acids; the sequence depends on the type of protein needed to be synthesized at the time.

Reference to Table 6–3 makes it immediately clear that some proteins have only a small amount of certain amino acids, or lack them completely. Gelatin, for example, contains no cysteine or tryptophan. It would appear that on a diet solely of gelatin the body would not be able to synthesize any protein that contains the amino acids tryptophan and cysteine (such as albumin). This problem, however, is not so serious as it first appears. The body does contain the necessary chemical machinery to make or interconvert one amino acid into another. Thus cysteine could be made from the other amino acids present in gelatin. Certain other necessary amino acids, including tryptophan cannot be prepared by the body and must be included in the daily diet, dramatizing the need for a balanced diet. These amino acids are called **essential amino acids** and are listed in Table 6–2 as EAA.

In Chapter 3 we found that carbon atoms containing four different substituents could exhibit a type of isomerism that we can call mirror-image isomerism (which concerns molecules that cannot be superimposed on their own mirror image). Now consider the amino acids in Table 6–2. None of them, with the exception of glycine, are superimposable on their mirror image (be sure to focus on the carbon bearing the hydrogen, the $CO_2H$ group, the $NH_2$ group, and the group in the rectangle). If we took a piece of steak, hydrolyzed the protein in it, and isolated all twenty amino acids, we would find amino acids with only one mirror-image form present (Figure 6–4, on the right). The other form would be completely lacking. In other words, each amino acid, when held up in space in the manner shown in the diagram on the right, would always have the substituent in the box pointing down and at you, the carboxyl group up, the amino group to the right and away, and the hydrogen to the left and away.

One of the most fascinating discoveries in science was the realization that molecules from living sources were almost always found in one mirror-image form, while the other mirror-image form was often completely missing. Consider the piece of steak again. Suppose this time the protein in the steak was composed of the amino acids shown on the left (the other mirror-image form). The steak would look the same, cook the same, and probably smell the same (or somewhat so). However, if you ate a piece, you would find that your body would probably not be able to digest it, and for that matter, the

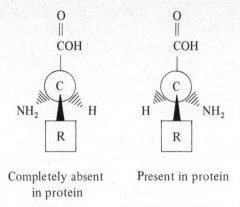

Completely absent          Present in protein
in protein

**Figure 6-4** "Mirror-image isomers" of the amino acids. Note that if the group in the box is H, the amino acid, glycine, is superimposable on its mirror image.

steak might kill you. Why? Well, probably for the same reason that a left-handed glove fits only a left hand. Will the mirror-image form of the keys in your pocket work the locks at home? The enzymes of your body have evolved to interact with molecules of a precise spatial geometry and if this condition is not met, they simply do not perform their task.

Why is it significant that only certain mirror-image forms are found in molecules from living sources? The same molecules (amino acids, for example) made synthetically from "inorganic" substances, are found to possess equal numbers of the two mirror-image forms. Since the two mirror-image forms have the same chemical and physical properties (except for one), they cannot be separated by conventional techniques. The problem that this fact poses is how and why did only one mirror-image form become involved in the life process when both forms are produced equally by natural processes (processes occurring at random on earth). The answer to this question (theory of prebiological evolution) at this time can only be based on speculation and is beyond the scope of this book.°

## LIPIDS

Of the three classes of organic compounds that make up the bulk of all living matter, lipids are probably the most diverse structurally.† Examples of types of lipids found widespread in nature are the waxes, fats, and steroids.

---

° The vital force associated with organic compounds at the time of Wöhler is not as ridiculous as it might appear if viewed in the following context. Most organic compounds from biological sources are identical to those made synthetically in all but one respect: the relative abundance of the two mirror-image forms present.

† In the discussion of lipids, carbohydrates, RNA and DNA that follows, bear in mind that only one mirror-image form is found in living systems (unless of course the molecule is superimposable on its mirror image).

The fats represent the most abundant class of lipids. Fats, or triglycerides, as they are often called, can be given by a general formula as shown.

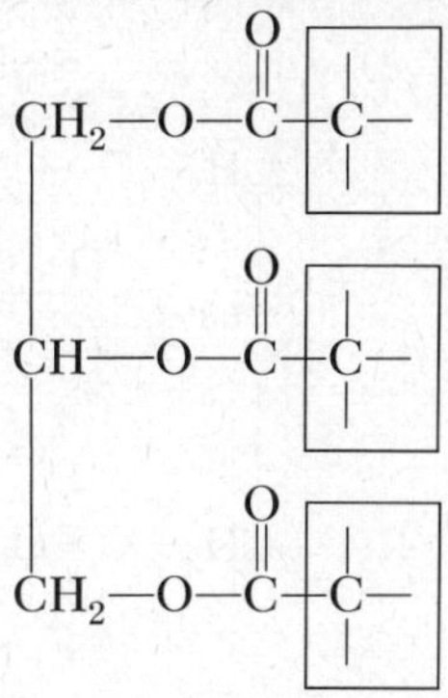

$$CH_2-O-\overset{\overset{\textstyle O}{\|}}{C}\boxed{-\overset{\textstyle |}{\underset{\textstyle |}{C}}-}$$

$$CH-O-\overset{\overset{\textstyle O}{\|}}{C}\boxed{-\overset{\textstyle |}{\underset{\textstyle |}{C}}-}$$

$$CH_2-O-\overset{\overset{\textstyle O}{\|}}{C}\boxed{-\overset{\textstyle |}{\underset{\textstyle |}{C}}-}$$

Triglycerides

The $\boxed{-\overset{|}{\underset{|}{c}}-}$ groups can be identical or different, generally consisting of only
carbon and hydrogen. Two typical $\boxed{-\overset{|}{\underset{|}{c}}-}$ groups are shown herewith.

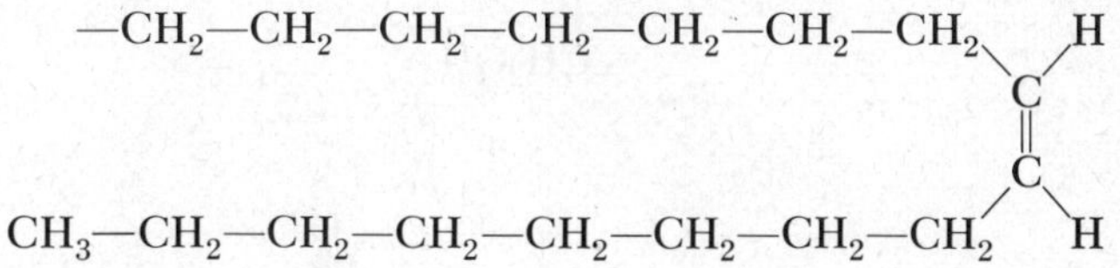

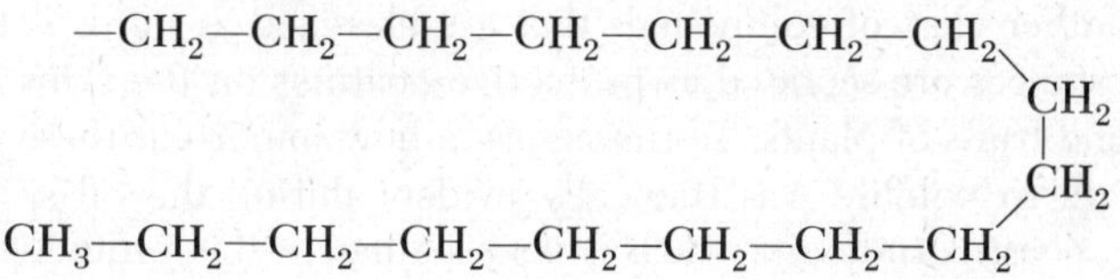

$$-CH_2-CH_2-CH_2-CH_2-CH_2-CH_2-CH_2 \diagdown \overset{H}{\underset{\displaystyle \|}{C}}$$
$$CH_3-CH_2-CH_2-CH_2-CH_2-CH_2-CH_2-CH_2 \diagup \underset{\displaystyle C}{} \diagdown H$$

$$-CH_2-CH_2-CH_2-CH_2-CH_2-CH_2-CH_2 \diagdown \underset{\displaystyle CH_2}{CH_2}$$
$$CH_3-CH_2-CH_2-CH_2-CH_2-CH_2-CH_2-CH_2 \diagup$$

They are abbreviated

$$-CH_2-(CH_2)_6-\overset{\overset{\textstyle H}{|}}{C}=\overset{\overset{\textstyle H}{|}}{C}-(CH_2)_7-CH_3$$

and

$$-CH_2-(CH_2)_{15}-CH_3,$$

respectively.

The triglyceride containing the carbon-carbon double bond would be
called an unsaturated fat.

If fats are treated with lye, they can be broken down into glycerol
and soap. If the reaction is carried out in a minimum of hot water, addition

of salt to the hot water will cause the sodium salt of the fatty acid to crystallize on cooling. The glycerol can be obtained by careful evaporation of the water. For centuries, the hydrolysis of fats was the method used to prepare soap.

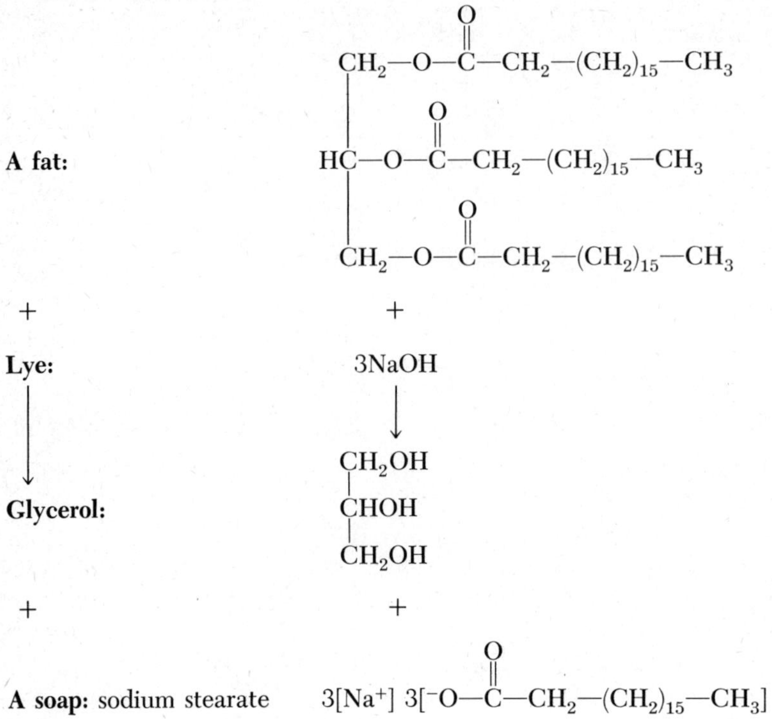

**A fat:**

$$CH_2-O-\overset{\overset{\displaystyle O}{\|}}{C}-CH_2-(CH_2)_{15}-CH_3$$

$$HC-O-\overset{\overset{\displaystyle O}{\|}}{C}-CH_2-(CH_2)_{15}-CH_3$$

$$CH_2-O-\overset{\overset{\displaystyle O}{\|}}{C}-CH_2-(CH_2)_{15}-CH_3$$

\+                                    +

**Lye:**                          3NaOH

$$\downarrow$$

                               $CH_2OH$
**Glycerol:**                   $CHOH$
                               $CH_2OH$

\+                                    +

**A soap:** sodium stearate    $3[Na^+]\ 3[^-O-\overset{\overset{\displaystyle O}{\|}}{C}-CH_2-(CH_2)_{15}-CH_3]$

Another class of compounds that are classified as lipids is the waxes. These substances are secreted as protective coatings on the skins and fur of animals, and fruits of plants. Testimony as to how important these protective coatings are to wildlife was tragically evident during the oil spills off the California coast. Crude oil, which is very similar to this protective coating, mixed with the coating and destroyed the insulating and water-repellent nature of feathers and fur. As a result, many birds and other wildlife depending on the sea for survival either drowned or perished from exhaustion.

Waxes are generally composed of esters of carboxylic acids and alcohols containing a larger number of carbon atoms. Beeswax, for example, is a complex mixture of many compounds. Myricyl palmitate is one important constituent. Lanolin, or wool fat, is another common wax. It is a very complex mixture of products, including steroids such as cholesterol (see below). Lanolin has the desirable property of absorbing a large amount of water without actually dissolving, making it useful as a medium in the preparation of ointments and cosmetics.

Many other types of lipids are necessary to the life processes. They are found widespread in the body and are important constituents of such organs as the brain and liver.

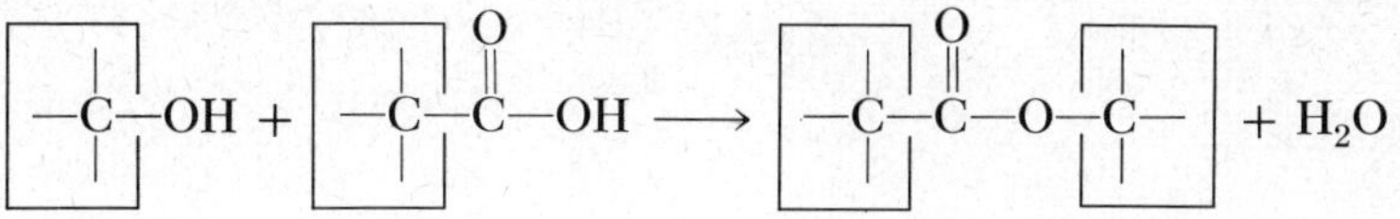

$$\left[-\overset{|}{\underset{|}{C}}-OH\right] + \left[-\overset{|}{\underset{|}{C}}-\overset{O}{\overset{\|}{C}}-OH\right] \longrightarrow \left[-\overset{|}{\underset{|}{C}}-\overset{O}{\overset{\|}{C}}-O-\overset{|}{\underset{|}{C}}-\right] + H_2O$$

An alcohol       An acid            An ester

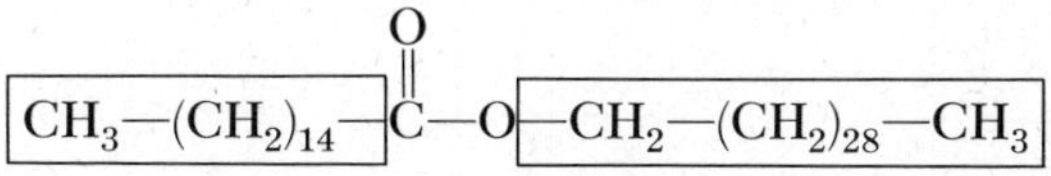

$$\boxed{CH_3-(CH_2)_{14}-\overset{O}{\overset{\|}{C}}-O}\boxed{-CH_2-(CH_2)_{28}-CH_3}$$

Myricyl palmitate

A further class of compounds that are considered lipids is the steroids. These compounds play a primary role in the digestive process, and as body regulators. Examples of this kind of compound are the sex hormones, vitamins of the D group, and the bile acids. The basic steroid skeleton is shown herewith. An abbreviated structure for this material is also given.

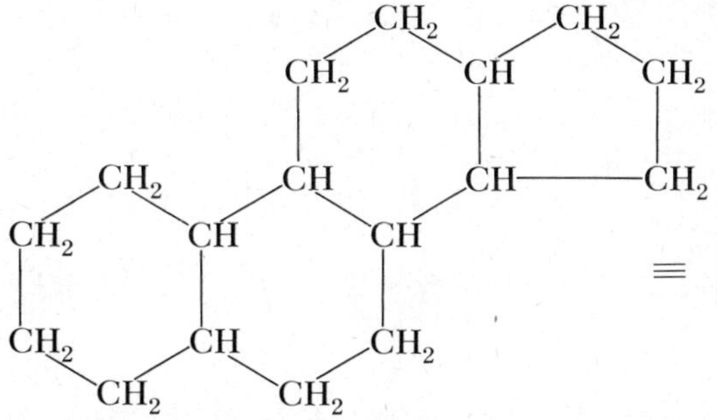

Steroid skeleton 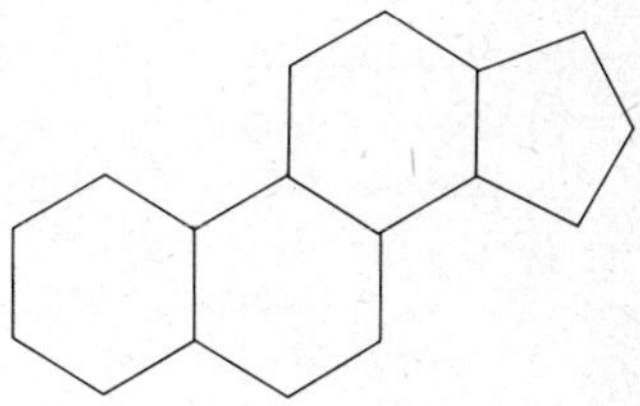       ≡       Abbreviated steroid nucleus

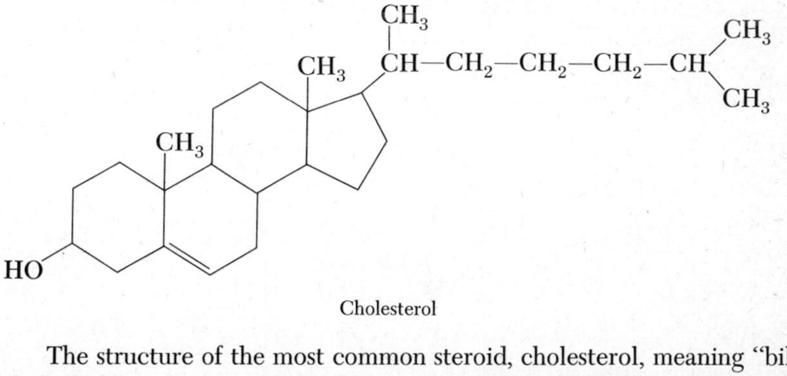

Cholesterol

The structure of the most common steroid, cholesterol, meaning "bile solid alcohol," is shown in the accompanying diagram. Cholesterol has been found in all animal tissues but appears to be completely lacking in plant tissue.

The bile acids are the most important end product of cholesterol metabolism. They are found in bile of most higher animals. The bile consists of a mixture of steroids. Two common examples, cholanic and cholic acids, are shown herewith. These steroids are generally found in the bile combined

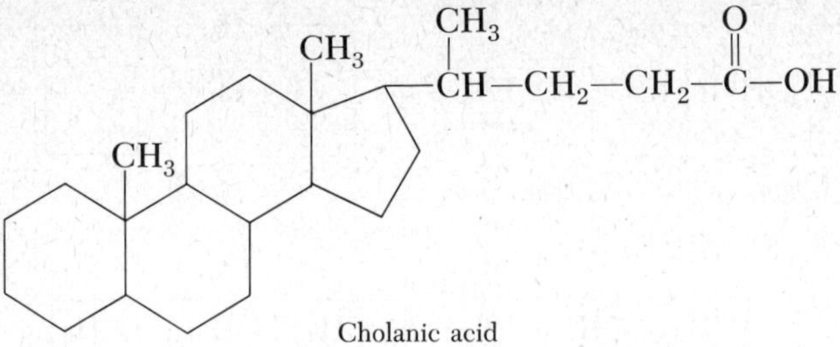

Cholanic acid

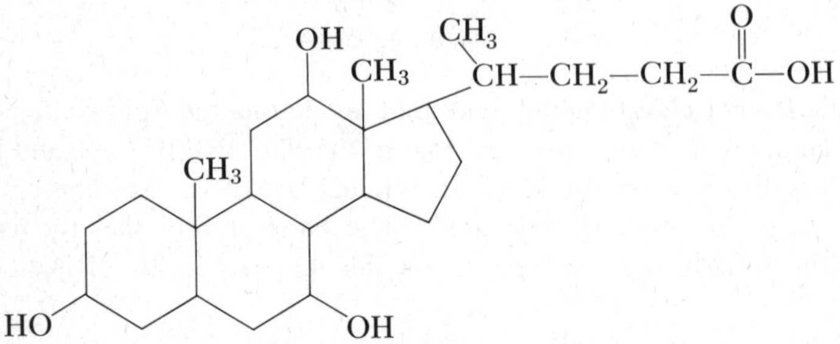

Cholic acid

with glycine. One of the primary functions of the water-soluble bile acids
is to increase the water solubility of fats by acting like a soap or detergent.
This permits fats and oils to travel in the bloodstream (which is primarily
water), after absorption through the intestine.

Vitamin D is another example of a series of compounds containing the

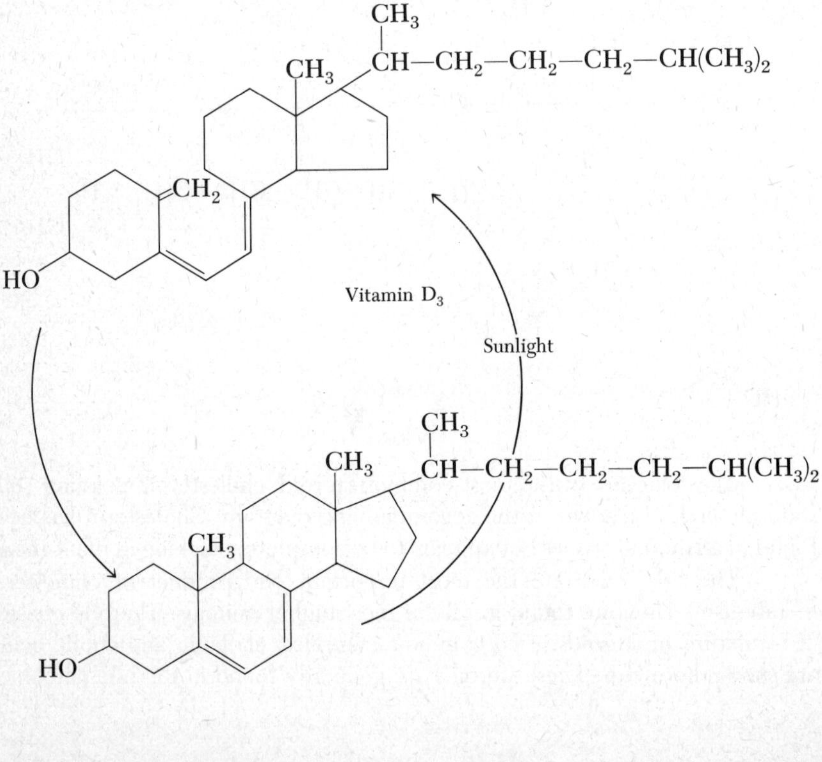

Vitamin D$_3$

Sunlight

steroid skeleton. The structure of vitamin $D_3$ is shown in the accompanying structural diagram. Most natural foods have practically no vitamin D activity. Milk, for example, is generally fortified with vitamin D concentrates. Vitamin D can be prepared by the absorption of sunlight by dehydrocholesterol, a reaction that takes place near the surface of the skin in sunlight. One function of vitamin D is to increase calcium and phosphorus absorption from the intestine into the body. This is particularly important for rapidly growing bone tissue, for example, in children.

## CARBOHYDRATES

Most of the organic matter in nature is composed of carbohydrates. Plant material is composed almost entirely of carbohydrates. Carbohydrates are even found to a limited extent in animals. Just as proteins can be broken down into amino acids, large complex carbohydrates can be broken down into simpler repeating units. Thus the carbohydrates, starch, sugar, and cellulose are related. The simplest sugars often have the formula $C_6H_{12}O_6$, but the general formula is $C_nH_{2n}O_n$, where $n$ is a very large number and varies with the type of carbohydrate. An example of a simple sugar with this formula is glucose. Fructose is another sugar with the same formula but a different arrangement of atoms. When fructose and glucose are connected, we obtain sucrose, or common table sugar. By combining each glucose unit in an orientation in which alternate glucose units are "flipped over" and by com-

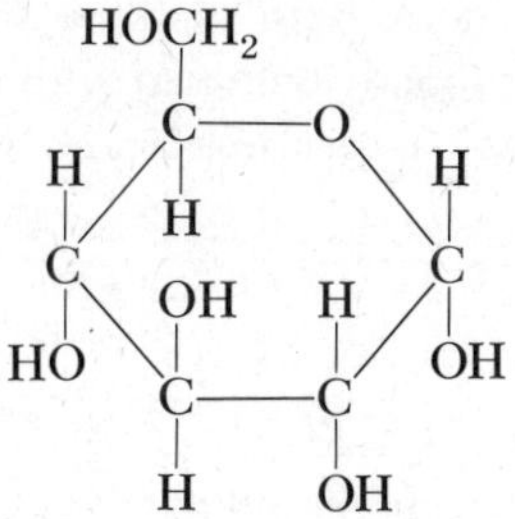

Glucose

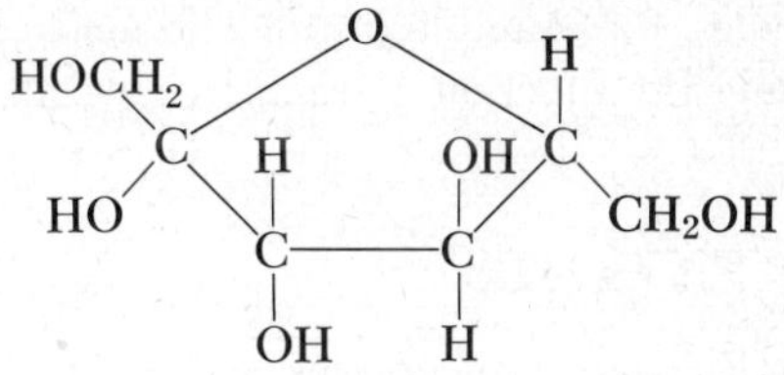

Fructose

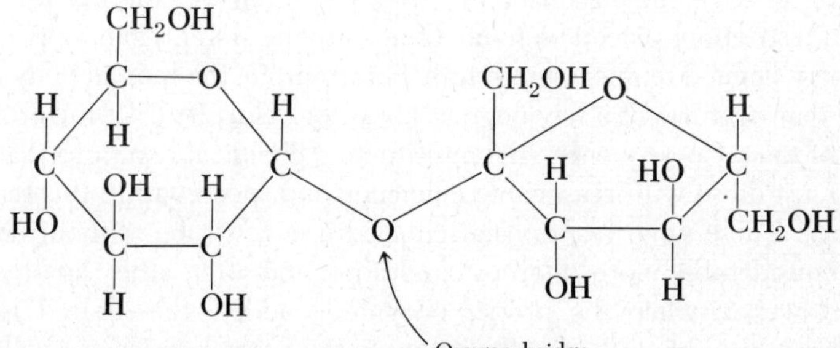

Oxygen bridge

Sucrose

bining about 2,000 such units per molecule, we would obtain cellulose, the main constituent of wood. If the glucose units are hooked up without "flipping," amylose, which is a starch, is obtained (see Figure 6–5).

It is quite remarkable that even though cellulose and amylose differ only slightly in the orientation of the glucose units, man is able to digest only the starch, amylose, whereas cellulose passes through the body unchanged. The reason for this is that the body possesses an enzyme, amylase, which is capable of cleaving the oxygen bridge between glucose units in amylose, but lacks the enzyme to cleave the glucose units in cellulose. Consider our enzyme amylase as a lock and amylose as the key. Given the proper fit, the enzyme surrounding the glucose units in amylose cleaves the oxygen bridge; if cellulose is used as the key, it does not fit the lock. This type of behavior illustrating the specificity of the reactions that occur in biochemical systems is generally the rule rather than the exception.

Two other sugars of considerable importance are ribose and deoxy-

Ribose                                   Deoxyribose

ribose. Note that both sugars contain only five carbon atoms and that deoxyribose contains one less oxygen than ribose does. Both of these sugars, although present in small amounts, are generally found combined with other molecules, for example, in nucleic acids, and play a very important role in directing the function of each living cell.


## ENZYMES

Enzymes are proteins, which are present in trace amounts and have the important function of regulating the countless number of reactions occurring in the body. They are often quite large and complicated in structure. The importance of enzymes becomes apparent when we consider the limitations and restrictions placed on living things. Living systems generally operate within very defined temperature limits. For example, the human body cannot tolerate temperatures of a few degrees above or below 98.6 °F for prolonged periods of time. Consequently, the multitude of chemical reactions that occur in cells must do so with reasonable efficiency and speed within this temperature range. These same reactions attempted in a test tube without enzymes require considerably more rigorous conditions, and often other side reactions occur. However, when the proper enzyme is added, the desired reaction occurs smoothly and with no other complications, as happens in cells.

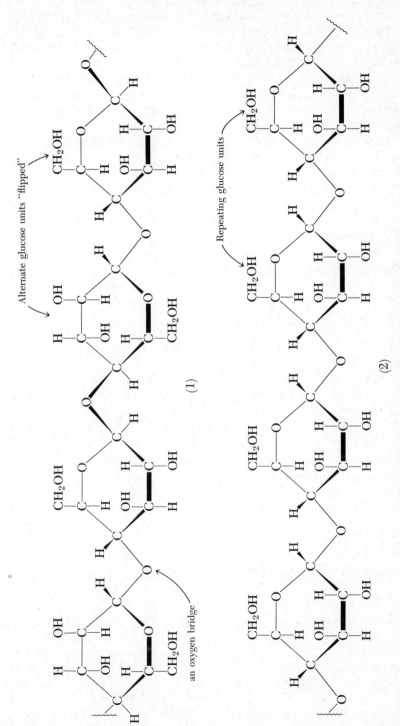

Alternate glucose units "flipped"

an oxygen bridge

(1)

Repeating glucose units

(2)

**Figure 6-5**  Structures of (1) cellulose and (2) amylose.

The role of enzymes in biochemical systems is to permit reactions to occur smoothly without the need to activate, or excite, molecules with heat. Furthermore, because the interaction between the enzyme and the substance it is attacking is specific, no other undesirable side reactions occur.

Enzymes can be considered the machinery a cell maintains to carry out its function. The cell is very tidy with its machinery, housing destructive enzymes in a protective covering (lysosome, Figure 6–12). Rupture of this sheath, from old age or from damage, leads to complete destruction of the cell. This is the manner in which dead or old cells are disposed of. Enzymes necessary for providing energy to the cell are housed in tiny cigar-shaped bodies, called mitochondria (Figure 6–12).

How do enzymes work? Although the intricate details of the mechanism of catalysis by enzymes is not fully known, the first step generally involves formation of an **enzyme-substrate complex,** probably by the lock-and-key concept. During the existence of the complex, a chemical reaction occurs, and it is followed by a breakup of the enzyme-product complex. This is shown diagrammatically in Figure 6–6 for the cleavage of the carbohydrate amylose by one of the amylase enzymes. This process continues until amylose is completely broken up. Often certain other molecules can combine to block the active site of the enzyme, particularly if the enzyme is not very specific. If this new enzyme substrate complex differs sufficiently so that no reaction occurs but the enzyme remains tied up, then the enzyme is removed from circulation.

Not much of a particular enzyme is present at any given time. The turnover time from substrate plus enzyme to products plus enzyme is generally very short. Hence a small amount of enzyme goes a long way. The presence of some new substrate that can successfully bind the enzyme for longer periods of time can virtually stop the intended function of the enzyme. The process of tying up an enzyme is called enzyme inhibition. The **pharmacologic** action of drugs is based largely on enzyme inhibition. "Nerve gas," insecticides, herbicides, and some antibiotics such as penicillin G have all been developed on the principle of enzyme inhibition. The toxicity of heavy metals such as mercury, lead, and cadmium, cyanide and other substances is also due to enzyme inhibition.

A great number of enzymes are in existence and assist in the numerous chemical reactions carried out by living systems. Over six hundred have been

An antibiotic, Penicillin G

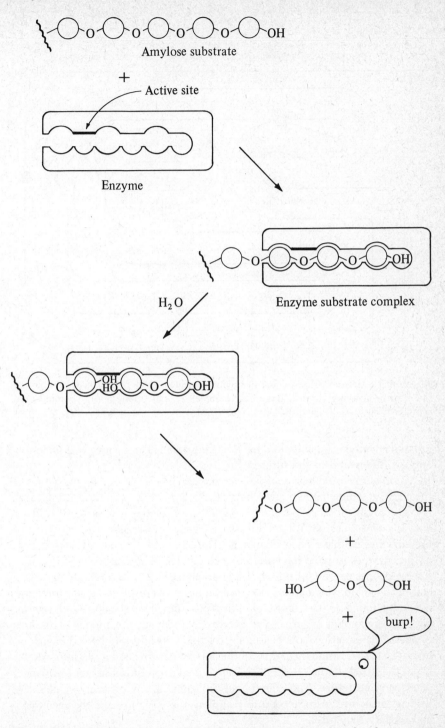

**Figure 6-6** Schematic diagram illustrating digestion of amylose by the enzyme amylase.

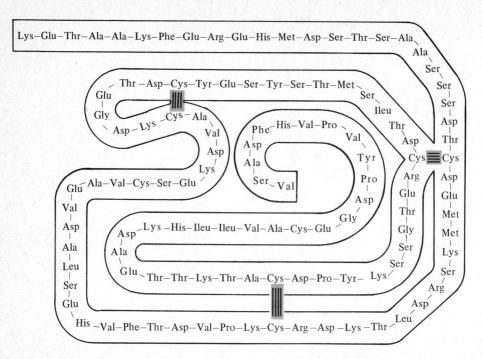

**Figure 6-7** Amino acid sequence in bovine pancreatic ribonuclease A, an enzyme involved in cleaving RNA, a large macromolecule, into smaller units. Shaded area represents sulfur-sulfur bonds.

isolated and about one hundred have been purified. Many more will undoubtedly be discovered in the future.

In our representation of the enzyme amylase (Figure 6–6), we depicted the enzyme somewhat as a jaw, chewing off parts of the substrate it is attacking. Obviously this is a gross oversimplification of the true structure of the enzyme. One question we might ask is, What is the chemical structure of the enzymes we have been discussing? Unfortunately, the structures of most of the enzymes isolated have not been elucidated. The reason for this is that since enzymes are found in only trace amounts, it is very difficult to isolate and purify them. Remember, enzymes themselves are proteins, and they are generally found in the midst of other proteins. The structures of several enzymes are known. Figure 6–7 illustrates the sequence of amino acid residues in bovine pancreatic ribonuclease A. To gain an insight into how the enzyme operates, to determine the nature of the active site, we need more information about the structure than is obtained from elucidating the amino acid sequence, or **primary structure.** The amino acid sequence of any protein is called the primary structure. Also needed to complete the structure are the **secondary** and **tertiary** structures.

It has been found that proteins are often twisted in a specific manner, as shown in Figure 6–8(a). The twisted structure that the peptide chain forms is called a helix and it represents one type of secondary structure of a protein.

The tertiary structure refers to the overall shape of the molecule and results from folding, or "wadding up," of the peptide chain [Figures 6–8(a) and 6–8(b)]. The amount of twisting and folding present in any protein is determined by the nature of the protein and by some external factors.

If the correct amino acid sequence is used in preparing a natural protein by synthetic means, the secondary and tertiary structures will often form unassisted, identical to the natural protein, provided that the molecule is given the opportunity to form the proper sulfur-sulfur and hydrogen bonds between chains. In complicated proteins this becomes more difficult. The proteins in a raw egg, for example, are very rigidly and specifically held in place by many sulfur-sulfur and hydrogen bonds. Boiling an egg simply ruptures these weak

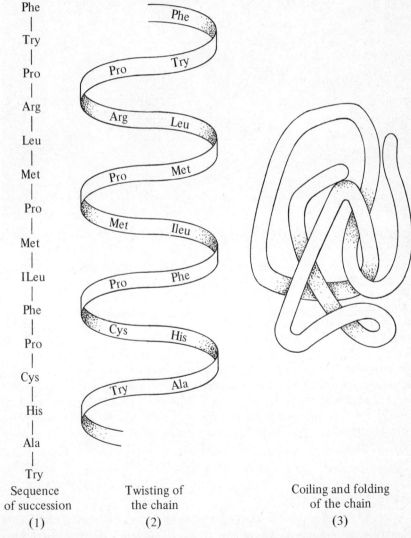

| Sequence of succession | Twisting of the chain | Coiling and folding of the chain |
|:---:|:---:|:---:|
| (1) | (2) | (3) |

**Figure 6-8(a)**  (1) Primary, (2) secondary, and (3) tertiary structures of proteins.

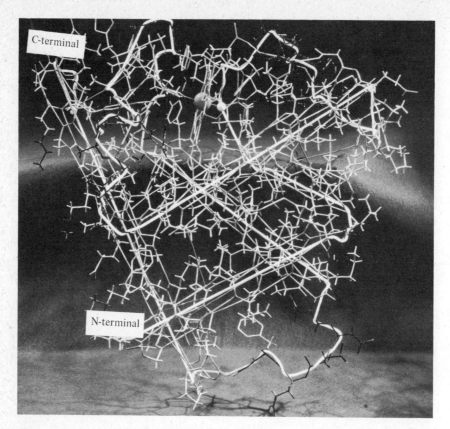

**Figure 6-8(b)** Photograph of model of muscle protein myoglobin from the sperm whale. Backbone of the protein,

$$\begin{array}{ccccc} & O & H & & O & H \\ & \| & | & & \| & | \\ -C & - & N & - CHR - & C & - N - \end{array}$$

can be followed by tracing the white cord. The folding and coiling of this cord represents the tertiary structure of the protein. (*Photograph courtesy of Professor John C. Kendrew.*)

bonds, resulting in a random uncoiling and unfolding of the protein. This causes the protein to **precipitate.** This process is irreversible partly because the protein has precipitated from solution and partly because of the complexity of the protein and the unlikelihood of there being a meeting of the same parts of the protein again at exactly the same time and place to form the proper sulfur-sulfur bonds.

## HORMONES

Like the enzymes, hormones are generally complex organic molecules. Often they are proteins. In animals they are secreted by specific tissues, traveling by way of the bloodstream to various organs and there exerting control over specific body processes. Hormones, like enzymes, are present in minute amounts. The specific role that hormones play is only now becoming known in detail. The profound effects that hormones exercise over body functions make endocrinology (the science dealing with the hormone-secreting glands) a very rapidly expanding field. The hormones secreted from the pituitary gland (attached at the base of the brain) are extremely important in the sexual development of adolescents, and are intimately involved in reproduction. These hormones are proteins in the small-to-medium-size range. The hormone oxytocin is routinely used in initiating labor at childbirth. This hormone causes uterine contractions. A very similar hormone, vasopressin, regulates the water balance in the body by regulating fluid excretion and thirst. The primary structure of these hormones is shown in Figure 6–9. Note that the shaded areas of these two small proteins (each a nona-peptide, meaning having nine amino acids present) are identical. Yet two simple substitutions, Ileu for Phe and Leu for Arg, convert the hormone for labor into the hormone for thirst. This transformation is truly remarkable. Other hormones with relatively simple structures are thyroxine, secreted by the thyroid gland, and adrenaline, secreted by the adrenal glands.

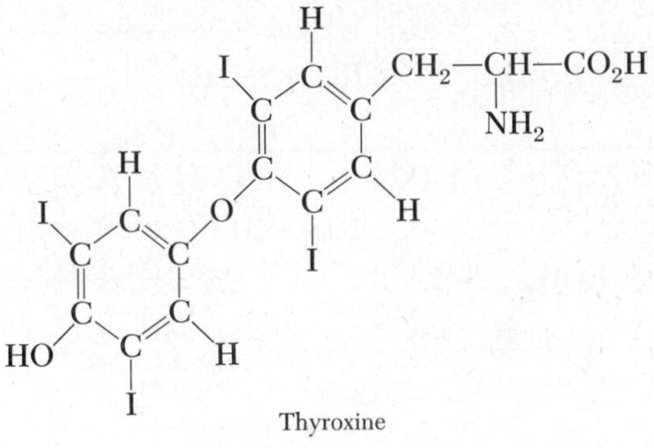

$$CH_2-CH-CO_2H$$
$$NH_2$$

Thyroxine

$$CH-CH_2-N-CH_3$$

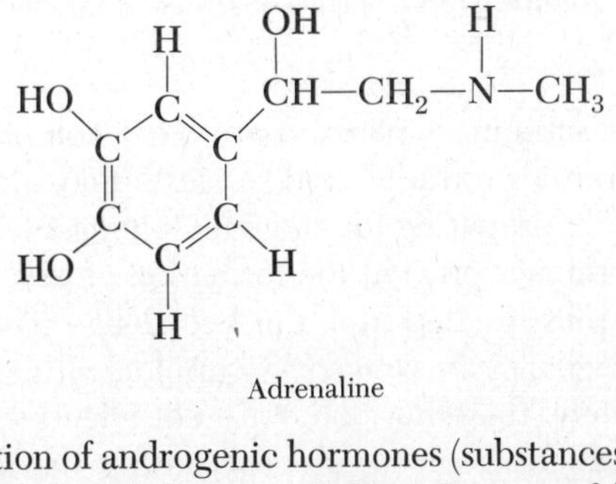

Adrenaline

The production of androgenic hormones (substances that promote male characteristics) occurs in the testes, and production of both androgen and estrogen (substances promoting ovulation) occurs in the ovaries. Examples of androgenic substances and estrogenic substances are shown in Figure 6–10. These substances are steroids. Note their resemblance to cholesterol. The

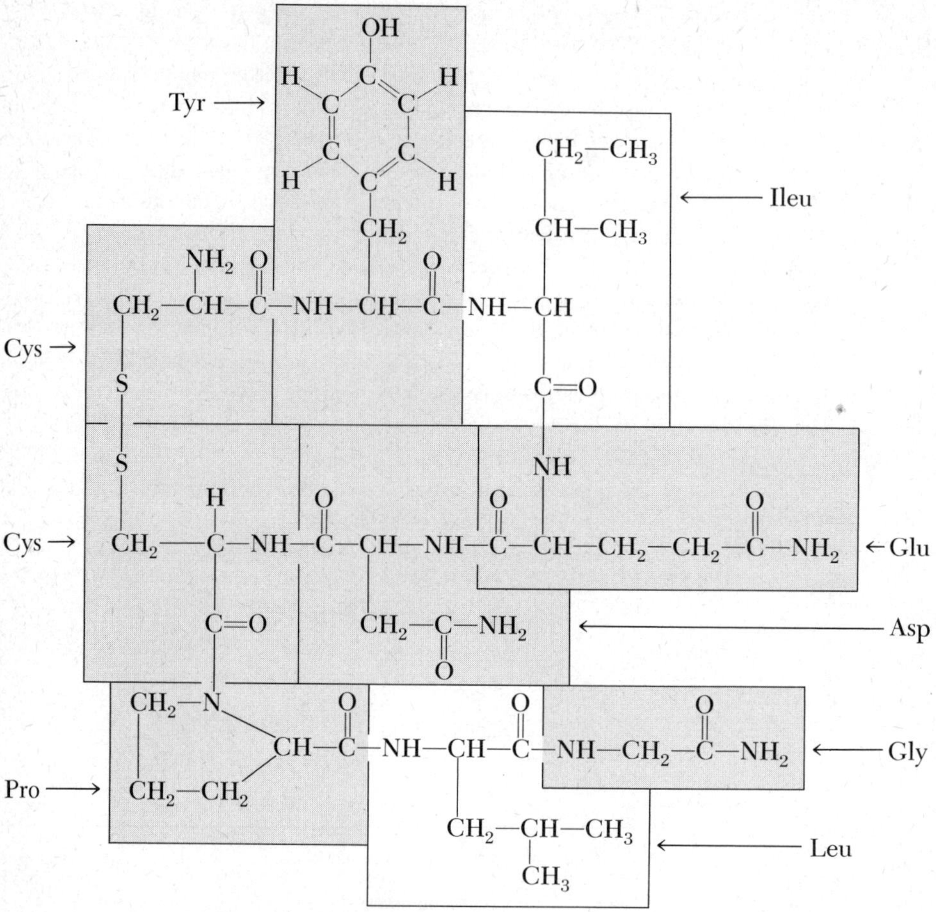

**Figure 6-9(a)** Primary structure of the hormone oxytocin.

functions that these substances perform are quite complex and interrelated. For example, the steroids estradiol and progesterone, along with other hormones, are involved in preparing for and maintaining a fertilized egg. During pregnancy these hormones prevent the formation of new eggs. Development of the birth control pills resulted from this knowledge. The idea was to create a state of pseudopregnancy in women by supplementing these substances in the diet. Unfortunately, these hormones are not effective when taken orally. Apparently some structural changes occur during the digestive process. A search was then initiated to find other substances that would not be affected when administered orally and that could mimic the effects of natural estradiol and progesterone. Four compounds out of a list of many that have been

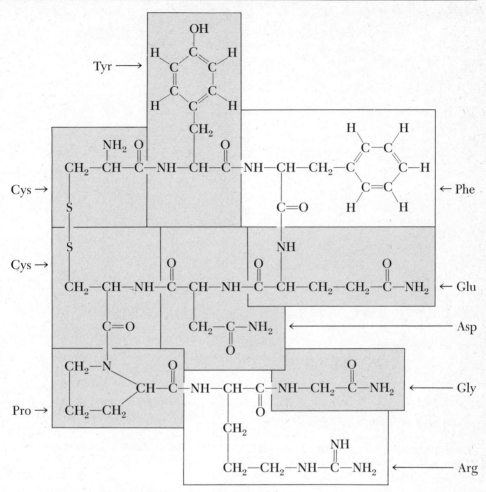

**Figure 6-9(b)** Primary structure of the hormone vasopressin.

developed as a result of this search are shown in Figure 6–11. It is quite likely that other types of birth control pills (possibly for use by the male) will also be developed.

## RNA AND DNA

We now come to the group of macromolecules responsible for transmitting information. At the earliest stage of our development, we all started as a single cell. From that cell, after a countless number of cell divisions, each of us emerged as a unique entity. The question we now ask is, How is the genetic information present in that first cell transferred during cell mitosis

## Androgenic Substances

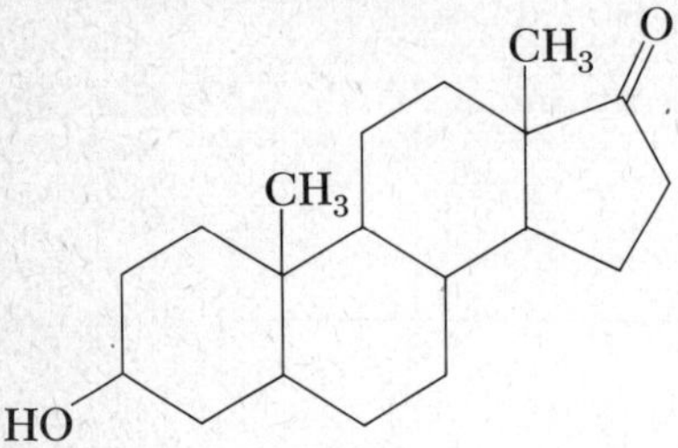

Androsterone

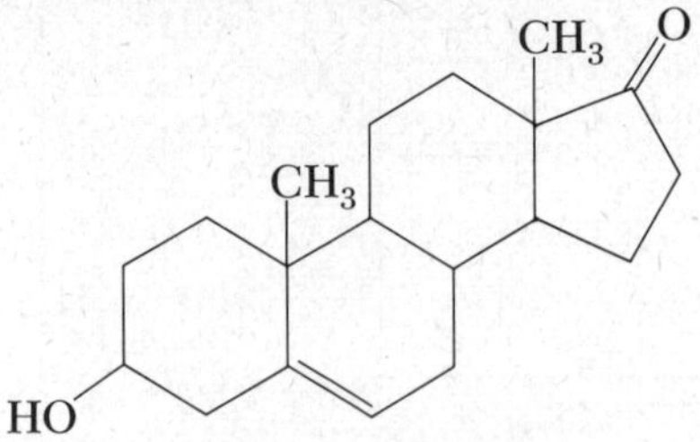

Dehydroisoandrosterone

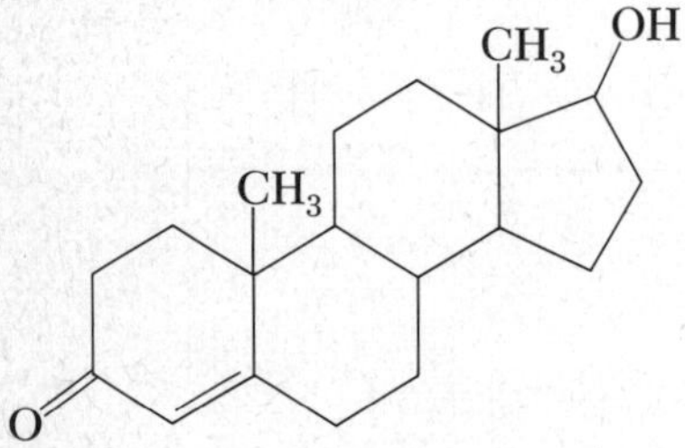

Testosterone

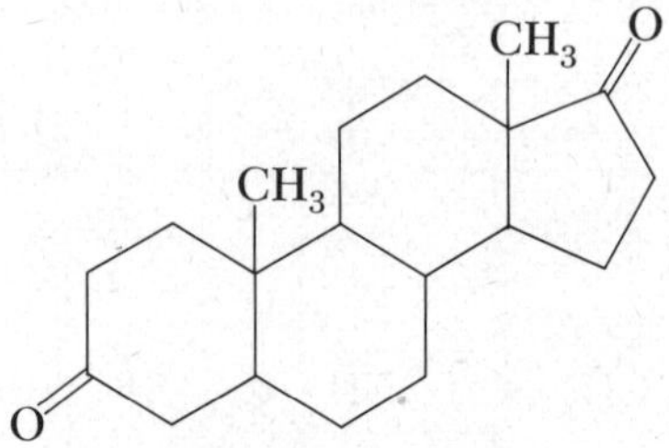

Androstene–3,17–dione

## Estrogenic Substances

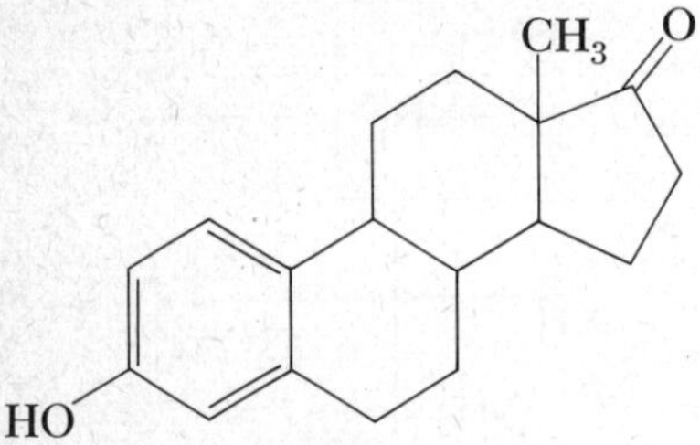

Estrone

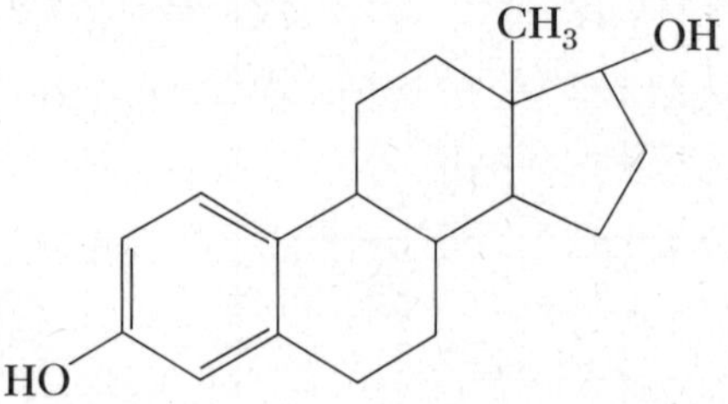

Estradiol

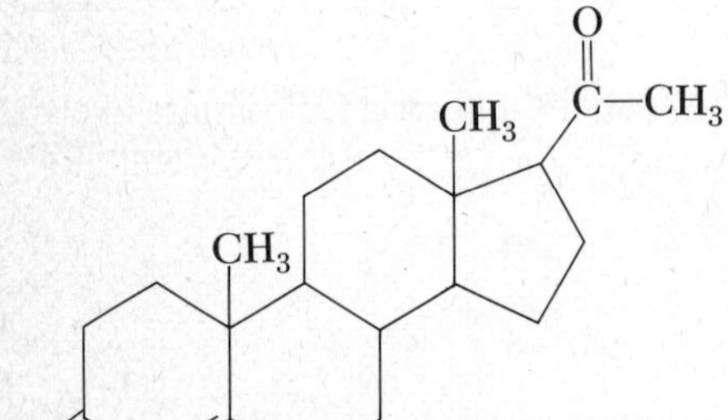

Progesterone

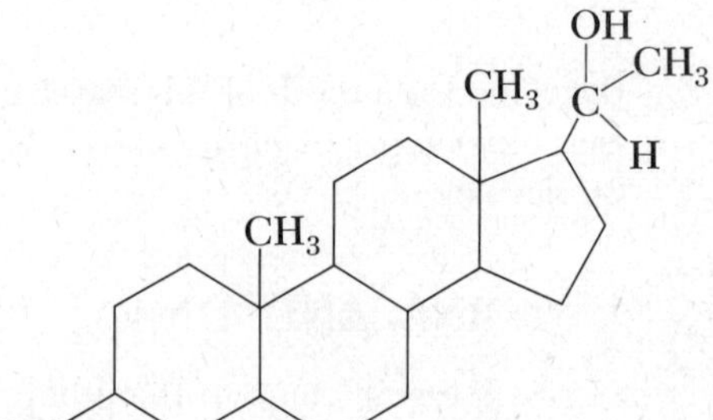

Pregnanediol

**Figure 6-10**  Androgenic and estrogenic substances.

**Synthetic Progesterones**

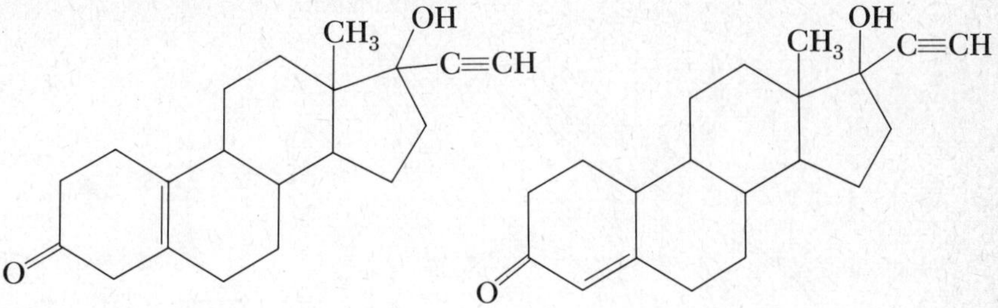

**Synthetic Estrogens**

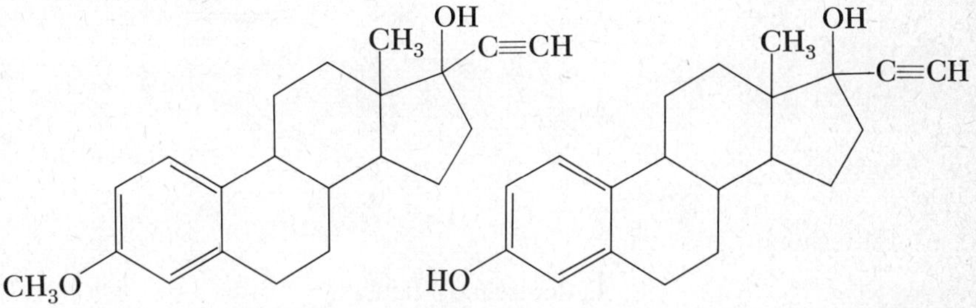

**Figure 6-11** Synthetic progesterones and estrogens.

(cell division)? The answer to this question is still incomplete. However enough information is available to permit us to sketch a rough picture of how the genetic information present in the chromosomes of a cell is reproduced and used to direct the operations of the cell. A diagram of a generalized animal cell is shown in Figure 6–12. The parts of the cell are listed as follows:

*Mitochondria* Sites of energy production for the cell.
*Cell membrane* Maintains integrity and individuality of cell.
*Ribosomes* Composed of RNA and protein, site of protein synthesis.
*Lysosome* Contains enzymes of the cell, sheathed in lipidlike casing.
*Nucleus* Governs operations of cell.
*Endoplasmic reticulum* Intricate series of canals for importing raw materials, exporting finished products.
*Nucleolus* Site of RNA synthesis.
*Chromatin* Makes up *chromosomes*.
*Chromosomes* Carriers of cell's hereditary material, DNA.

Just prior to mitosis, considerable activity is observed in the nucleus. First the chromatin content of the nucleus is doubled and then the scattered chromatin is condensed to form the chromosomes. The main constituent of

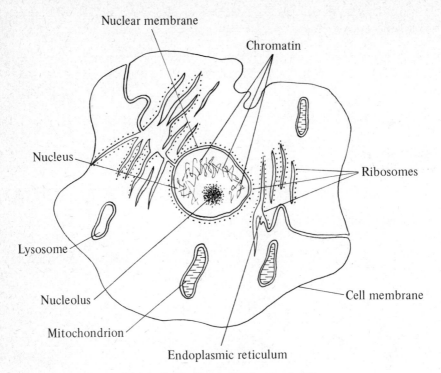

**Figure 6-12**  Generalized animal cell.

the chromosomes (and chromatin) is a substance called deoxyribonucleic acid (DNA). Each DNA molecule is very large, being composed of many smaller subunits, called **nucleotides.** Nucleotides in turn can be broken down into three constituent parts, a sugar, a nitrogen-containing base, and a phosphate (in the

Deoxyribose

Phosphate
(phosphoric acid)

Base: Adenine, A

Base: Guanine, G

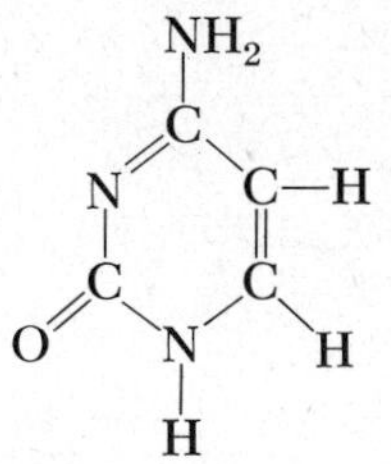

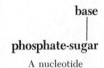

Base: Cytosine, C 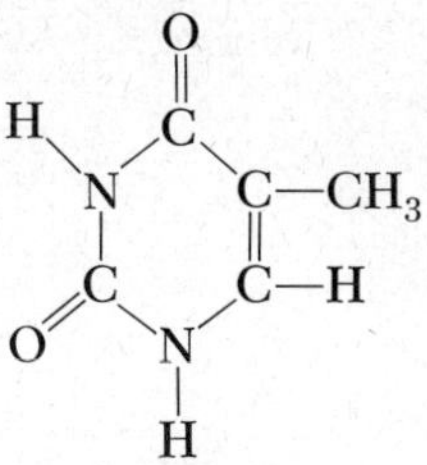                    Base: Thymine, T

form of phosphoric acid). For DNA the sugar is always deoxyribose, while four bases are usually found in DNA. They are adenine, guanine, cytosine, and thymine. These constituents of the nucleotide are combined in the following manner:

**base**

|

**phosphate-sugar**

A nucleotide

Structures of the nucleotides are shown in Figure 6–13(a). The structure of the DNA molecule, as in proteins, depends on the sequence of the repeating units (or nucleotides). This is shown in Figure 6–13(b). The number of units per molecule varies, going as high as a hundred million. The backbone of the DNA molecule can be considered to be composed of alternating phosphate and deoxyribose molecules. Through some very elegant X-ray studies, using information gathered by many other workers, Watson and Crick found that two strands of the backbone of DNA were twisted in a helical fashion similar to the twisting found in the secondary structure of some proteins (Figure 6–14). Furthermore they found that the bases attached to the phosphate-sugar backbone were always oriented toward the inside of the helix and that whenever the base adenine was found, thymine was always found opposite to it and attached to a different backbone (see Figure 6–16). A similar relationship existed between guanine and cytosine. It is known that a positive interaction exists between A and T and between G and C. This positive interaction, shown in Figure 6–15, is called hydrogen bonding and it helps to keep the helical structure of the DNA molecule from unraveling; its most important function, however, is to allow precise copying of the DNA molecule.

The double-stranded nature of DNA and the specificity of the bases of adenine for thymine and guanine for cytosine suggest a mechanism for the transmission of genetic information. You will recall that prior to mitosis (cell division) the chromatin content (DNA) of the cell is doubled. This can be interpreted, on a molecular basis, as a replication of each strand of DNA. The *complementary* strand of DNA can be synthesized by the cell using a single strand of the original DNA as a **template.** The specific base interactions between A and T, G and C mentioned earlier, effect the complete reproduction of the original DNA helix. This process is shown in Figures 6–16 and 6–17.

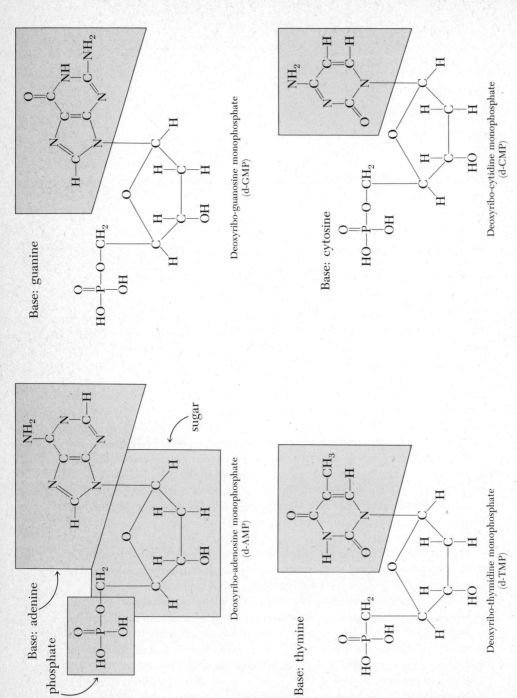

**Figure 6–13(a)** Structures of the subunits of the DNA molecule.

Base: guanine

Deoxyribo-guanosine monophosphate
(d-GMP)

Base: cytosine

Deoxyribo-cytidine monophosphate
(d-CMP)

Base: adenine

phosphate

sugar

Deoxyribo-adenosine monophosphate
(d-AMP)

Base: thymine

Deoxyribo-thymidine monophosphate
(d-TMP)

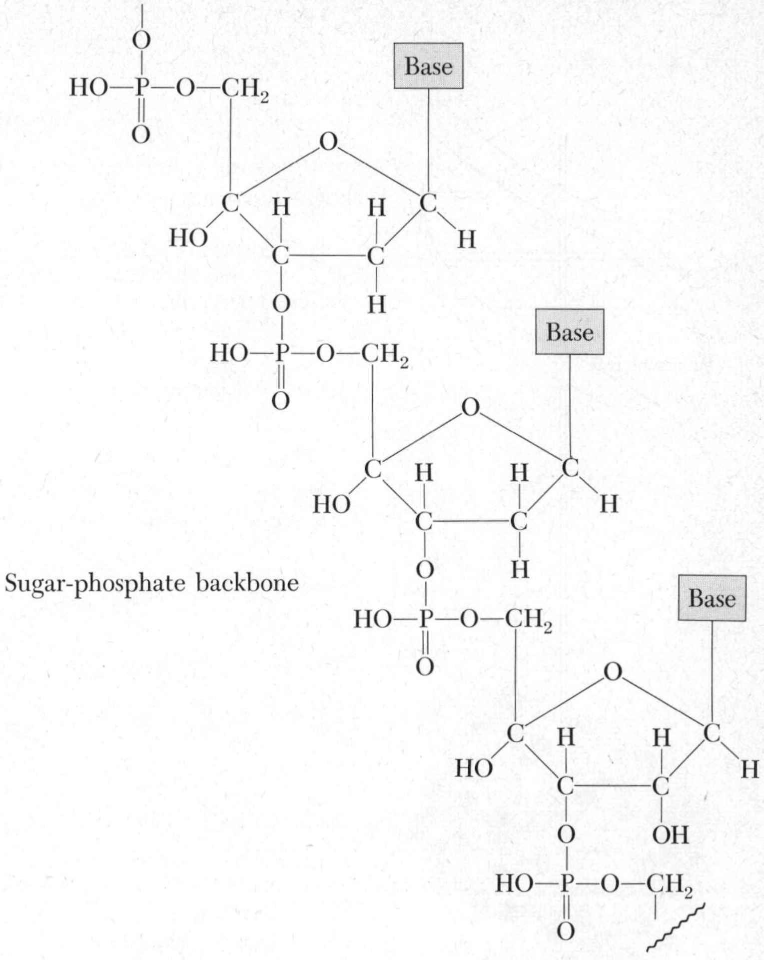

**Figure 6-13(b)**   Structure of the DNA molecule.

On a macroscopic level, we can use the following as an analogy to DNA duplication. Consider a hand and a snugly-fitting glove. If we remove the hand from the glove, we can use the hand as a model to make another glove or the glove to make another hand. Using either mold, or template, we can reproduce the essential features of both hand and glove. Replication of DNA is thought to occur in such a manner.

The role that DNA plays in cells is as an informational molecule. It contains directions, in the form of a template, for the production of protein. The chain of events that the DNA in the nucleus of cells initiates is the synthesis of ribonucleic acid (RNA). Ribonucleic acid in turn leads to the production of protein by the cell. Any random change in the DNA, particularly

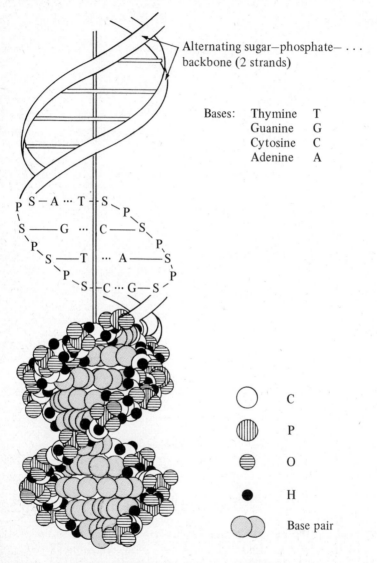

Alternating sugar–phosphate– . . . backbone (2 strands)

Bases:  Thymine  T
        Guanine  G
        Cytosine C
        Adenine  A

○ C

◉ P

⊖ O

● H

◯◯ Base pair

**Figure 6-14** The double-stranded helical nature of DNA drawn to dramatize the helical nature of the sugar–phosphate–sugar– . . . backbone (top); the hydrogen bonding between base pairs (middle); and the detailed structure showing all the atoms of C, H, O, P, and N (lower). (*Adapted from Bennett Frieden*, Modern Topics in Biochemistry (*New York: Macmillan, 1969*).)

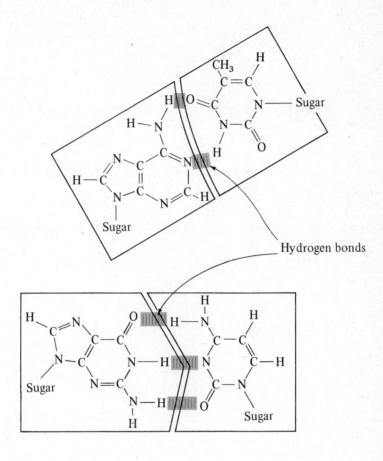

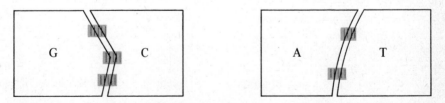

**Figure 6-15** Hydrogen bonding between adenine and thymine, and guanine and cytosine.

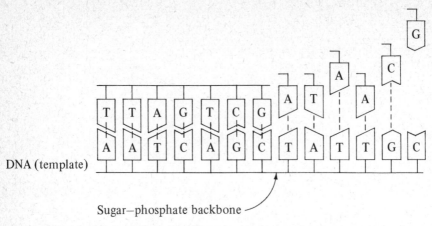

DNA (template)

Sugar–phosphate backbone

**Figure 6-16**  Biosynthesis of DNA (two-dimensional representation).

in the sequence of nucleotides, leads to a synthesis of protein quite different from that intended. Thus the protein needed never is synthesized. This situation gives rise to mutations, which in most circumstances are harmful to the organism.

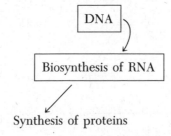

DNA

Biosynthesis of RNA

Synthesis of proteins

RNA molecules are remarkably similar to those of DNA with the exceptions that ribose replaces deoxyribose in the sugar–phosphate–sugar–· · · backbone and uracil replaces thymine as a nitrogen base. Otherwise, the gross structural features of RNA are identical to those of DNA (Figure 6–18).

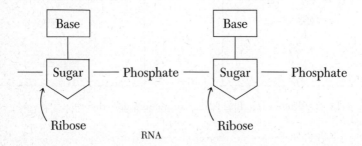

RNA

A typical virus chemically consists of ribonucleic acid intermeshed with a protective protein coat. Viruses act by penetrating the host cell and then

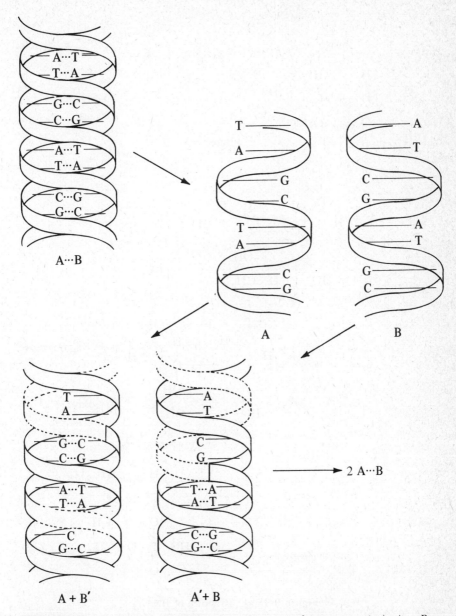

**Figure 6-17** Biosynthesis of DNA (three-dimensional representation): A···B = original double-stranded α helix; A + B = two single-stranded α helices; A + B′ = A + partially synthesized B helix; A′ + B = B + partially synthesized A helix; two A ··· B = double-stranded α helix.

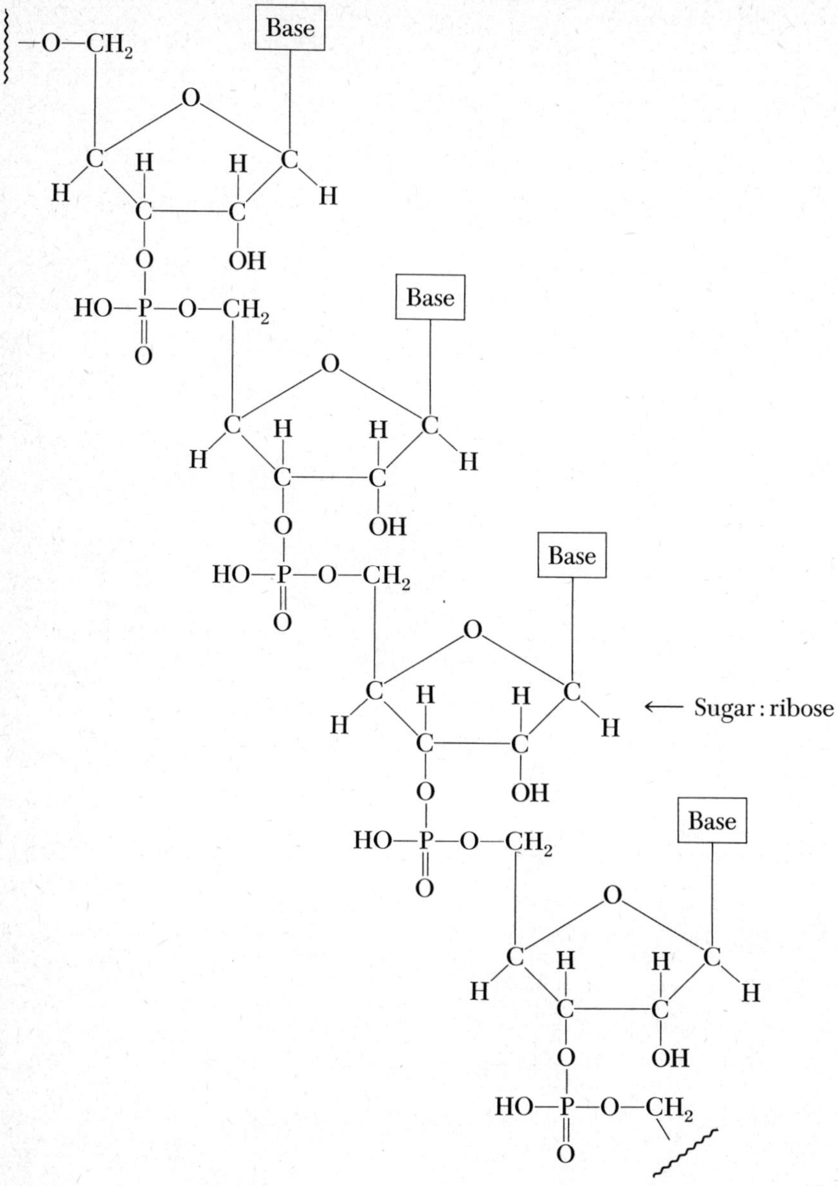

**Figure 6-18(a)**   Structure of RNA.

using the machinery of the host cell, ribosomes, mitochondria, and so on, to produce more virus at the expense of the host. Tobacco mosaic virus (Figure 6–19), for example, can be isolated, crystallized (like sugar or salt), and stored indefinitely. Dispersed in water, it again actively infects the host.

A question we might ask ourselves now is, How is RNA synthesized

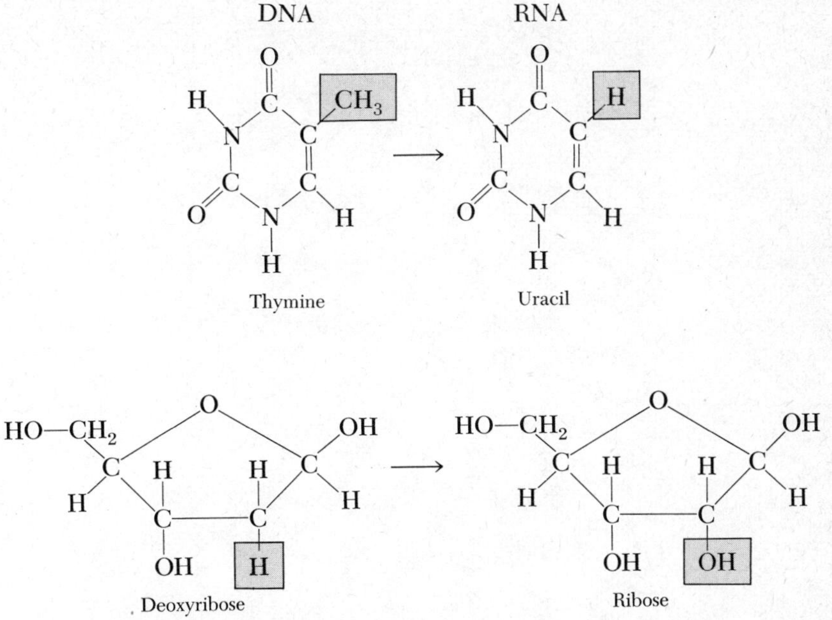

DNA    RNA

Thymine    Uracil

Deoxyribose    Ribose

**Figure 6-18(b)** Substitutions in conversion of DNA to RNA.

from DNA, and how does the information stored in the sequence of the DNA nucleotides get transferred to the RNA molecule? The answer to this question can easily be understood as an extrapolation of the DNA biosynthesis we have just discussed. If the DNA molecule were to unravel either partially or completely, then the DNA could serve as a template for RNA synthesis. This is shown in Figure 6-20. Recalling that one of the chief forces involved in keeping the two strands of DNA together was the hydrogen-bond interactions of the base pairs, we see that the RNA complement to DNA could easily be prepared by allowing RNA nucleotides to bind to the DNA. The similarity in structure and shape between ribose and deoxyribose and between uracil and thymine (see Figure 6-18) permits substitution of ribose and uracil for deoxyribose and thymine in the replication process. Thus in a manner similar to DNA biosynthesis, the synthesis of RNA is achieved using DNA as the template. Ribonucleic acid biosynthesis occurs in the nucleolus of the cell (Figure 6-12).

The RNA molecules that are produced vary widely in size and function, depending on their role. Some of the RNA is used to transfer to the ribosomes the amino acids needed for protein synthesis (transfer RNA).[*] Transfer RNA usually consists of seventy to eighty nucleotides.

[*] The ribosomes are attached to the endoplasmic reticulum (Figure 6-12) and serve as an assembly bench for protein synthesis.

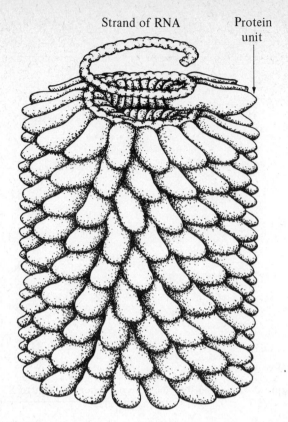

Strand of RNA     Protein unit

**Figure 6-19**  Diagram of tobacco mosaic virus (TMV). The coat of this virus is of protein, which can be subdivided into 2,200 units, each with the same amino acid sequence. The interior of TMV consists of a single strand of coiled RNA. (*Redrawn from J. W. Baker and G. E. Allen,* Matter, Energy and Life (*Reading, Mass.: Addison-Wesley, 1964*).)

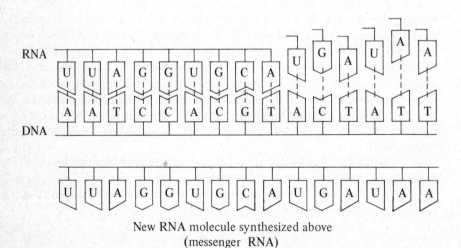

New RNA molecule synthesized above
(messenger RNA)

**Figure 6-20**  Biosynthesis of messenger RNA.

238

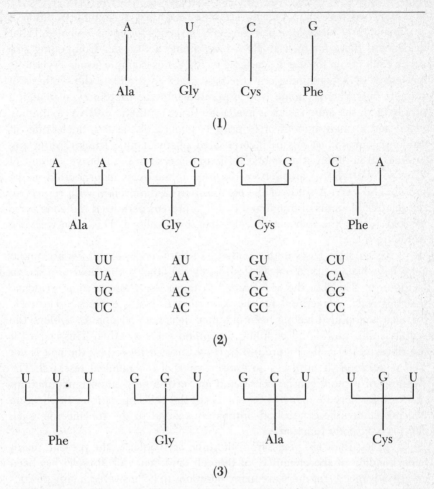

**Figure 6-21** Possible nucleic base-amino acid codes. Ala = alanine; Cys = cysteine; Gly = glycine; Phe = phenylalanine. (1) One nucleic base, one amino acid: four possible combinations. (2) Two nucleic bases, one amino acid: sixteen possible combinations. (3) Three nucleic bases, one amino acid: sixty-four possible combinations.

Messenger RNA, the biosynthesis of which is shown in Figure 6–20, has a primary structure complementary to a portion of DNA. It is believed that the code determining the sequence of amino acids in protein synthesis is transferred from DNA by messenger RNA (m-RNA). Let us briefly consider how this might be possible. Recalling that all protein consists of twenty amino acids, some manner must be found in which each amino acid can be specifically called for during protein synthesis. Clearly one base on m-RNA cannot code for one amino acid since there are only four different bases and twenty amino acids (see Figure 6–21). Unless the amino acid sequence in protein synthesis is nonspecific, this is impossible. A similar argument holds for the suggestion that two bases code for one amino acid, since there are only sixteen ways in which two base pairs can be arranged. The first acceptable solution is obtained if three bases, attached consecutively, code for one amino acid. What evidence is there for such an RNA base–amino acid code? Compelling evidence comes from the use of synthetic m-RNA. For example, using a synthetic messenger RNA containing only the base uracil to stimulate the synthesis of protein, only phenylalanine is incorporated into the protein even though a supply of all the amino acids is available. Using synthetic m-RNA containing uracil and a small amount of cytosine (sequence of the two nucleotides at random), four amino acids are incorporated, phenylalanine, leucine, serine and proline, all in different amounts. With an increase in the relative amount of cytosine, the relative amounts of the four amino acids incorporated in the protein also vary. A study of the composition of the protein with respect to the relative amounts of uracil and cytosine present permitted the code to be worked out for the four amino acids mentioned above. This code is shown in Figure 6–22.

Using other bases in the synthetic m-RNA, codes for all twenty amino acids have been worked out. Codes for several other amino acids are shown in Figure 6–23. Note the redundancy in the code. There is some evidence that this code, worked out for the bacterium *E. coli* is universal. The code for each amino acid having been obtained, it is next possible to achieve the synthesis of a protein by coupling the amino acids together. This occurs at the ribosomes of cells (Figure 6–12). How this is achieved by the cell is not well understood; it furnishes an exciting area of biochemical research. The synthesis of protein can be considered one of the most important functions of a cell, for here lie many of the secrets of the cell's operations. Replication and protein syntheses are both intimately related to the specific role each cell plays in body functions.

In less than 150 years, from the time of Wöhler to the present, man's understanding of the chemistry of the cell (and hence of himself) has been greatly advanced. Yet there are many questions to be answered, many puzzles still to be unraveled. In the future lie further revelations of the fascinating operations of the cell, and an opportunity to correct its disorders and perhaps to uncover the secrets of life itself.

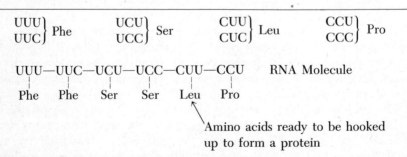

$$\left.\begin{array}{l}\text{UUU}\\\text{UUC}\end{array}\right\}\text{Phe} \qquad \left.\begin{array}{l}\text{UCU}\\\text{UCC}\end{array}\right\}\text{Ser} \qquad \left.\begin{array}{l}\text{CUU}\\\text{CUC}\end{array}\right\}\text{Leu} \qquad \left.\begin{array}{l}\text{CCU}\\\text{CCC}\end{array}\right\}\text{Pro}$$

UUU—UUC—UCU—UCC—CUU—CCU      RNA Molecule

Phe    Phe    Ser    Ser    Leu    Pro

Amino acids ready to be hooked up to form a protein

**Figure 6–22**  Code for the amino acids Phe, Ser, Leu, and Pro, using messenger RNA containing uracil and cytosine, in the bacterium *E. coli.*

$\left.\begin{array}{l}\text{UUU} \\ \text{UUC}\end{array}\right\}$ Phe   $\left.\begin{array}{l}\text{CUU} \\ \text{CUC} \\ \text{CUA} \\ \text{CUG}\end{array}\right\}$ Leu   $\left.\begin{array}{l}\text{AUU} \\ \text{AUC} \\ \text{AUA}\end{array}\right\}$ Ileu   $\left.\begin{array}{l}\text{GUU} \\ \text{GUC} \\ \text{GUA} \\ \text{GUG}\end{array}\right\}$ Val

$\left.\begin{array}{l}\text{UCU} \\ \text{UCC} \\ \text{UCA} \\ \text{UCG}\end{array}\right\}$ Ser   $\left.\begin{array}{l}\text{CAU} \\ \text{CAC}\end{array}\right\}$ His   AUG$\}$ Met   $\left.\begin{array}{l}\text{GCU} \\ \text{GCC} \\ \text{GCA} \\ \text{GCG}\end{array}\right\}$ Ala

$\left.\begin{array}{l}\text{UAU} \\ \text{UAC}\end{array}\right\}$ Tyr   $\left.\begin{array}{l}\text{CAA} \\ \text{CAG}\end{array}\right\}$ Glutamic Acid   $\left.\begin{array}{l}\text{ACU} \\ \text{ACC} \\ \text{ACA} \\ \text{ACG}\end{array}\right\}$ Thr

$\left.\begin{array}{l}\text{UGU} \\ \text{UGC}\end{array}\right\}$ Cys   UGG$\}$ Try   $\left.\begin{array}{l}\text{GGU} \\ \text{GGG}\end{array}\right\}$ Gly

**Figure 6-23** Summary of the m-RNA codes for a representative number of amino acids for the bacterium *E. coli.*

## POSTSCRIPT

One of the most compelling reasons for introducing biochemistry to all students, science majors or not, is to expose them to the current level of scientific competence in an area of immense moral concern. Man may in the near future learn how to tamper with the genetic code. Should this possibility be realized, in addition to the significant scientific achievement engendered, the benefits to man may be immense and/or the horrors unbelievable. Benefits to man in the area of correcting disorders transmitted genetically would be realized, while the possibility of creating human robots out of test tubes, devoid of any human emotion, also exists. The manner in which scientific information is used should be of concern to everyone. Perhaps you will be interested in Aldous Huxley's classic novel *Brave New World* and George Orwell's *Nineteen-Eighty-Four* as an introduction to this possibility.

## *PROBLEMS*

1. Why is it essential to eat a balanced diet?
2. List some cases where lipids act as protective coatings.
3. Why is the fact that DDT and other insecticides are soluble in lipids and insoluble in water significant?
4. How are dead, damaged, or old cells removed in living systems?

5. Explain in simple terms how the birth control pills are thought to work.

6. Explain how viruses work.

7. Describe what is meant by enzyme specificity.

8. Describe what is meant by enzyme inhibition.

9. (a) How many dipeptides that contain the amino acids glycine and alanine can be formed? Write the structures.

   (b) How many dipeptides that contain the amino acids methionine and serine can be formed? Write the structures.

   (c) How many tripeptides containing glycine and alanine can be formed? Write the structures using abbreviations.

10. How many tripeptides that contain the amino acids glycine, alanine, and serine can be formed? Write the structures using abbreviations.

11. (a) Predict the primary structure of a protein synthesized by the following messenger RNA.

    UUU–UUC–UCU–UCC–UAU–ACU–GCU–GUU–AUU–CAA–UUC

    (b) The following portion of the DNA molecule is used to synthesize m-RNA. Predict the sequence of amino acids coded by this DNA.

    GAA–AGG–GGG–GTG–GTT–GTC     DNA

12. Suppose you used a synthetic m-RNA to stimulate protein synthesis. What protein would you synthesize if the only base in your m-RNA was uracil (U). Suppose you now were able to incorporate cytosine as well. If the composition of your m-RNA was 33% cytosine, 67% uracil and the sequence of these bases were incorporated at random:

    (a) What possible base sequences could you get in your m-RNA? List them (there are 8), UUU, etc.

    (b) Which sequences should be present in the greatest amount based on the composition of U and C?

    (c) Using Figures 6–22, 6–23, what proteins should be incorporated in the m-RNA stimulated synthesis of protein?

    (d) Which amino acids should be present in greatest amount in the protein?

13. Repeat Problem 12 using guanine (33%) and uracil (67%). The code was deciphered by similar types of experiments. Once it was known which amino acids were incorporated and also how much was incorporated as a function of base composition, it was possible to determine which three bases code for each amino acid.

14. Suppose you have isolated an enzyme that will *only* hydrolyze peptide bonds between *ala*nine and *ser*ine. What must the sequence of amino acids be in a protein containing seven amino acids if you isolate the following peptides after using your enzyme? (There are three possible answers.)

    ala–pro–ala + ser–glu–cys–ser

Suppose you now took the same protein and cooked it up with stomach acid for a short time and isolated the following. What must the sequence of the protein be?

> pro–ala–ser
> cys–ser–ala + 5 amino acids
> glu–ser–ala
> ser–glu–cys
> glu–cys–ser

15. A hexapeptide (six amino acids) is chewed up by the stomach to produce only 3 amino acids. If the reaction time in the stomach is reduced, one can isolate the following dipeptides. Omit cyclic possibilities.

protein (hexapeptide) ⟶ stomach ⟶ ala, pro, his

protein (hexapeptide) ⟶ stomach ⟶ ala–pro
(short period    his–his
of time)      his–als
            pro–his
            pro–pro

Suggest a sequence for this protein consistent with the data.

16. Draw the mirror images of serine and phenylalanine. Are the mirror images superimposable on the original structure? Use the molecular model provided as an aid.

17. Hardening of the arteries, or arteriosclerosis, is caused in part by a deposit of cholesterol inside the arteries. Gallstones are also almost pure cholesterol. Is it unusual to expect cholesterol to be deposited by blood? Why?

## BIBLIOGRAPHY

1. T. Bennett and E. Frieden. *Modern Topics in Biochemistry.* New York: Macmillan, 1969.

2. E. West and W. Todd. *Textbook of Biochemistry,* 3rd ed. New York: Macmillan, 1961.

3. J. Fruton and S. Simmonds. *General Biochemistry,* 2nd ed. New York: Wiley, 1960.

4. I. Liener. *Organic and Biological Chemistry.* New York: Ronald Press, 1966.

5. J. Baker and G. Allen. *Matter, Energy and Life.* Reading, Mass.: Addison-Wesley, 1970.

6. G. Simpson and W. Beck. *Life. An Introduction to Biology,* shorter ed. New York: Harcourt, Brace & World, 1969.

7. Albert L. Lehninger. *Biochemistry.* New York: Worth, 1971.

8. V. Petrow. "Current Aspects of Fertility Control." *Chemistry in Britain* 167 (1970).

9. "Biochemistry of the Pill Largely Unknown." *Chemical and Engineering News*, March 27, 1967, p. 44.

# SUGGESTED READING (BIOCHEMISTRY)

1. "Man into Superman: The Promise and Perils of the New Genetics." *Time*, April 19, 1971, pp. 33–52.

2. J. D. Watson. *The Double Helix*. New York: New American Library, 1969.

3. R. F. Steiner and H. Edelhoch. *Molecules and Life*. D. Van Nostrand, Princeton, N.J., 1965.

# Population

# 7

*It was the best of times, it was the worst of times, it was the age of wisdom, it was the age of foolishness, it was the epoch of belief, it was the epoch of incredulity, . . .*

Charles Dickens

In November 1970, Pakistan was battered by a cyclone that destroyed 500,000 lives. By January 1971, the loss in population had been recovered. This tremendous growth in the population of a country over such a short period of time is not restricted to Asian countries. In America, enough new people are added every year to equal the combined populations of Rhode Island, Delaware, Idaho, Nevada, and Montana, approximately three million. A question we should ask ourselves is, What sort of pressure does population growth place on our environment? Consider the resources and raw materials necessary to maintain an individual according to present American standards. Each individual in the United States can expect to consume directly or indirectly 21,000 gallons of gasoline, 10,000 pounds of meat, 28,000 pounds of milk and cream, and countless other articles in their lifetime. Prosperity, to all of us, is directly or indirectly related to these statistics. But at what cost? What are the tangible effects in producing an additional 200 million pounds of steel, 6.3 million electric refrigerators, or 11 million automobiles? A trip to any large industrial city will reveal some of the most obvious costs. The urban sprawl, the maze of highways, adulterated color and texture of the surrounding air and water, crowding, noise, garbage, and odors are just a few of the distressing conditions that everyone shares. Everyone involved in urban living would gladly welcome the removal of these irritating and noxious contaminants.

Pollution today is an important moral, social, and political issue. A considerable amount of publicity has been given to efforts to curb pollution. Devices to control automobile emissions, restrictions on the incineration of garbage, and the addition of electrostatic precipitators to remove particulates from smokestacks are just a few of the means introduced to control pollution of the air. Technological improvements in the treatment of water, such as updating of sewage disposal systems and addition of tertiary treatment plants, are also being made. However, these efforts might be compared with our efforts in treating the common cold; they concentrate on the symptoms rather than the disease. Such an approach in medicine is justified on the basis that while the symptoms are easily treated, the cure is often unknown. With pollution, the approach is successful so long as the pollutants are small in amount relative to the environmental systems that have to remove them. We

**Little boxes on a hillside.** Affluence can't cure the dense treeless subdivisions that are part of the urban sprawl. (*Reprinted, by permission, from* St. Louis Post-Dispatch. *Photograph by David Gulick.*)

might ask ourselves what the net effect is of encouraging the reduction of automobile emissions on one hand while stimulating the production of more and more automobiles on the other?

Is pollution simply a symptom of a more complex problem? We know that population pressures and technology have contributed toward creating pollution problems and that these problems all appear in a cyclic relationship. This chapter will attempt to dramatize this vicious cycle by focusing on some of the new and existing applications of technology that are being, or have been, introduced to cope with ever increasing social, environmental, and demographic pressures.

## HISTORY OF POPULATION

*We have met the enemy and he is us.*

**Pogo**

Any historical development of population must begin with a discussion of man's past, estimated at approximately two million years. It is believed that man's first 1,700,000 years were primarily concerned with learning how to make and use tools. The next 150,000 years were invested in the art of initiating, maintaining, and transporting fire. Then for the next 140,000 years, man explored the earth and domesticated plants. Approximately 8,000 years were spent in the development of field agriculture and village life. Man's historical development during the last 2,000 years is well known. It is estimated that during most of the two million years the world's population totaled eight million. The growth of population during this time is shown in Figure 7–1 and could, for most of this time, be approximated by a straight line.

For the longest part of the time that man has existed on earth, his lifetime has been short and full of hazards. At first, he barely managed to replace himself. As long as man was nomadic, the human population fluctuated with climatic changes, availability of food, and so forth. About 10,000 years ago a significant change occurred. Man domesticated plants and invented farming, and in doing so began a change that clearly differentiated him from the rest of the animal kingdom. Dependent on farming, man, now attached to the soil, was intimately concerned with the fertility of the land, rainfall, temperature, and the like. It is believed that many of the primitive religions have their roots in this stage of man's development, not only because farming afforded additional leisure time but also because it developed a total dependence on uncontrollable and unpredictable factors, such as the weather. High human fertility was necessary if the race was to survive and prosper, and in time fecundity was often incorporated into the religion. Infant mortality was high. Even in the last century in some of the underdeveloped countries, 25 percent of the children born did not survive their first year. Parents who

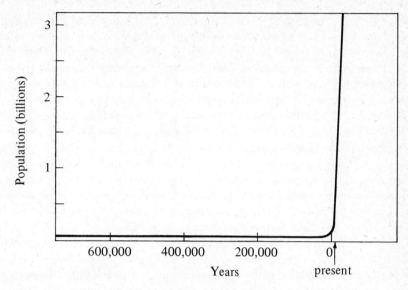

**Figure 7-1** World population as function of time.

wanted to be assured of having children to take care of them in their old age, tried to have as large a family as possible. The "steady-state" concentration of people (birthrate equal to death rate) of several million people existed for thousands of years, fluctuating somewhat during famine and drought. The majority of the population during this time was located in rural areas. Since the total world population was small, man's impact on the environment was negligible. Although occasional plagues of locusts and other insects occurred, such events were the result of factors quite independent of man's presence. Ignorant of methods of farming, such as crop rotation and contour plowing that we know today, the farmer often had to move to other fertile ground periodically. Although his effect on the land in many cases was devastating, the availability of virgin land and his limited numbers kept constantly in check by natural processes caused little irreversible damage to the environment.

The last 2,000 years, however, have changed most of this. The Dark Ages, the Renaissance, the Reformation, the industrial revolution and the technological revolution have created an environment so different from our evolutionary past that it is questionable whether we are biologically and socially equipped to handle it. The first step in this series of events was the development of equipment that would bring larger areas of land under cultivation. This freed some people from the soil, and the development of the modern city ensued. The availability of more food affected the population, but in an indirect way. The traditional birthrate remained essentially constant but what was effected through advances in medicine and increases in food production was a decrease in the death rate so that it no longer equaled the birthrate. Life expectancy was also increased. Although the need for a large birthrate was no longer necessary, many people, particularly in the under-developed countries, have continued with the old traditions to the extent that the dimensions of the problem have become staggering. An estimated 3.5 billion people are now alive, 55 percent in Asia alone. This is more than double the number of people who inhabited the earth at the start of the twentieth century. The 120 million people born each year far outbalance the 50 million who die. Seventy million people are added annually to the world's population and this number is increasing.

Obviously population cannot continue to expand indefinitely. Some experts insist that even our present world population of 3.5 billion is too great for the world to sustain far into the future. Others believe that if population were stabilized at 6 or 7 billion, no insurmountable problems would result.

There are three factors that determine the rate of population growth. Two of them, the birthrate and the death rate, have been mentioned. The third factor is often completely ignored. This factor, which has played an important role in stabilizing the European population, is related to the average age at which couples become parents. Consider the following two couples, both having two children apiece. If Ted and Carol have 2 children at age 20 and their children behave similarly, at age 65 they will have 8 great grandchildren, 4 grandchildren and 2 children, a total of 14 descendants. Bob and Alice, who waited until age 30 to have children (and whose children behaved similarly), would at age 65 have 4 grandchildren and 2 children, a total of 6 descendants. Note the difference that a decade makes.

The remarkable increase in population over the past few decades can mathematically be *approximated* by a geometrical progression (for example, $2^n$: 2, 4, 8, 16, . . .). The result of any process described by a geometrical progression is basically the same. Uncontrolled growth of bacteria, insects, or people leads to the same result, an "explosion." Population growth continues in an ever increasing rate until at some point a new steady-state concentration is again achieved.

Recently, John Miller, a venerable Ohio farmer passed away at age 95 and was survived by 410 descendants, perhaps the largest number descended from any living American. John Miller was actually a part of the population explosion of the twentieth century. His family started with just seven children and during most of his life, his family could not be considered large. He simply survived long enough to see how a geometrical progression works. At the time of his death, the mailman was bringing word of a new descendant *every ten days*. A foremost thought in John Miller's mind was summarized in a question he constantly kept repeating, "Where will they all find good farms?"

Decrease in population brought about by an increase in death rate over birthrate can also be approximated mathematically by a geometrical progression. We might describe this process as a "population implosion." For example, if the death rate of a species for some reason exceeds the birthrate for some length of time, the species may become extinct. The dinosaur, passenger pigeon, and dodo are some of the countless examples of species that have followed this path.

## BIRTH CONTROL

What are our alternatives to overpopulation? Do we have any means at our disposal to deal with increasing birthrates? One possible solution is to maintain the current laissez-faire attitude. At some time a new steady-state concentration, where the birthrate and death rate balance by natural factors, will again be reached. There is some evidence to support the contention that we may soon be approaching a new steady-state concentration in America. Recent economic expansion and increases in the standard of living appear to be lowering the average life expectancy of Americans. The result is not surprising considering that a large portion of the world's goods are produced and consumed (about 40 percent) by Americans. The tons of $SO_2$, CO, NO, $NO_2$ spewed into the air of a city are bound to have some adverse effects on its inhabitants. Widespread use of tobacco can be considered another nail in the coffin. Lung disorders such as lung cancer and emphysema are on the rise. Those most affected by these disorders, however, are the ones least able to bring about reform, the very young and the very old.

Another possible solution in bringing the birthrate into balance with the death rate is worldwide birth control. Probably no medical development has had greater impact on birth control programs in the underdeveloped world than the intrauterine device. Although the exact mode of action is still uncertain, it is believed that the intrauterine device operates by preventing zygote (fertilized egg) implantation in the uterus following ovulation. Normally, the egg passes into the funnellike end of the uterine tube and if fertilized, it is implanted in the uterine wall. Otherwise, the egg disintegrates in passing through the uterus. The intrauterine device represents a foreign object in the vagina and thus triggers the natural abortive mechanisms present in the body. There are many different kinds and shapes of intrauterine devices and their mode of operation may vary somewhat. Although devised over a hundred years ago, the intrauterine loop has only in the past few years achieved widespread use.

In 1960, oral contraceptives were introduced and the contraceptive revolution ensued. A decade and 700 million more people later, public opinion concerning the pill remains divided, both on medical and religious grounds. There remains caution and concern over the pill's possible long-range side effects. It remains to be determined whether there is any connection between prolonged use of the pill and diabetes, cancer, and diseases of the circulatory system. Crude estimates indicate that about 14 million women (about 7 million American women) use oral contraceptives.

Some efforts in the birth control area are being directed toward men.

The most effective method presently available is the vasectomy. Vasectomies are operations that are easily performed and require no hospitalization. Only a small incision is required to sever and seal the sperm duct, which connects the testis with the seminal vesicle and prostate gland. Since the sperm duct is severed "upstream" from the seminal vesicle and prostate gland, which provide the fluid in which the spermatozoa are transported, a vasectomy only removes the spermatozoa from the seminal fluid. Completely normal sexual activity is still possible.

Inexpensive and possibly reversible, this method of maintaining family size is gaining wide acceptance throughout the world. Those people concerned with the morality of abortion or worried about the possible negative effects of long-term use of the pill find this method of birth control particularly attractive. In addition, sperm banks have become available; they permit long-term storage of sperm at very low temperatures. Thus, if reasons should develop for which a male who has had a vasectomy should want more off-spring, his wife can be artificially impregnated with his own sperm. In 1970, 750,000 Americans submitted themselves to this 25-minute surgical procedure. India appears to have the most ambitious program in this area. However, the situation in parts of Asia has been termed hopeless by some **demographers.** They feel that enough technicians could not be trained in time to perform the necessary vasectomies. They predict mass starvation in certain over-populated regions of the world by the end of this decade.

Abortion, particularly in the United States, has never been officially intended or allowed as a form of birth control until recently. Instead its primary function has been directed toward preserving the health of the living. This attitude is reflected in the Family Planning Services and Population Research Act of 1970, which provides funds for research in biomedical con-traceptive development and for program implementation related to family planning. This act specifically prohibits funds from being used in areas con-cerning abortion as a method of family planning.

A glimpse at Table 7–1 will show some data that indicate that this attitude on the part of government is not necessarily prevalent in other countries. The abortion rate in Rumania for 1967–68 (3.9 abortions/1 live birth) bears mentioning. The remarkably high rate of abortions in conjunction with a decreasing birthrate caused the Rumanian government in 1966 to restrict abortions to cases of special circumstance or to women over 45 who already had a moderate-size family. As a result, the birthrate doubled in 1967, but again dropped in the succeeding year, indicating that other methods, including illegal abortion, were being used.

A consideration of the estimated number of illegal abortions demon-strates that many countries are not far behind Rumania. The data in Table 7–2 suggest that abortion is being used as a prominent form of birth control.

The moral issues regarding abortion are often heatedly debated. One thing appears certain, if man on a worldwide basis does not voluntarily limit his population, natural factors will limit it for him. These factors may take

**Table 7-1**
**Incidence of Therapeutic Abortions in Selected Countries,**
**Data from 1967 and 1968**

| Country | Population (Millions) | Number of Live Births (Thousands) | Number of Abortions (Thousands) | Ratio of Abortions to Live Births |
|---|---|---|---|---|
| Rumania | 19.4 | 278.4 | 1,100 | 3.9 |
| USSR | 230 | 4,190 | 6,000 | 1.4 |
| Hungary | 10.3 | 148 | 200 | 1.3 |
| Bulgaria | 8.4 | 124.7 | 98 | 0.7 |
| Japan | 101 | 1,935 | 747 | 0.4 |
| Yugoslavia | 20.1 | 388 | 150 | 0.4 |
| Czechoslovakia | 14.4 | 216 | 85 | 0.4 |
| England and Wales | 55 | 700 | 100 | 0.14 |
| Sweden | 7.9 | 122 | 96 | 0.08 |

SOURCE: Adapted from H. Rudel, F. A. Kinel, and M. R. Henzl, *Textbook of Birth Control*, (Macmillan).

the form of starvation, disease of epidemic proportion, or complete collapse of mental and social well-being. The 3.5 million people, mostly children, in the underdeveloped countries who will starve to death this year and the countless others who suffer irreversible damage due to malnutrition are merely a handful compared to what is expected for the future.

Recently, an international team at the Massachusetts Institute of Technology completed a computer-simulated study in which they tried to determine, using best guesses, the effect of the world's natural resources on population. Their results, which correlated data on current physical, economic, and social relationships, forecast a dramatic drop in population following exhaustion of the world's nonrenewable natural resources (oil, coal, and so

**Table 7-2**
**Estimated Number of Illegal Abortions (1967)**

| Country | Number of Abortions (Thousands) | Ratio of Abortions to Live Births |
|---|---|---|
| Austria | 200–300 | 1.5–2.0 |
| France | 200–400 | 1.5–3.0 |
| West Germany | 1,000–3,000 | 1.0–3.0 |
| Chile | 100–200 | 0.4–0.8 |
| Uruguay | 160 | 3.0 |
| USA | 300–3,000 | 0.1–0.6 |

SOURCE: Adapted from A. Klinger, *Demografia y Economia* 12 (1969): 479 and cited in C. Djerassi, *23rd International Congress of Pure and Applied Chemistry, Special Lectures*, vol. 3 (London: Butterworth's, 1971).

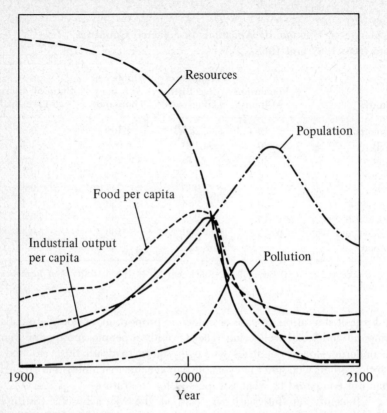

**Figure 7-2**  Exhaustion of resources cuts industrial output. Predicted relationships between several social and economic variables as a function of time. Vertical scales are combined simply to demonstrate the interrelation of the variables, and are not of equal value. Horizontal scale: time. (*From* The Limits to Growth: *Project on the Predicament of Mankind, by Donella H. Meadows, Dennis L. Meadows, et al. A Potomac Associates Book published by Universe Books, New York, 1972. Cited in* Chemical and Engineering News, *March 6, 1972.*)

on). The results of their study are summarized in Figure 7–2. The model this team used was based on the interrelationship among population, industrial output, food, nonrenewable resources, and pollution. Even in the most optimistic system—one having a 75 percent recycling of materials (iron, paper, and so on), use of nuclear power, reduction of pollution to 25 percent of the 1970 value, doubled food yield, and worldwide birth control—a significant drop in population is predicted before the twenty-first century as a result of pollution and exhaustion of natural resources.

Man's numbers cannot increase indefinitely. At some point a steady state must be reestablished. If the MIT calculations are correct, man's numbers must decrease to avert what must surely be considered a world catastrophe of immense proportions. The only choice we have is in the manner in which man's numbers are stabilized.

# INSECTICIDES AND PESTICIDES

The development of modern technology has permitted an ever increasing amount of land to be cultivated. This has allowed feeding of an ever expanding population. Population pressures in turn have forced us to become increasingly dependent on technology. In contrast with earlier farms that grew a variety of crops, modern technological farms as a matter of economics have to grow a single crop over vast expanses of land. The small diversified farm kept the heterogeneous insect population in check by attracting natural predators and by providing only a limited food supply. The advent of the modern farms completely upsets this delicate balance. Growing of a single crop over acres and acres of land, necessary in the feeding of a nonagricultural society, offers an unlimited food supply to a relatively few species of insects. Any insect attracted by this supply prospers and there is a population explosion of the species. To solve this problem the farmer, often encouraged by chemical companies, has initiated widespread use of cheap chemical agents as a control. Furthermore, since the public has been willing to pay more for the perfect plump fruit or vegetable, further encouragement of the traditionally conservative farmer was not necessary. Extensive use and often misuse of these chemicals and indifference or ignorance about their long-range consequences have created a very serious ecological crisis.

The advent of DDT in 1946 ushered in the widespread use of insecticides. They have been of considerable benefit to man in the areas of disease control and increased crop production. The incidence of tropical diseases such as malaria and yellow fever has been drastically reduced. Upon application, chemical insecticides act quickly and isolated treatments are relatively inexpensive. However, as is often the case, there is also another side to the story. Consider the chlorocarbon insecticides (compounds of carbon, chlorine, and hydrogen), which because of their broad spectrum of activity toward insects have been extensively used. DDT is probably the best known of the many chlorinated hydrocarbons used (Table 7–3). What happens to the DDT after it has been sprayed? Unfortunately, nothing much. DDT is not readily biodegradable. That is to say, bacteria and other living organisms that degrade most carbon-containing compounds into $CO_2$ and $H_2O$ are not capable of metabolizing DDT. Consequently, the amount of DDT in the environment has been steadily increasing. DDT and the other chlorocarbon insecticides are stored in the fatty tissue of living things. Recent findings show that alarming concentrations of these substances are present in most living species. The effects of DDT, for instance, are not necessarily limited to insects. This substance, when present in sufficient concentration, is lethal to many other forms of life. The success in using DDT (and most other pesticides) in controlling insects rests on the fact that the lethal dosage for insects is considerably smaller than for animals, partly because of the insects' relatively small size. However, the shorter life span and prolific nature of most insects allows considerable opportunity for adaptation through natural selection. With

## Table 7-3
## Some Common Chlorocarbon Insecticides

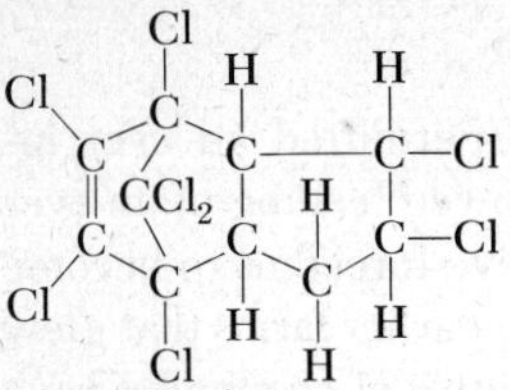

Chlordane

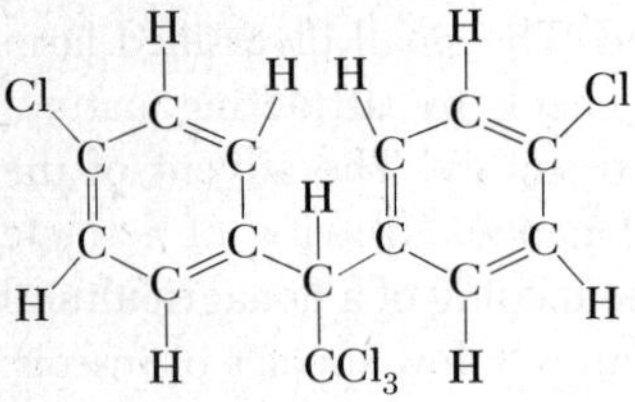

DDT

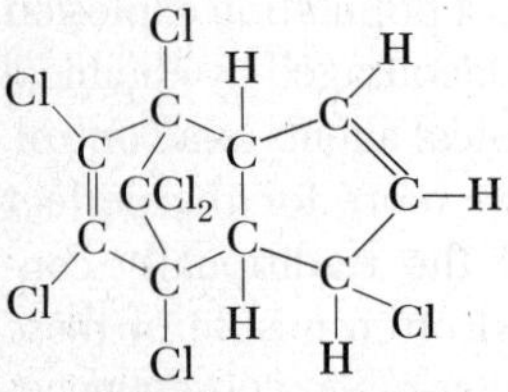

Heptachlor

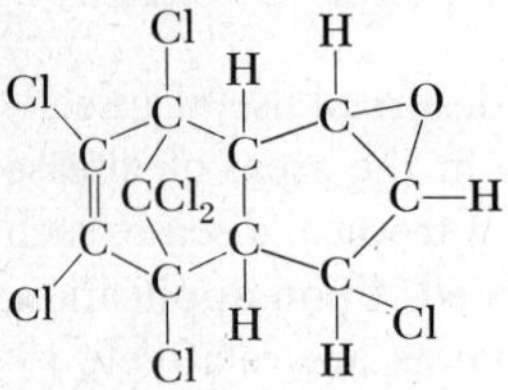

Heptachlor epoxide

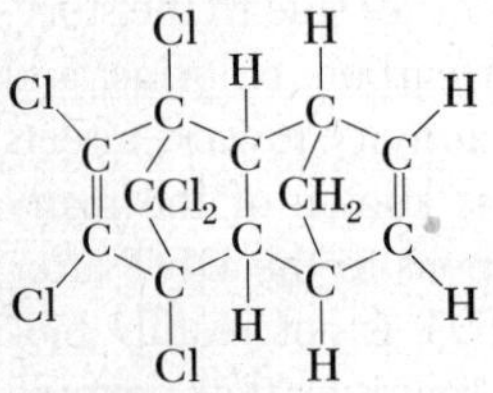

Aldrin

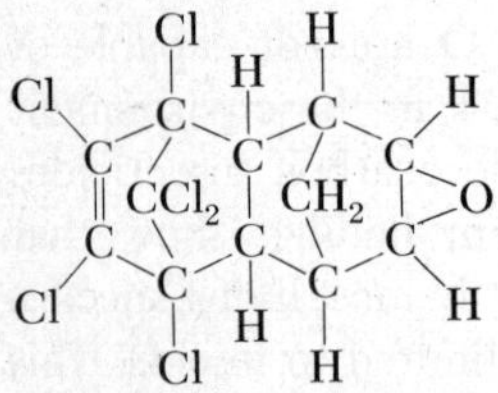

Dieldrin

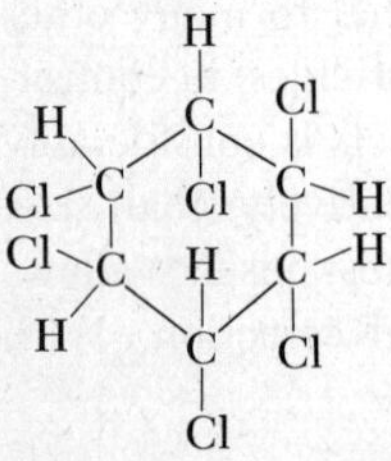

Lindane

chemical pesticides, it generally takes ten to twenty generations to develop a resistant strain of insects. Often larger dosages of insecticide are necessitated by the insects' resistance to the drug. In the Canite Valley in Peru, DDT and similar pesticides were used to control an insect pest. Natural predators were also killed, resulting in a temporary population explosion of this pest when a strain of the insect became resistant to the insecticide.

Approximately 150 million pounds of chlorinated hydrocarbon insecticides and an additional 700 million pounds of other pesticides are "consumed" annually. The chlorinated hydrocarbons do slowly degrade through the action of certain microorganisms and sunlight. However, the degradation products are often as toxic as the insecticide itself.

There are approximately one million different insect species that have survived and flourished for some 250 million years. Insects are probably the most successfully adaptable creatures on earth. Only a tiny fraction of these insects are as yet considered pests by man (6,000). To control them, Americans practice an art termed overkill by some. For example, to control the lawn moth, dosages as high as 40 lb per acre of DDT have been used whereas 11 lb per acre would have sufficed. In addition, some programs aimed at eliminating some insects have in the process inadvertently killed many species of insects of definite benefit. The differential susceptibility of various species of insects is of serious concern. For example, the use of DDT in apple orchards to control pests eliminated populations of a certain ladybird beetle, the prime predator of the pest, red mite. The population of the red mite exploded, causing significant damage to the orchard. Apparently bees are extremely susceptible to many insecticides (such as carbaryl, Table 7–4). As a result, dosages used to control other insects have a drastic effect on the population of this useful insect. In California, the bee population was decimated to such an extent that bees had to be imported from the State of Washington to pollinate the orchards.

What can replace DDT and the other chlorocarbon insecticides? Our current economic structure is based on the premise that a few people can supply the nutritional needs of the many. It relies heavily on the use of **herbicides, insecticides, nematocides, rodenticides, fungicides,** and the like.

Prior to the development of the large technological farms and insecticides, a rather sophisticated method of controlling insects had been developed. The knowledge that certain plants were capable of repelling particular insects was passed from generation to generation. By a judicious choice of plants surrounding a farmer's crops and home, insects could be controlled and damage minimized. With the advent of cheap chlorocarbon insecticides, this practice was discontinued. Now that the danger from the widespread use of insecticides is receiving attention, interest in the active ingredients present in these plants has been renewed.

Pyrethrin, a compound obtained from pyrethrum flowers (related to chrysanthemums), has long been known to be an active insecticide. The main drawbacks of using natural insecticides are their relatively high cost. Advantages of pyrethrum (composed of a mixture of compounds called pyrethrins,

## Table 7-4
## Differential Susceptibility of Insects to Some Common Insecticides

| | Housefly | Bee | Grasshopper | Cockroach | Rat (oral) |
|---|---|---|---|---|---|
| *Carbamates* (used on agricultural crops) | | | | | |
| Pyrolan | 0.0032 | 0.013 | ..... | ..... | 0.090 |
| Carbaryl | >0.500 | 0.0023 | ..... | >0.133 | 0.540 |
| Isolan | 0.025 | 0.013 | ..... | ..... | 0.013 |
| *Chlorinated Hydrocarbons* (insecticides) | | | | | |
| DDT | 1008–0.21 | 0.114 | 9.380 | 0.010 | 0.400 |

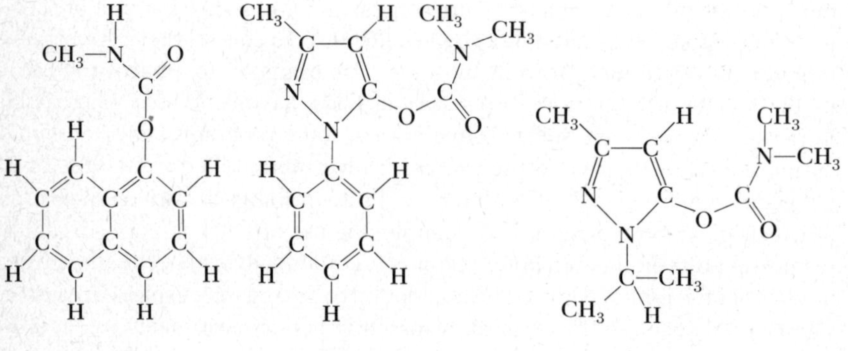

Carbaryl                          Pyrolan                          Isolan

NOTE: Lethal dose in milligrams (0.001 gram) of pesticide per gram of insect (topical dose unless otherwise noted). The broad spectrum of activity of chlorocarbons (toxic to most insects), shown here for DDT, made them particularly attractive as insecticides.

Figure 7–3) are that this mixture is readily biodegraded, and nontoxic to mammals and fish, and that no case of insect resistance has been documented (contrast this with DDT). Perhaps the pyrethrum story can teach us a lesson about our environment. The rapidity with which these compounds degrade may contain the secret to their success. Their short lifetime in the environment greatly reduces the probability of insects' developing resistance to them by minimizing the contact time between the chemical and the insect. Recently a synthetic pyrethrin has been developed that is about five times as effective as the natural pyrethrins.

Rotenone and some related compounds found in roots of many plants have also been used as insecticides. The advantages and disadvantages of rotenone are basically the same as for pyrethrum, with the exception that rotenone is toxic to fish.

Another technique used to control pests involves the use of "trap crops." Apparently insects are differentially attracted by crops. Alfalfa for example preferentially attracts insects that are troublesome to cotton. The

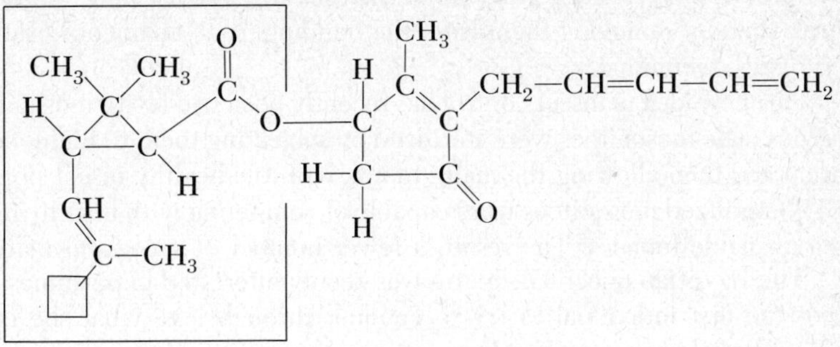

**Figure 7-3** (1) Structure of natural pyrethrins: Pyrethrin I if □ contains a $CH_3$;

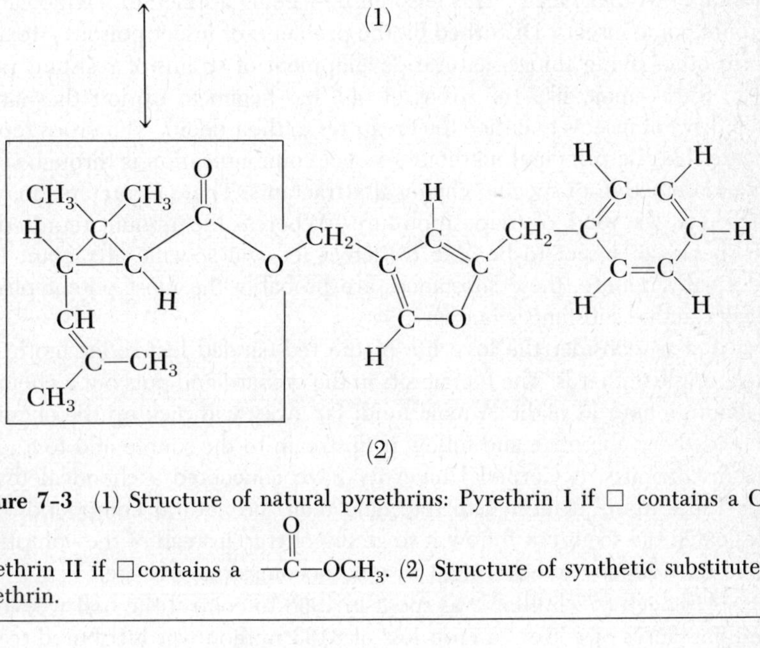

Pyrethrin II if □ contains a $-\overset{O}{\overset{\|}{C}}-OCH_3$. (2) Structure of synthetic substitute for pyrethrin.

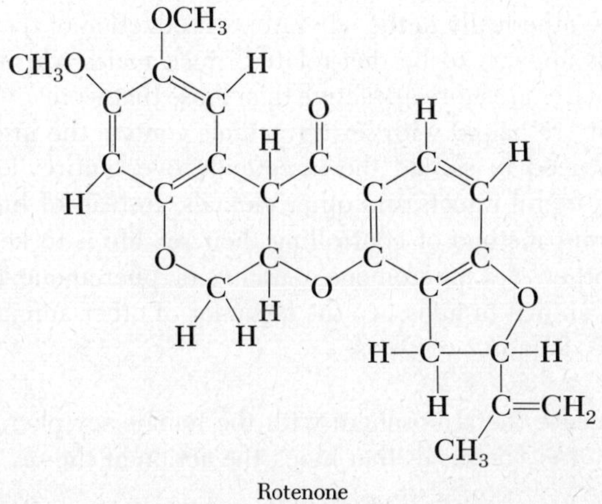

Rotenone

age-old practice of releasing guinea pigs into flea-infested buildings to attract the fleas, thereby removing them from the building, is an analogous example of this "bait" technique.

One new idea in insect control has recently been tried with mosquitoes. Numerous male mosquitoes were sterilized by subjecting them to high-energy radiation and then allowing the males to mix with the healthy insect population. The sterilized mosquitoes were capable of competing with healthy males for the available females. The result, a fewer number of successful matings.

The Egyptian queen Cleopatra was keenly interested in perfumes. She was not the first individual to try to conquer through love what she could not obtain by other means. This lesson is now being applied to a wide variety of troublesome insects. Disturbed by the problems of insect poisons—toxicity toward other living things, natural development of strains of resistant pests, and so forth—man, like the sirens of old, has begun to exploit the natural sexual drive of insects to entice the creatures to their doom. Numerous reports indicate that the principal method of insect communication is through secretion and detection of specific chemical attractants. These attractants may be classified as sex, food, or oviposition lures. Whereas the amount required for each species of insect to be able to detect its own specific attractant is so fantastically minute, these substances are probably the most potent **physiologically** active substances known.

Let us consider the love life of the red-banded leaf-roller moth that attacks apple orchards. The female sits in the orchard and puts out a chemical for about an hour at night. A male moth far away will pick up the chemical on his feathery antennae and follow it upstream to the source and to a night of woo. Chemists at Cornell University have concocted a chemical that is thirty times more attractive to the male than the natural compounds. The males catch the scent but follow it to an insect trap instead of the voluptuous female they expect and the night of woo becomes one of woe.

Although $75 million was spent in 1968 to control the boll weevil on 10 million acres of cotton, a crop loss of $133 million was attributed to this insect. Recently, the male boll weevil's sex attractant was isolated, identified, and produced synthetically in the laboratory. Production of the sex attractant by boll weevils appears to be diet-related, since males who have been fed fresh cotton squares are more attractive than those that were given an artificial diet. Traps that are baited with sex attractants contain the insecticides. This eliminates the need to spread the insecticide over entire fields and helps prevent killing useful insects and other animals. Instead of luring insects to a trap, a different method of controlling their sex life is to keep them from finding each other. Sex pheromone masking (a pheromone is a substance secreted by an animal to influence the behavior of other animals of the same species) can be done in two ways:

1. Permeate the atmosphere with the female sex pheromone.
2. Use other chemicals that block the action of the sex attractant, for

example, both formaldehyde and Chanel No. 5 (a popular lure for another species) have been shown to be effective.

Some idea of the work involved in isolating and identifying a sex attractant is obtainable from the case of *Porthetria dispar,* the gypsy moth. Study on the gypsy moth was culminated in 1960 after thirty years of work on the isolation, identification, and synthesis of the sex attractant. To isolate this sex pheromone it was necessary to clip the last two abdominal segments of 500,000 virgin females, extract and separate the many components present, and finally purify the active component. A total of only 0.02 grams was isolated from all these moths. The structure of the sex pheromone was shown to be 10-acetoxy-*cis*-7-hexadecen-1-ol by chemical studies. Note that this material contains a carbon atom with four different substituents. It is consequently capable of existing in two nonsuperimposable mirror-image forms. Only one of these mirror-image forms is produced by the moth.

Carbon atom bearing 4 different substituents

$$CH_3-(CH_2)_5-\overset{\overset{\displaystyle H}{|}}{\underset{\underset{\displaystyle O-\overset{\overset{\displaystyle }{\|}}{\underset{\displaystyle O}{C}}-CH_3}{|}}{C}}-CH_2-CH=CH-(CH_2)_5CH_2OH$$

The common housefly is a danger to all of us. Its involvement in the spreading of typhoid, dysentery, diarrhea, cholera, yaws, trachoma and other diseases has long been known. The mating behavior of the housefly has been studied extensively since the time of Aristotle. Recently, the sex pheromone has been isolated from the cuticle and the feces of the female housefly and identified as 9-tricosene. To control this pest in the past, it was necessary to spray with chlorocarbon pesticides, generally in enclosed areas. For this reason, using the insect's own sex pheromone to control the insect has considerable appeal. This substance, called Muscalure,° has recently become commercially available.

$$CH_3-CH_2-CH_2-CH_2-CH_2-CH_2-CH_2-CH_2-CH=CH-\underset{\displaystyle |}{CH_2}$$
$$CH_3-CH_2-CH_2-CH_2-CH_2-CH_2-CH_2-CH_2-CH_2-CH_2-\overset{\displaystyle CH_2}{|}$$

9-Tricosene

Not all insects deploy sex attractants. Other possible methods of insect control have been found by investigating their fascinating life cycles. In the normal course of events, many insect larvae form a pupa within which they quietly rearrange most of their molecules to the adult, or moth, form. When

---

° Trademark.

a normal pupa splits open, the adult emerges intent on sexual activity and the continuation of the life cycle. The *intrapupal* molecular rearrangement is controlled and directed by a family of hormones. One of them, the juvenile hormone, will cause a pupa to molt into a second smaller pupa by shifting the control mechanism acting within the pupa. Synthetic compounds, more active than the insect's own juvenile hormone, have been produced for yellow mealworms and milkweed bugs. As little as 10 nanograms (0.00000001 g) applied to the abdomen produces defective **metamorphosis** and incapacity to mature sexually. The compounds are not insecticides since doses of up to 10 mg (0.01 g) are not fatal.

Insect communication by means of chemical agents appears to be very general. In addition to attractants, defensive secretions are also widely used by insects. Chlorinated hydrocarbons recently were found in the defensive secretions of one species of insect that had been collected in an area sprayed by chemical insecticides. The same chemical was entirely missing in insects of the same *genus* not exposed to the insecticide. Apparently this species had developed pathways for detoxifying the insecticide while at the same time utilizing it for its own defense. The main constituent of the defensive secretion from a bee or wasp sting is irritating but not fatal for most people. Wouldn't it be a turn of fate if wasps and bees, constantly exposed to insecticides, modified and incorporated these insecticides into their own defensive secretions and then proceeded to give us a taste of our own medicine?

Another new method in insect control that has enormous potential is the use of insect diseases as a method of insect control. Like man, insects suffer from disease. Aristotle noted a mysterious illness that killed honeybees, and numerous other insect afflictions were noted by other careful observers. Insect pathology, the study of insect diseases, developed in the nineteenth century, partly because of a particular silkworm fungus that interfered with the profitable production of silk. Since that time, a considerable amount has been learned about insect pathology. Insect viruses, for example, are very effective in controlling insect population. They are quick-acting, very specific, completely harmless to plants and animals, and are already a natural part of the environment. Several reasons exist for the lack of widespread use of these pathogens, however. First, there is a fear of opening another Pandora's box. It took twenty years to find out about the ill effects of chlorinated hydrocarbons, and the amount of irreversible damage they have caused is still unknown. Another obstacle, one which may soon be solved, is the cost of commercial production of these pathogens.

Recently VIRON/H° became the first virus to be approved for widespread use as a pesticide. This virus is specific for the genus *Heliothis*, in which there are two economic pests, *H. virescens*, the tobacco budworm, and *H. zea*, commonly known as the tomato fruitworm, corn earworm, or cotton

---

° Trademark.

bollworm, its name depending on the crop it attacks. There are over 300 other known viruses that can be used to control insects.

Another biological pesticide, Thuricide° (active ingredient, *Bacillus thuringiensis*) has been field-tested recently. This bacterium is active against the gypsy moth, which has been blamed for defoliation of more than 1.5 million acres of trees and shrubs in the northeastern United States in 1971. We are likely to see extensive use of this pesticide in the near future.

What is the likelihood that widespread use of biological pesticides will cause such changes that these agents must then be classifiable as a menace to other forms of life? No one really knows. However, according to the United States Department of Agriculture, without the use of substantial amounts of pest killers, the nation would be unable to feed its people.

"Now we can go back to organic agriculture if we must . . . but before we do that somebody must decide which fifty million Americans we'll let starve." (Secretary of Agriculture Butz (1972).) This statement is a clear indication of the commitment to technology we have made as a result of population pressures.

It is quite evident that knowledge gained through the study of insects will lead to better ways of controlling them. We must remember, however, that our survival is directly dependent on their survival. Selective control of harmful pests is clearly preferable to indiscriminate destruction of the insect community. Consider the alternatives: a few mosquito bites or spending each summer under the hot sun pollinating plants.

## FOOD ADDITIVES

*Poison is in everything and nothing is without poison.*

Paracelsus

Economics and the world's growing population have greatly influenced world food production. In the underdeveloped countries, considerable effort has been directed at increasing food production by using new methods of farming and by growing special hybrid crops. In the developed countries, the emphasis in food production has been in providing the consumer with convenience, sensuous appeal, and nutrition. To achieve this, a host of chemical additives have been introduced, particularly in processed foods. To anyone who reads labels, terms like butylated hydroxyanisole, monosodium glutamate, saccharin, and ammonium chloride are common household words. Some 2,600 chemicals are used in food production, as preservatives, emulsifiers, artificial sweeteners, and flavoring, and as bleaching, brightening, thickening, and softening agents. The use of food additives constitutes a $485 million operation.

---

° Trademark.

What is a food additive? Can sugar be considered one? The distinction between food additives and ingredients is not a clear one. Most food additives used to condition food are of no significant nutritional value. Others are used as dietary supplements. The difficulty in distinguishing between "food additives" and "ingredients" lies in the fact that foods (ingredients), like additives, are chemicals. The distinction between the two is often an artificial one, since many additives—ascorbic acid (vitamin C), for example—may have been present in the original food, but either destroyed or lost during processing. There are, however, a series of chemicals, preservatives for example, that are not generally found in natural foods but are "additives" in the strictest sense of the word. A list of common additives "generally recognized as safe" is shown in Table 7–5.

Are these food additives safe? There is nothing new in the idea of food additives. Thousands of years ago the Chinese used ethylene and propene produced by the combustion of kerosene to ripen bananas and peas, and the ancient Egyptians used colorings for their food. Pickling in salt and the fermentation processes resulting in the production of lactic acid, ethyl alcohol, or acetic acid are other methods of food preservation, dating back many years. Curing ham with brine containing nitrites and nitrates is another long-established process. What is new is the multitude of new chemical agents that have been introduced only recently. Even though these additives have been found "safe" to use, there is no guarantee that in combination with some other substance, some medicine or other food additive, side effects will not result.

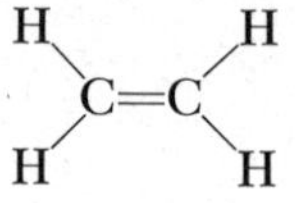

Ethylene

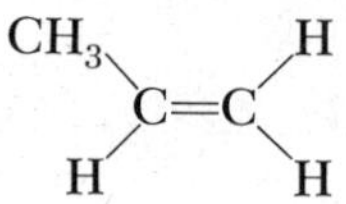

Propene

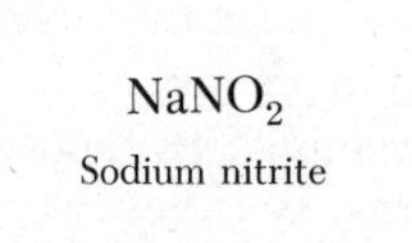

Lactic acid

$CH_3CH_2OH$

Alcohol
(ethyl alcohol)

$NaNO_2$

Sodium nitrite

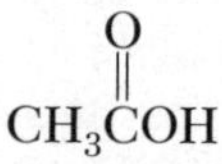

$CH_3\overset{\displaystyle O}{\overset{\|}{C}}OH$

Acetic acid
(vinegar)

$NaNO_3$

Sodium nitrate

What is safe? How are food additives, and medicines for that matter, pronounced safe. By applying Paracelsus's law, "Poison is in everything and nothing is without poison," the ground rules are established. It is the dosage that makes a drug either a poison or a remedy, whether the reference is to a food coloring, alcohol, water, aspirin, DDT, or lead. What constitutes a moderate dose? The answer to this question is extremely difficult and for the most part can only be answered by reaching some sort of compromise between the therapeutic value of the drug and the drug's undesirable side effects. Monosodium glutamate, for example, is a flavor enhancer; with liberal use it produces profuse sweating, a condition called *Kwok's* disease. Determination

$$\text{HO}-\overset{\overset{\text{O}}{\|}}{\text{C}}-\underset{\underset{\text{NH}_2}{|}}{\text{CH}}-\text{CH}_2-\text{CH}_2-\overset{\overset{\text{O}}{\|}}{\text{C}}-\text{O}^-\ \ \text{Na}^+$$

Monosodium glutamate (MSG)

of a safe level of consumption for a drug requires getting a considerable amount of information about the drug's fate in the body: whether it is metabolized or excreted, the site or sites it acts on, and the biological response that it initiates. This is just a portion of the information that should be available before a safe dosage can be prescribed. Unfortunately, answers to these questions are generally not known and the mode of action of most chemicals on the body is not well understood. To circumvent this problem, compound testing is carried out on a more empirical basis. In the case of new food additives and drugs, for example, animals are fed large doses of the prospective additive for consecutive generations and the effects of the drug are noted. The dosages may be 10 to 100 times the amount designed to be used for human consumption. Furthermore, honest and thorough efforts are made to determine what happens to the additive after it is ingested, and to determine what the effects of this additive are in combination with other additives frequently used. After some further testing, safe dosages are estimated and the substance is released for its final testing by the public.

All of the information man has learned about drugs in the past has ultimately been derived from using himself as a guinea pig. The present is no different. Perhaps the most dangerous aspect about the present is that we are all guinea pigs. In our technological society, we all participate in a "chemical feast" whether we want to or not. We can only hope that we are similar enough biologically to the animals we have tested, so that in the long run we shall not be adversely affected by the new products. Most of the additives in foods are new, having been introduced during the past twenty years. This constitutes less than one generation on a human timetable, and hence these drugs can hardly be considered as having withstood the test of time. Yet we are consuming these food additives at an unprecedented rate. It is estimated that the average American consumes more than five pounds of these nonnutritive additives every year, an amount that is expected to rise.

Food additives in general can be categorized in two general groups, as food preservatives and as additives used to improve the appearance, texture, and flavor of food. Refrigeration, canning, and drying constitute the other

# Table 7-5
## Partial Listing of Additives in USDA's "Generally Recognized as Safe (GRAS)" List.

*Anticaking Agents*
Aluminum calcium silicate
Calcium silicate
Magnesium silicate
Sodium aluminosilicate
Sodium calcium aluminosilicate
Tricalcium silicate

*Chemical Preservatives*
Ascorbic acid
Ascorbyl palmitate
Benzoic acid
Butylated hydroxyanisole
Butylated hydroxytoluene
Calcium ascorbate
Calcium propionate
Calcium sorbate
Caprylic acid
Dilauryl thiodipropionate
Erythorbic acid
Gum guaiac
Methylparaben
Potassium bisulfite
Potassium metabisulfite
Potassium sorbate
Propionic acid
Propyl gallate
Propylparaben
Sodium ascorbate
Sodium benzoate
Sodium bisulfite
Sodium metabisulfite
Sodium propionate
Sodium sorbate
Sodium sulfite
Sorbic acid
Stannous chloride
Sulfur dioxide

Thiodipropionic acid
Tocopherols

*Emulsifying Agents*
Cholic acid
Desoxycholic acid
Diacetyl tartaric acid esters
 of mono- and diglycerides
Glycocholic acid
Mono- and diglycerides
Monosodium phosphate
 derivatives of above
Propylene glycol
Ox bile extract
Taurocholic acid

*Nutrients and Dietary
Supplements*
Alanine
Arginine
Ascorbic acid
Aspartic acid
Biotin
Calcium carbonate
Calcium citrate
Calcium glycerophosphate
Calcium oxide
Calcium pantothenate
Calcium phosphate
Calcium pyrophosphate
Calcium sulfate
Carotene
Choline bitartrate
Choline chloride
Copper gluconate
Cuprous iodide
Cysteine
Cystine

Ferric phosphate
Ferric pyrophosphate
Ferric sodium pyrophosphate
Ferrous gluconate
Ferrous lactate
Ferrous sulfate
Glycine
Histidine
Inositol
Iron, reduced
Isoleucine
Leucine
Linoleic acid
Lysine
Magnesium oxide
Magnesium phosphate
Magnesium sulfate
Manganese chloride
Manganese citrate
Manganese gluconate
Manganese glycerophosphate
Manganese hypophosphite
Manganese sulfate
Manganous oxide
Mannitol
Methionine
Methionine hydroxy analogue
Niacin
Niacinamide
Phenylalanine
D-pantothenyl alcohol
Potassium chloride
Potassium glycerophosphate
Potassium iodide
Proline
Pyridoxine hydrochloride
Riboflavin
Riboflavin-5-phosphate

Serine
Sodium pantothenate
Sodium phosphate
Sorbitol
Thiamine hydrochloride
Thiamine mononitrate
Threonine
Tocopherols
Tocopherol acetate
Tryptophane
Tyrosine
Valine
Vitamin A
Vitamin A acetate
Vitamin A palmitate
Bitamin $B_{12}$
Vitamin $D_2$
Vitamin $D_3$
Zinc chloride
Zinc gluconate
Zinc oxide
Zinc stearate
Zinc sulfate

*Sequestrants*
Calcium acetate
Calcium chloride
Calcium citrate
Calcium diacetate
Calcium gluconate
Calcium hexametaphosphate
Calcium phosphate, monobasic
Calcium phytate
Citric acid
Dipotassium phosphate
Disodium phosphate
Isopropyl citrate
Monoisopropyl citrate

Potassium citrate
Sodium acid phosphate
Sodium citrate
Sodium diacetate
Sodium gluconate
Sodium hexametaphosphate
Sodium metaphosphate
Sodium phosphate
Sodium potassium tartrate
Sodium pyrophosphate
Sodium pyrophosphate, tetra
Sodium tartrate
Sodium thiosulfate
Sodium tripolyphosphate
Stearyl citrate
Tartaric acid

*Stabilizers*
Acacia (gum arabic)
Agar-agar
Ammonium alginate
Calcium alginate
Carob bean gum
Chondrus extract
Chatti gum
Guar gum
Potassium alginate
Sodium alginate
Sterculia (or karaya) gum
Tragacanth

*Miscellaneous Additives*
Acetic acid
Adipic acid
Aluminum ammonium sulfate

Aluminum potassium sulfate
Aluminum sodium sulfate
Aluminum sulfate
Ammonium bicarbonate
Ammonium carbonate
Ammonium hydroxide
Ammonium phosphate
Ammonium sulfate
Beeswax
Bentonite
Butane
Caffeine
Calcium carbonate
Calcium chloride
Calcium citrate
Calcium gluconate
Calcium hydroxide
Calcium lactate
Calcium oxide
Calcium phosphate
Caramel
Carbon dioxide
Carnauba wax
Citric acid
Dextrans
Ethyl formate
Glutamic acid
Glutamic acid hydrochloride
Glycerin
Glyceryl monostearate
Helium
Hydrochloric acid
Hydrogen peroxide
Lactic acid
Lecithin
Magnesium carbonate

Magnesium hydroxide
Magnesium oxide
Magnesium stearate
Malic acid
Methylcellulose
Monoammonium glutamate
Monopotassium glutamate
Nitrogen
Nitrous oxide
Papain
Phosphoric acid
Potassium acid tartrate
Potassium bicarbonate
Potassium carbonate
Potassium citrate
Potassium hydroxide
Potassium sulfate
Propane
Propylene glycol
Rennet
Silica aerogel
Sodium acetate
Sodium acid pyrophosphate
Sodium aluminum phosphate
Sodium bicarbonate
Sodium carbonate
Sodium carboxy-
    methylcellulose
Sodium caseinate
Sodium citrate
Sodium hydroxide
Sodium pectinate
Sodium phosphate
Sodium potassium tartrate
Sodium sesquicarbonate
Sodium tripolyphosphate

Succinic acid
Sulfuric acid
Tartaric acid
Triacetin
Triethyl citrate

*Synthetic Flavoring Substances*
Acetaldehyde
Acetoin
Aconitic acid
Anethole
Benzaldehyde
N-butyric acid
d- or l-carvone
Cinnamaldehyde
Citral
Decanal
Diacetyl
Ethyl acetate
Ethyl butyrate
Ethyl vanillin
Eugenol
Geraniol
Geranyl acetate
Glycerol tributyrate
Limonene
Linalool
Linalyl acetate
l-Malic acid
Methyl anthranilate
3-Methyl-3-phenyl
    glycidic acid ethyl ester
Piperonal
Vanillin

NOTE: To be on this list an additive must have been in use before 1958. Saccharin, which was originally on this list and still in use has been removed. Additives introduced after 1968 (not on this list) must be introduced individually and must meet certain specifications of safety.

main methods of preserving foods. **Antioxidants,** such as butylated hydroxyanisole and butylated hydroxytoluene, are routinely used to prevent food oils, shortening, and many other foods from turning rancid. Calcium propionate is used in breads and baked goods to prevent spoilage. Sodium benzoate, citric acid, and lactic acid are a few of the many other preservatives commonly used. Many of these substances have been in use for long periods of time and can be considered safe. Others are relatively new.

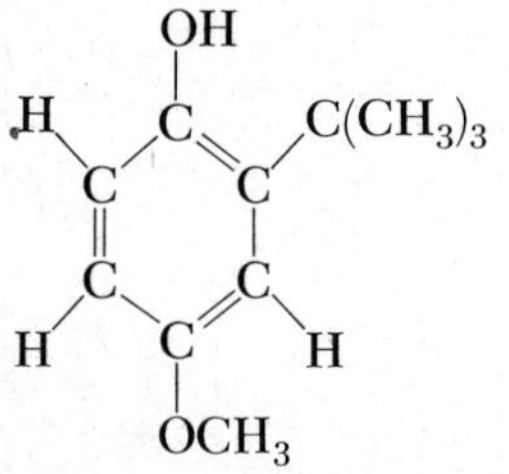

Butylated hydroxyanisole

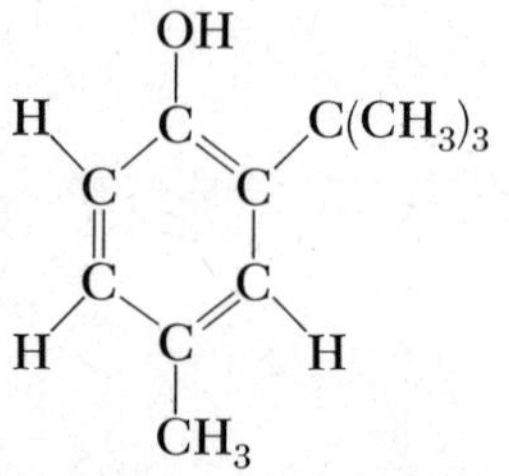

Butylated hydroxytoluene

$[CH_3—CH_2—CO_2^-]_2Ca^{+2}$

Calcium propionate

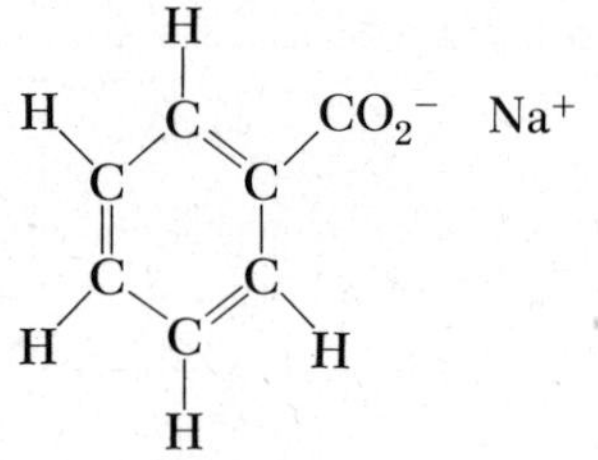

Sodium benzoate

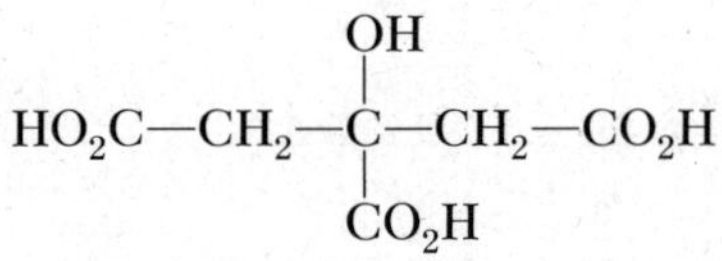

$$HO_2C—CH_2—\overset{\displaystyle OH}{\underset{\displaystyle CO_2H}{\overset{|}{\underset{|}{C}}}}—CH_2—CO_2H$$

Citric acid

Many people object to bland food and drink, so flavorings are often added to confectionery, canned foods, and beverages. Next time you are in a supermarket, note the added wax on cucumbers. Although this improves the appearance of the cucumber and keeps it fresh longer by reducing the rate of water loss, it also provides an ideal place for bacteria and other microbes to collect. Peeling the cucumber, of course, removes the wax and much of the nutritive value of the cucumber.

In a society that is becoming increasingly weight-conscious, many people want a sweetening agent that will not add calories to their diet.

Saccharin has long been used for this purpose; it is also cheaper than sugar. (For a comparison of the relative sweetness of these substances, see Table 7–6.) For a time cyclamates were used, but their use was suspended as a precaution after tests showed that rats that were fed large doses of a 10:1 mixture of cyclamate and saccharin developed bladder cancers. It was subsequently reported that rats that were fed cyclamate without saccharin, at one-sixth of the dose that led to the original ban, also developed bladder cancer. The adverse effect of cyclamates in rats may be due to its conversion to cyclohexylamine, a known **carcinogen.** Saccharin is now being critically reviewed. There is no question that for those who must restrict their intake of sweets, such as diabetics, artificial sweeteners are very important. The widespread use of these drugs in foods and drink, however, is another matter.

$$
\begin{array}{c}
NH_2 \\
|\\
CH \\
CH_2 \quad CH_2 \\
CH_2 \quad CH_2 \\
CH_2
\end{array}
$$

Cyclohexylamine

The most dangerous kinds of food additives are those that are introduced to food either by accident or as residues. These additives are not added to food at the time of processing but are incorporated earlier in the food chain either by accident or design. Examples of these unintentional additives range from the chlorocarbon insecticides (DDT, endrin, dieldrin) and antibiotics in animal feed (penicillins, sulfonamides, neomycin, streptomycin) to other miscellaneous additives such as diethyl stilbestrol (substance suspected of possessing carcinogenic properties in man and used to fatten cattle), and organomercurials (used to prevent mildewing of seeds) and so on.

Some of the additives used to "condition" animals prior to sacrifice have recently been criticized. Antibiotics used in humans, for example, will slowly be phased out of animal feed because of the possibility that low-level use of the drugs in feed may be linked to establishing drug-resistant bacteria in man. Other drugs however, will likely be substituted in their place.

The use of food additives has become a necessity in modern food processing. The extent to which we have become committed to these additives because of an increasing population, however, is not clear. The use of additives simply for convenience is no excuse for indiscriminate use of these drugs. Yet the economics of the marketplace often achieves this result. It is quite certain that food additives are here to stay. The increasing consumption of these nonnutritive substances is obviously a condition necessitating careful surveillance and control.

**Table 7-6**
**The Relative Sweetness of Some Artificial Sweeteners**

| Name | Structure | Relative Sweetness |
|---|---|---|
| Glycerol (glycerin) | 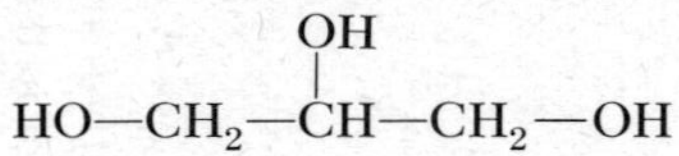 | 1.0 |
| Sucrose (table sugar)[a] | 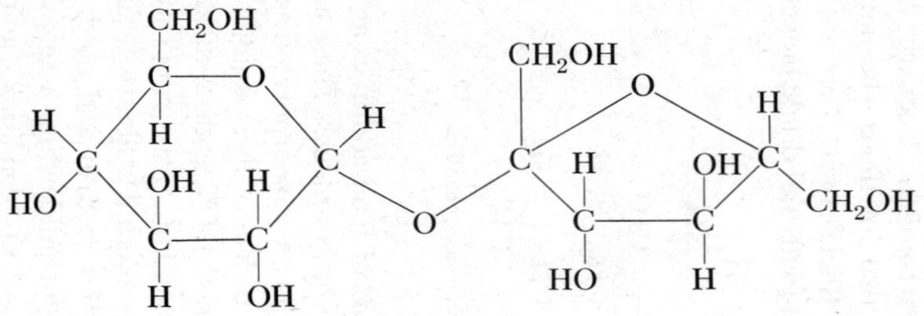 | 1.0 |
| Ethylene glycol (permanent antifreeze; *poisonous!*) | $HO-CH_2-CH_2-OH$ | 1.3 |
| Sodium cyclamate (banned in U.S.) | 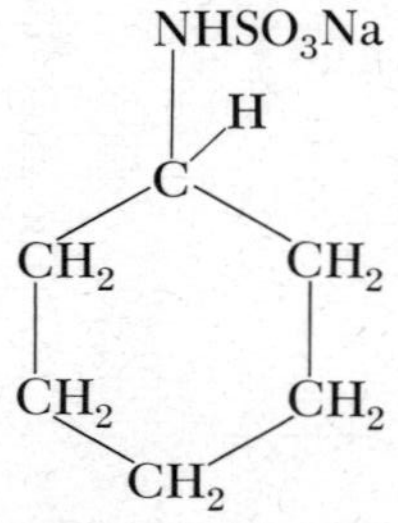 | 110. |

Dulcin (Sucrol) 200.

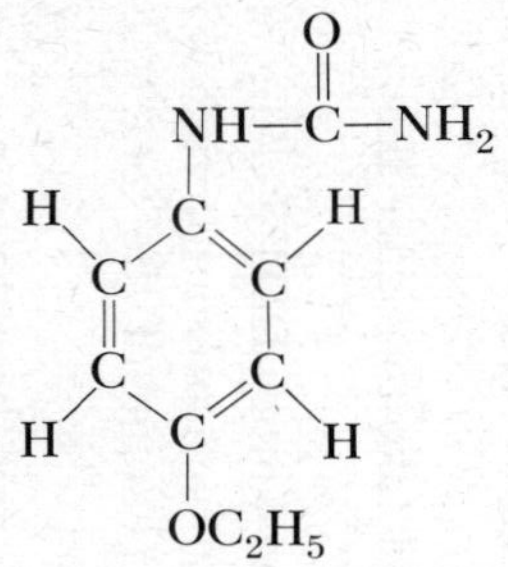

$O$
$\parallel$
$NH—C—NH_2$

H

C

C H

NH C C

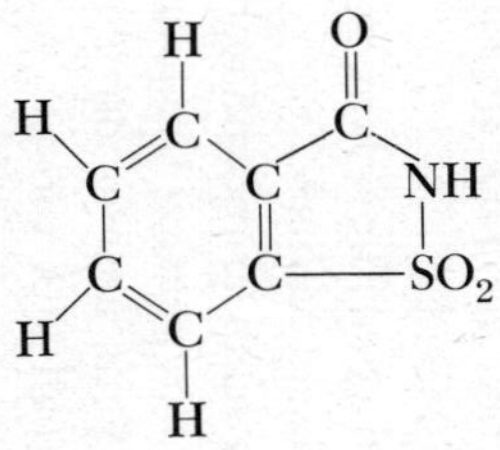

C OC₂H₅

C C

H H

H H

Saccharin 550.

$O$
$\parallel$
$NH—SO_2$

C C

H C C

C H

C C

H H

5-Nitro-2-propoxyaniline (Ultrasüss) 2,000.
(banned in U.S.)

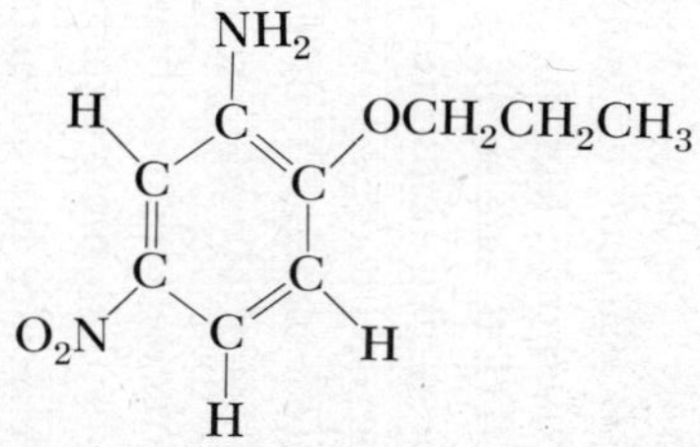

OCH₂CH₂CH₃

H C C

NH₂ C C H

C C

H H

O₂N

SOURCE: Data from *The Merck Index*, 7th ed. (Rahway, N.J.: Merck and Co., 1962).
[a] Note the position of table sugar relative to the last four entries on the list. The low value of sugar in relation to these nonnutritive food sweeteners is consistent with sugar's role as a food rather than a pacifier.

271

## *PROBLEMS*

1. (a) If the world's population doubles every twenty years, what will it be after 100 years?
   (b) If the world's population is reduced by one-half every ten years, what will it be after a century?

2. Table 7–2 lists some estimated numbers of illegal abortions. Compare this number to the abortion rate in countries where abortion is legal. What do these data seem to imply?

3. Discuss the statement: Zero population growth can be achieved if every family restricts itself to two offspring.

4. Under what conditions would the statement in Problem 3 be correct in this decade?

5. (a) If the half-life of DDT in the environment is twenty-five years, estimate how long it would take for DDT to be reduced by a factor of ten assuming no more is added.
   (b) Based on a half-life of ten years, how long will it take to reduce the DDT to one tenth its present value?
   (c) Assuming a half-life of ten years, approximately how much DDT could be used each year without increasing the total amount present in the environment? Calculate your answer on a percentage of the total amount already present (call it $X$ tons).

6. Why do the chlorocarbon insecticides pose a more serious threat to living systems than most other types of insecticides?

7. (a) What are the advantages of using biological pesticides?
   (b) What are some of the possible disadvantages?

8. How are drugs and food additives tested?

9. What are desirable characteristics for test animals used as "guinea pigs" in drug testing?

10. If automobile emissions were reduced to 20 percent of the current levels and if we assume the population doubles every twenty years, approximately how long would it take for the total automobile emissions to reach present-day levels?

11. (a) According to the text, what are some of the most serious problems concerning the use of food additives?
    (b) Even though the additives today are probably safer than those used twenty or more years ago, why do these drugs pose a greater threat to *Homo sapiens* as a species?

12. In the earliest days of civilization there was war, famine, disease, and natural disaster (fire, flood, earthquake). Science was not well developed.

Religion and superstition were usually employed to account for the unexplainable. Most of us would agree that science has advanced to a more sophisticated level yet there is still war, famine, disease and natural disaster. Religion is still popular and superstition and witchcraft are still rampant. Question: Has science really affected society?

# BIBLIOGRAPHY

1. *Report of the Secretary's Commission on Pesticides and Their Relation to Environmental Health. Parts 1 and 2.* U.S. Department of Health, Education and Welfare, 1969.

2. *XXIIIrd International Congress of Pure and Applied Chemistry, Special Lectures, Vol. 3, Boston, Mass., 1971.* C. Djerassi. London: Butterworth's, 1971.

3. *Organic Pesticides in the Environment.* Advances in Chemistry Series 60, American Chemical Society Publications.

4. David M. Kiefer. "Population." *Chemical and Engineering News*, October 7, 1968, p. 118.

5. V. Petrow, "Current Aspects of Fertility Control." *Chemistry in Britain* 167 (1970).

6. B. C. L. Weedon. "Food Additives." *Chemistry in Britain* 242 (1970).

7. D. A. Carlson, M. S. Mayer, D. L. Silhacek, J. D. James, M. Beroza, and B. A. Bierl. "Sex Attractant Pheromone of the Housefly: Isolation, Identification, and Synthesis." *Science* 174 (1971):76.

8. "EPA Approves Biological Pesticide." *Chemical and Engineering News*, February 14, 1971, p. 13.

9. F. Greer, C. M. Ignoffo, and R. F. Anderson. "The First Viral Pesticide: A Case History." *Chemical Technology*, June 1971, pp. 342–47.

10. Frank Carey. "A Projection of Disaster for Mankind. . . ." *St. Louis Post Dispatch*, March 19, 1972, p. 60; *Chemical and Engineering News*, March 6, 1972, p. 3.

11. Patricia McCormack. "Male Sterilization: A Popular Alternative." *St. Louis Post-Dispatch*, August 11, 1971, p. 4F.

12. "Food Additive Makers Face Intensified Attack." *Chemical and Engineering News*, July 12, 1971, p. 16.

13. James S. Turner. "Chemical Feast." New York: Grossman, 1970.

14. Martin Jacobson. "Insect Sex Attractants." *Interscience* 1965, chap. 9.

# SUGGESTED READINGS

1. Garrett Hardin. *Population, Evolution and Birth Control*. San Francisco: W. H. Freeman, 1969.

2. Garrett DeBell. *The Environmental Handbook*. New York: Ballantine Books, 1970.

3. Paul R. Ehrlich. *The Population Bomb*. New York: Ballantine Books, 1970.

4. Martin Brown. *The Social Responsibility of the Scientist,* selected topics New York: Free Press, 1971.

5. Robert Rienow and Leona T. Rienow. *Moment in the Sun*. New York: Ballantine Books, 1967.

6. Rachael Carson. *Silent Spring*. Boston: Houghton Mifflin, 1970.

7. Frank Graham, Jr. *Since Silent Spring*. Boston: Houghton Mifflin, 1970.

8. Barry Commoner. "The Environmental Cost of Economic Growth." *Chemistry in Britain* 8 (February 1972):52.

9. G. O. Kermode. "Food Additives." *Scientific American* 226, (3) 1972:15.

10. P. R. Erhlich. "Eco-Catastrophe." *Ramparts*, September 1969, pp. 62–64.

11. J. S. Holt. "Food Resources of the Ocean." *Scientific American*, September 1969, p. 178.

12. T. Whitesides. *Defoliation*. New York: Ballantine Books, 1970.

13. L. Pauling. *Vitamin C and the Common Cold*. San Francisco: W. H. Freeman, 1970.

14. "Man's Control of the Environment." *Congressional Quarterly*, August 1970. Washington, D.C.

15. Don Widener. *Timetable for Disaster*. Los Angeles: Nash, 1970.

# APPENDIX A  *Stoichiometry*

*If you can measure that of which you speak,*
*and can express it by a number, you know something of your subject.*

Lord Kelvin (1824-1907)

**1(a)  CGS Units**  Most concepts in science can be expressed as a function of three variables, length, time, and mass. In this text, all discussion will refer to one standard system, the metric system.

Early systems of measurement were not uniform. For example, the inch at one time was defined as the sum of the lengths of three grains of barley. In 1790, Louis XVI issued a decree for the development of a uniform standard and asked the French Academy of Science to suggest some standards. The metric system was developed as a response to this decree.

**1(b)  Length**  The basic unit of length in the metric system is the meter. This unit of length was defined as one ten-millionth of the distance between the equator and the north pole. The distance was then marked on a bar of platinum and iridium and kept in a triply locked vault at the International Bureau of Weights and Measures at Sevres, France. In 1960, this length was redefined as a function of the wavelength of a particular kind of light. The new definition was established because of the need for a more accurate standard of measure. The actual length of the meter for most purposes however remains the same.

Listed herewith are the most common prefixes used in connection with fractions of a meter. For reference purposes, the English system equivalents (foot, inch) are also listed.

| *milli-* | one thousandth |
|----------|----------------|
| *centi-* | one hundredth  |
| *kilo-*  | one thousand   |

| Length | | |
|--------|--|--|
| Meter (m) | | 39.4 inches |
| Millimeter (mm) | one thousandth of a meter | 0.0394 inch |
| Centimeter (cm) | one hundredth of a meter | 0.394 inch |
| Kilometer (km) | one thousand meters | 3,280 feet |

**1(c)  Mass**  The concept of mass historically has been an area of intense philosophical discussion. For our purposes it is sufficient to say that mass refers to the number of "fundamental" particles that compose matter (that is, neutrons, protons, and electrons): but for a deeper insight into this concept, the relationships between mass, energy, and time must be considered.

The basic unit of mass is the gram. A penny has a mass of about 3 grams. Very closely related, but different, is the concept of weight. Weight is effectively caused by the attractive forces between two masses. An excellent illustration of the difference between weight and mass is the weightlessness experienced in space. Although the mass of an object in space remains essentially constant, the absence of any celestial bodies nearby causes a condition of weightlessness. The smaller mass of the moon (as compared with earth) allows astronauts to walk easily on the moon, even as they carry massive backpacks. Since the effects of gravity are essentially constant throughout the surface of the earth, we often use weight interchangeably with mass. Weight is directly proportional to mass in a constant gravitational field (for example, the earth's gravity).

| | |
|---|---|
| Gram (g) | |
| Milligram (mg) | one thousandth of a gram |
| Kilogram (kg) | one thousand grams |

**1(d) Volume**  The basic unit of volume is the liter. The liter is now defined as the volume of 1,000 cubic centimeters. Originally the volume of a liter was defined as the volume of 1,000 grams of pure water at one atmosphere pressure and 3.98 °C. Under the old definition, a liter equaled 1000.028 cm³. *Milliliter* and *cubic centimeter* can be used interchangeably.

**1(e) Temperature**  What is temperature? We are all familiar with the thermometer and the fact that a thermometer measures temperature, but what is the exact property of matter that we are measuring? To gain an insight into the concept of temperature, let us consider a small particle in motion. This particle would have some kinetic energy ($\frac{1}{2}$ mass of the particle $\times$ (velocity of the particle)$^2$) associated with its motion. If in some fashion we could increase the velocity of this particle, we should find that the property that we call temperature would also increase. Temperature measures the average kinetic energy of particles called molecules. We are all familiar with the following effect. On a cold sunny day the inside of an automobile is often much warmer than its environment. Sunlight passing through the car window is absorbed by the interior surface of the car and the sun's energy is ultimately converted to thermal energy. Air molecules colliding with the car's interior surface pick up some of this kinetic energy and thus dissipate the energy. As a result the temperature of the surrounding air increases.

**2(a) Exponential Notation**  Most of the numbers we deal with in daily life are generally conveniently handled by our common decimal notation. However in science, we often deal in numbers that are either extremely large or infinitesimally small. For example, the distance of the earth from the sun is 13,000,000,000,000 cm. The size of this number makes it very cumbersome to handle. Exponential notation has been developed to allow us to easily handle numbers such as this. This notation

| | |
|---|---|
| $1.0 = 1 \times 10^0$ | $1 = 1 \times 10^0$ |
| $10.0 = 1 \times 10^1$ | $0.1 = 1 \times 10^{-1}$ |
| $100.0 = 1 \times 10^2$ | $0.01 = 1 \times 10^{-2}$ |
| $1000.0 = 1 \times 10^3$ | $0.001 = 1 \times 10^{-3}$ |
| $4000.0 = 4 \times 10^3$ | $0.004 = 4 \times 10^{-3}$ |

can be summarized by stating that the exponent $n$ in $10^n$ is simply the number of places the original decimal point is moved. If the decimal point is moved to the left, the exponent is positive; if it is moved to the right, the exponent is negative. Thus the distance of the earth from the sun can be stated simply as $1.3 \times 10^{13}$ cm.

**2(b)** *Addition, Subtraction* The main thing to bear in mind in adding or subtracting numbers expressed in scientific notation is that the real position of the decimal point is controlled by the exponent. Therefore as the first step prior to addition or subtraction, you should convert your numbers to the same power of ten. Subtraction or addition can then be carried on in the normal fashion. Thus:

$$
\begin{aligned}
3{,}000 &= 3 \times 10^3 \longrightarrow 3.0 \times 10^3 = 3.0 \times 10^3 \\
300{,}000 &= 3 \times 10^5 \longrightarrow 3.0 \times 10^5 = 300 \times 10^3 \\
\hline
303{,}000 & \qquad\qquad\qquad\qquad\qquad\quad 303 \times 10^3
\end{aligned}
$$

$$
\begin{aligned}
4 \times 10^{-4} &\longrightarrow 4.0 \times 10^{-4} \longrightarrow 40 \times 10^{-5} \\
-2 \times 10^{-5} &\longrightarrow -2.0 \times 10^{-5} \longrightarrow -2 \times 10^{-5} \\
\hline
& \qquad\qquad\qquad\qquad\qquad\qquad 38 \times 10^{-5}
\end{aligned}
$$

$$
\begin{aligned}
3 \times 10^3 &\longrightarrow 3.0 \times 10^3 \longrightarrow 3.0 \times 10^3 \\
3 \times 10^{-1} &\longrightarrow 0.0003 \times 10^3 \longrightarrow 0.0003 \times 10^3 \\
\hline
& \qquad\qquad\qquad\qquad\qquad 3.0003 \times 10^3 = 3{,}000.3
\end{aligned}
$$

or

$$
\begin{aligned}
3 \times 10^3 &\longrightarrow 3.0000 \times 10^3 \longrightarrow 30{,}000 \times 10^{-1} \\
3 \times 10^{-1} &\longrightarrow \qquad 3 \times 10^{-1} \longrightarrow 3 \times 10^{-1} \\
\hline
& \qquad\qquad\qquad\qquad\quad 30{,}003 \times 10^{-1} = 3{,}000.3
\end{aligned}
$$

**2(c)** *Multiplication* Multiplication of two numbers expressed in this manner simply requires multiplying the preexponential component in normal fashion and then *adding* the exponents.

$$
\begin{array}{rr}
1.3 \times 10^{13} & 1.3 \times 10^{13} \\
\times\; 2 \times 10^3 & \times\; 2 \times 10^{-3} \\
\hline
2.6 \times 10^{16} & 2.6 \times 10^{10}
\end{array}
$$

**2(d)** *Division* Division of two numbers expressed in this manner requires division of the preexponential components in normal fashion followed by *subtraction* of the exponent in the denominator from the exponent in the numerator.

$$
\frac{6 \times 10^2}{3 \times 10^4} = 2 \times 10^{-2} \qquad\qquad \frac{5.0 \times 10^6}{4.0 \times 10^2} = 1.25 \times 10^4
$$

An alternative way of accomplishing this is shown below. First all the exponents in the denominator are converted to the numerator by changing the sign of the exponent ($1/10^4 = 10^{-4}$). Addition of these exponents completes this operation.

$$
\frac{6 \times 10^2}{3 \times 10^4} = \frac{6 \times 10^2}{3 \times 10^4} \times \frac{10^{-4}}{10^{-4}} = \frac{6 \times 10^2 \times 10^{-4}}{3 \times 10^0} = \frac{6}{3} \times 10^{-2}
$$

$$
\frac{6 \times 10^2}{3 \times 10^{-4}} = \frac{6 \times 10^2}{3 \times 10^{-4}} \times \frac{10^{+4}}{10^{+4}} = \frac{6 \times 10^6}{3} = 2 \times 10^6
$$

3(a) *Atomic Weights*  All matter is composed of small particles, which are called protons, neutrons, and electrons. Two of these types, the electron and the proton, bear electrical charges. These charges are equal in magnitude but opposite in sign. The masses and consequently the sizes of two such particles are not equal, however. The electron is negatively charged, considerably smaller, and lighter than the proton. Neutrons, as the name suggests, bear no formal charge. Through some elegant experiments, which we shall not discuss here, scientists have been able to measure the masses of these three types of particles. As you can see, the mass of a proton is about 1,800 ($1.8 \times 10^3$) times as large as the mass of an electron. The mass of a neutron can be considered to be essentially the same as the mass of a proton.

| | |
|---|---|
| Mass of a proton | $1.67252 \times 10^{-24}$ gram |
| Mass of an electron | $0.00091 \times 10^{-24}$ gram |
| Mass of a neutron | $1.67482 \times 10^{-24}$ gram |

All the elements in the periodic table are made up of varying numbers of protons, neutrons, and electrons. Since all elements in the periodic table are electrically neutral, they must contain the same number of protons ($+$) and electrons ($-$). Furthermore, the differences in chemical behavior of the elements can be directly related to the numbers of protons and electrons found in the atom. The number of neutrons in the nucleus does not affect the chemical behavior of an element. Neutrons can be considered as nuclear glue, keeping the nucleus intact by insulating the positively charged protons in the nucleus.

3(b)  Let us now consider the mass of various elements in the periodic table. Hydrogen is an element with one proton and one electron. Therefore the mass of an atom of hydrogen is approximately $1.67 \times 10^{-24}$ gram.

**Table A-1**

| Element | Constituent Parts | Mass (grams) |
|---|---|---|
| H | 1 proton + 1 electron | $1.67 \times 10^{-24}$ |
| He | 2 protons + 2 neutrons + 2 electrons | $6.68 \times 10^{-24}$ |
| C | 6 protons + 6 neutrons + 6 electrons | $20.1 \times 10^{-24}$ |
| B | 5 protons + 5 neutrons + 8 electrons | $16.7 \times 10^{-24}$ |

The masses of single atoms of H, He, C, and B are given in Table A–1. The masses of these elements are approximately equal to the sum of the masses of the individual protons, neutrons, and electrons.

3(c)  The size of an atom is so small that chemists cannot deal with a single atom directly. To establish workable conditions, the chemist therefore deals with large numbers of atoms at one time, such as $6 \times 10^{23}$ atoms. This number is called the Avogadro number in honor of Amadeo Avogadro, who first interpreted the behavior of gases in terms of the number of reacting atoms. The Avogadro number has the same relationship to atoms and molecules as the term *dozen* has to *eggs;* namely that it is a convenient number to deal with when referring to submicroscopic particles. When a chemist is referring to $6 \times 10^{23}$ atoms, he calls this a gram atom. Thus 0.1 gram atom of sodium would refer to $0.1 \times 6 \times 10^{23}$ atoms of sodium. If instead of atoms as the particles in question, we were dealing with molecules (a molecule being composed of two or more atoms chemically bonded together and therefore acting as a single particle), we should refer to the Avogadro number of molecules as a mole. These two terms are important ones:

$$6 \times 10^{23} \text{ atoms} \equiv 1 \text{ gram atom;}$$

$$6 \times 10^{23} \text{ molecules} \equiv 1 \text{ mole.}$$

One of the problems associated with the convenience of using exponential notation is the tendency to lose perspective of the magnitude of a number. For example, suppose an Avogadro number of cigarettes were placed end to end to create a "super" king-size cigarette. It would take the light from the glowing tip about 240,000 years to reach the other end (light travels at the speed of 186,000 miles/sec or $3 \times 10^{10}$ cm/sec).

**3(d)** Having defined the gram atom as $6 \times 10^{23}$ atoms, let us now calculate the mass of a gram atom of H, He, C, and O atoms. The mass of this number of atoms

| Element | Mass $\left( \dfrac{\text{gram/atom} \times}{\text{atoms/gram atom}} \right)$ | Gram-atomic weight (grams/gram atom) |
|---|---|---|
| H | $1.67 \times 10^{-24} \times 6 \times 10^{23}$ | $= 1.0$ |
| He | $6.68 \times 10^{-24} \times 6 \times 10^{23}$ | $= 4.0$ |
| C | $20.0 \times 10^{-24} \times 6 \times 10^{23}$ | $= 12.0$ |
| B | $16.7 \times 10^{-24} \times 6 \times 10^{23}$ | $= 10.0$ |

is called the gram-atomic weight. As you can see, these calculations in the table simply generate the gram-atomic weights listed in the periodic table.

**3(e)** On closer inspection of these numbers, you will notice slight deviations of our numbers with those in the periodic table. Boron (B) is the most marked deviant. Why is the atomic weight of boron 10.8 when we calculate 10.0 for this element?

To answer this question we must return to a statement made earlier. We mentioned that the chemical properties of an element depend only on the number of protons in the nucleus and the corresponding number of electrons surrounding that nucleus. Implied in this statement is that any nucleus that contains the same number of protons but a different number of neutrons would behave *chemically* as the same element. Such is the case. The element boron contains two nuclear species, one containing 5 protons and 5 neutrons ($^{10}_{5}B$) and one containing 5 protons and 6 neutrons ($^{11}_{5}B$). These two species are called isotopes. Since isotopes cannot be separated by ordinary chemical means, an average atomic weight has been defined. In any handful of atomic boron, 80 percent of the atoms are $^{11}_{5}B$ and 20 percent are $^{10}_{5}B$. If boron were composed only of $^{11}_{5}B$, then one gram atom of boron would weigh 11 grams. Similarly, one gram atom of $^{10}_{5}B$ weighs 10 grams. Naturally occurring boron is a mixture of both these isotopes in a ratio of 8 atoms of $^{11}_{5}B$ (80 percent) to 2 atoms of $^{10}_{5}B$ (20 percent). In $6 \times 10^{23}$ atoms of naturally occurring boron, $4.8 \times 10^{23}$ atoms are $^{11}_{5}B$ and $1.2 \times 10^{23}$ atoms are $^{10}_{5}B$. Their combined weight is equal to 10.8 grams, as shown below.

$$80\% \times 6 \times 10^{23} \text{ atoms} = 4.8 \times 10^{23} \text{ atoms } ^{11}_{5}B$$

$$20\% \times 6 \times 10^{23} \text{ atoms} = 1.2 \times 10^{23} \text{ atoms } ^{10}_{5}B$$

$$\begin{array}{l} 4.8 \times 10^{23} \text{ atoms} \times 18.4 \times 10^{-24} \text{ grams}/^{11}_{5}B \text{ atom} = \phantom{0}8.8 \text{ grams } ^{11}_{5}B \\ 1.2 \times 10^{23} \text{ atoms} \times 16.7 \times 10^{-24} \text{ grams}/^{10}_{5}B \text{ atom} = \phantom{0}2.0 \text{ grams } ^{10}_{5}B \\ \hline \phantom{0000000} 6 \times 10^{23} \text{ atoms of boron (as a mixture)} = 10.8 \text{ grams} \end{array}$$

**Exercise**   Show that the mass of a $_5B^{11}$ atom is $18.4 \times 10^{-24}$ grams.

Isotopes are generally the rule rather than the exception in the periodic table. Most elements contain nuclei with different numbers of neutrons in the nucleus. The atomic weights of these elements are simply averages of the atomic weight of each individual isotope times the relative concentration in which they occur naturally. One interesting aspects of defining atomic weights in this manner (as an average) is the requirement that the relative abundance of isotopes of a given element remain constant regardless of the source of the element or the geographical location where it is obtained. This is generally the case. Exceptions, however, do exist. One example is the isotope deuterium, which contains 1 neutron and 1 proton in its nucleus. The natural abundance of deuterium is about 0.8 percent (that is, about 8 out of 1,000 atoms of hydrogen are deuterium). In the Dead Sea, an isolated body of water, evaporation and condensation of water over a period of millions of years has led to an abnormally high enrichment of deuterium (as HOD). This has occurred because of slight differences in the vapor pressure of HOD as compared with $H_2O$. HOD has a slightly lower vapor pressure at a given temperature than $H_2O$ (vapor pressure is a measure of a substance's ability to escape from solution) so that the water evaporating from the Dead Sea contained less deuterium than the water condensing and falling as rain. Over long periods of time this has led to a significant enrichment.

**4 Mole and Molecules**   Up to now we have considered the masses of individual atoms. Let us now expand our considerations to include molecules as well. Methane is a simple molecule containing five atoms, one central carbon atom and four hydrogens attached to it. This material is the principal constituent of natural gas (85 percent) and the main constituent of the atmospheres of Jupiter, Saturn, Uranus, and Neptune. From the masses of carbon and hydrogen, we can calculate that the mass of one methane molecule is $26.68 \times 10^{-24}$ gram. A mole of methane gas ($CH_4$) would thus weigh 16 grams.

$$\begin{array}{l} \phantom{4}C = 20.0 \times 10^{-24} \text{ gram} = 20.0 \times 10^{-24} \text{ gram} \\ 4H = 4(1.67 \times 10^{-24}) \text{ gram} = \phantom{0}6.68 \times 10^{-24} \text{ gram} \\ \hline \phantom{4H = 4(1.67 \times 10^{-24}) \text{ gram} = } 26.68 \times 10^{-24} \text{ gram/molecule} \end{array}$$

$$26.68 \times 10^{-24} \text{ gram/molecule} \times 6 \times 10^{23} \text{ molecules/mole} = 16 \text{ grams}$$

This number is called the molecular weight of methane. A simpler way of calculating the molecular weight of methane is to add the atomic weight of carbon (12) with the atomic weight of the appropriate number of hydrogens ($4 \times 1$). This sum (16) gives us the weight of an Avogadro number of methane molecules. The molecular weights (mw) of $SO_2$ and $CO$, two notorious air pollutants, are 64 and 28.

$$S \text{ gram} = \text{atomic weight} = 32$$
$$O \text{ gram} = \text{atomic weight} = 16$$
$$C \text{ gram} = \text{atomic weight} = 12$$

Molecular weight of $CO = 12 + 16 + = 28$ grams
Molecular weight of $SO_2 = 32 + 16 + 16 = 64$ grams

Suppose after sampling a large volume of air from some metropolitan center, we isolated 6.4 grams of sulfur dioxide. To how many moles and how many molecules of $SO_2$ would this correspond?

$$SO_2 \text{ molecular weight} = 64 \text{ grams/mole}$$
$$6.4 \text{ grams} \div 64 \text{ grams/mole} = 0.1 \text{ mole } (Ans.)$$
$$0.1 \text{ mole} \times 6 \times 10^{23} \text{ molecules/mole} = 6 \times 10^{22} \text{ molecules } (Ans.)$$

It is important to remember that the mole concept is identical to the concept of a dozen or a gross. Anyone dealing with a large number of articles (or particles) quickly learns to deal in collective terms rather than in terms of the individual articles (3 dozen as opposed to 36).

**5 *Equations and Calculations*** The role of mathematics has been the main distinguishing feature between the physical and the social sciences. Historically, the chemical equation arose from the early chemists' concern about the quantitative nature of matter and its reactions. Applications of mathematics to these reactions was a natural consequence of these investigations.

**5(a) *Balancing chemical equations*** The combustion of gasoline is a familiar chemical reaction. The equation that represents this reaction is given herewith.

| | Reactants | | Products |
|---|---|---|---|
| Eq. (1) | $C_8H_{18} + O_2$ | $\longrightarrow$ | $CO_2 + H_2O.$ |
| Eq. (2) | $C_8H_{18} + 12.5O_2$ | $\longrightarrow$ | $8CO_2 + 9H_2O.$ |
| Eq. (3) | $2C_8H_{18} + 25O_2$ | $\longrightarrow$ | $16CO_2 + 18H_2O.$ |

Before trying to balance this equation, let us first agree on what the various symbols in the equation mean. The arrow in this equation (Equation 1) designates the direction in which the reaction proceeds and so allows a distinction to be made between reactants and products. Mathematically it can be replaced by an equals sign since every reactant atom must be found somewhere in the products. The subscript 2 in carbon dioxide or oxygen indicates that each molecule contains two oxygen atoms combined in some manner. Changing this number would indicate a completely different molecule. Balancing this equation is achieved by equating the number of atoms of each element on each side of the arrow by means of numbers. Thus the eight carbons in $C_8H_{18}$ are balanced by $8C$ in $CO_2$; eighteen hydrogens in $C_8H_{18}$, by $18H$ in water ($9H_2O$); and so on. These numbers precede the molecular formula as shown in Equation 2 and indicate the number of discrete molecules formed or consumed in the reaction. These numbers must, however, always be whole numbers. Equation 2 is mathematically balanced but physically impossible because half-molecules of oxygen do not exist under normal conditions. We can remedy this situation by multiplying both sides of Equation 2 by 2. This leads to a balanced and physically meaningful equation (Equation 3). Some other balanced equations are given herewith.

$$\text{Eq. (1)} \qquad H_2 + O_2 \longrightarrow H_2O$$

$$H_2 + \tfrac{1}{2}O_2 \longrightarrow H_2O$$

$$2H_2 + O_2 \longrightarrow 2H_2O$$

$$\text{Eq. (2)} \qquad NaOH + HCl \longrightarrow NaCl + H_2O$$

$$\text{Eq. (3)} \qquad 2Na + 2H_2O \longrightarrow 2NaOH + H_2$$

$$\text{Eq. (4)} \qquad 6CO_2 + 6H_2O \longrightarrow C_6H_{12}O_6 + 6O_2$$

**5(b) Chemical Calculations** Having learned to balance chemical equations, we might now wonder why we bothered. Suppose we consider the oxidation of acetylene, a typical combustion reaction welders run when operating an oxyacetylene torch:

$$2C_2H_2 + 5O_2 \longrightarrow 2H_2O + 4CO_2.$$

Because a welder is usually interested in operating at the highest obtainable temperature, he must adjust his reactants accordingly, since at the two extremes (no oxygen or acetylene) combustion ceases. Our chemical equation requires 5 molecules of oxygen for every 2 molecules of acetylene. Alternatively, we can say that 5 dozen molecules of oxygen react with 2 dozen molecules of acetylene—but more conveniently, 5 moles of oxygen react with 2 moles of acetylene. Converting from moles to grams, we find that 52 grams of acetylene ($2 \times$ molecular weight of acetylene, 26) and 160 grams of oxygen ($5 \times$ molecular wt. of $O_2$, 32) combine to produce 36 grams of water ($2 \times$ molecular wt. of water, 18) plus 176 grams of carbon dioxide ($4 \times$ molecular wt. of $CO_2$, 44). Note that the mass of the reactants, 212 grams (160 grams + 52 grams), equals the mass of the products, 212 grams (176 + 36).

The combustion of natural gas is another familiar reaction:

$$CH_4 + 2O_2 \longrightarrow CO_2 + 2H_2O.$$

Suppose you were interested in determining the amount of oxygen that would be required to burn 1.6 grams of methane. Our balanced equation requires 2 moles of oxygen for every mole of methane, and since 1.6 grams is 0.1 mole of methane (molecular wt. 16), we should require 0.2 mole of oxygen (molecular wt. 32) or 0.2 mole $\times$ 32 grams/mole = 6.4 grams of oxygen. If we had used less oxygen, we should have experienced incomplete combustion, perhaps with formation of toxic carbon monoxide.

**5(c) Percentage Composition** One question you may have asked yourself by now is: How do we know that the molecular formula of methane (for example) is $CH_4$? Before trying to answer this question, let us determine what the percentage composition of carbon and hydrogen in $CH_4$ would be. We know that one mole of this material would weigh 16 grams and that there is 1 gram atom of carbon or 12 grams of carbon per mole of $CH_4$. This means that $\frac{12}{16} \times 100\% = 75.0\%$ of this material by weight is carbon. The remainder therefore must be hydrogen, or 4g/16 × 100 = 25.0% of this material is hydrogen. These two figures (75.0% C, 25.0% H), therefore, represent the percentage composition of $CH_4$.

**5(d) Empirical Formulas**   The empirical formula of a substance is simply the ratio of atoms of the different elements in a molecule. Suppose we were given the percentage composition of a substance as 75.0% C and 25.0% H. This tells us that 100 grams of this substance contains 75.0 grams C and 25.0 grams of H. How many gram atoms of carbon and hydrogen do we have?

$$75.0 \text{ grams C} \div 12 \text{ grams/gram atom} = 6.25 \text{ gram atoms C}$$

$$25.0 \text{ grams H} \div 1 \text{ gram/gram atom} = 25.0 \text{ gram atoms H}$$

Since the number of gram atoms is directly proportional to the number of atoms, the ratio of H atoms to C atoms is 4. Our substance must contain four atoms of H for each atom of C.

$$H/C = \frac{25.0 \text{ gram atoms H} \times 6 \times 10^{23} \text{ atoms/gram atom}}{6.25 \text{ gram atoms C} \times 6 \times 10^{23} \text{ atoms/gram atom}} = \frac{4 \text{ H atoms}}{\text{C atoms}}$$

This is as far as we can go without knowing the molecular weight of this substance. We don't know whether our substance is $CH_4$ or $C_2H_8$, $C_3H_{12}$, . . . Numerous methods for determining molecular weights have been developed. However, for the present, let us assume the molecular weight of this substance is 16. Since $CH_4$ (16) is the molecular weight, our molecular formula must be $CH_4$.


# PROBLEMS

1. Convert the following numbers into exponential notation: 1,050, 0.002, 73,000, 0.05.

2. Convert the following numbers to the normal decimal system: $3 \times 10^3$, $3 \times 10^{-4}$, $4 \times 10^7$, $4 \times 10^{-7}$.

3. Add the following: $3 \times 10^2 + 4 \times 10^3$; $6 \times 10^{-2} + 4 \times 10^{-3}$; $6 \times 10^{-2} + 4 \times 10^3$.

4. Multiply the following: $3 \times 10^3 \times 4 \times 10^3$; $3 \times 10^2 \times 3 \times 10^{-2}$; $4 \times 10^{-2} \times 4 \times 10^{-4}$.

5. Divide the following: $4 \times 10^{-3} \div 2 \times 10^2$; $6 \times 10^2 \div 3 \times 10^3$; $6 \times 10^{-2} \div 3 \times 10^{-2}$.

6. Calculate the mass of a single atom each of oxygen, lithium, and deuterium, and the mass of $6 \times 10^{23}$ atoms of these elements.

7. The element chlorine has two isotopes, $^{35}_{17}Cl$ and $^{37}_{17}Cl$, in an abundance of 66.7% and 33.3% respectively. What is the average gram-atomic weight of chlorine?

8. How many gram atoms are there in 3 grams of Li, 6 grams of C, 4 grams of Al, and 14 grams of silicon?

9. Calculate the molecular weights of $NH_3$, HCN, $CO_2$, NaCl.

10. How many moles and how many molecules are there in 2.8 grams of CO, 3 grams of NO, 260 grams of $C_2H_2$?

11. Balance the following equations.

(a) $Na_2CO_3 + HCl \longrightarrow NaCl + CO_2 + H_2O$.

(b) $C_3H_6 + O_2 \longrightarrow CO_2 + H_2O$.

(c) $H_2O \longrightarrow O_2 + H_2$.

(d) $Zn + HCl \longrightarrow ZnCl_2 + H_2$.

12. How many grams of oxygen are needed to produce two moles, or 88 grams, of $CO_2$ in the following reaction?

$$C_2H_4 + 3O_2 \longrightarrow 2CO_2 + 2H_2O$$

13. How many grams of sodium hydroxide are necessary to neutralize 6 grams of acetic acid? How many grams of $H_2O$ are formed?

$$C_2H_4O_2 + NaOH \longrightarrow NaC_2H_3O_2 + H_2O.$$

14. What is the percentage composition of the elements in $CO_2$, $H_2O$, $CH_3OH$?

15. A compound consists of 50 percent sulfur and 50 percent oxygen by weight. What is its empirical formula?

# APPENDIX B  Nomenclature for Organic Compounds (A Limited Presentation)

There are different ways to name compounds. We use one form for clarity, but on occasion, another commonly accepted name may be written underneath in parenthesis. We use a prefix and suffix in devising the name. For alkanes, the suffix is *-ane*. The prefixes are listed for compounds from $C_1$ through $C_{10}$. We assume that the carbons are bonded in a straight chain, that is, as

$$C-C-C-C \quad \text{and not as} \quad C-\underset{\underset{C}{|}}{C}-C.$$

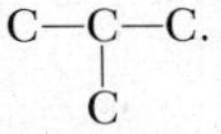

When one hydrogen atom is removed, we have a group that has the same prefix but a new suffix, *-yl*.

| Molecular Formula | Name | Group | Name of Group |
|---|---|---|---|
| $CH_4$ | Methane | $CH_3\cdot$ | Methyl |
| $C_2H_6$ | Ethane | $C_2H_5\cdot$ | Ethyl |
| $C_3H_8$ | Propane | $C_3H_7\cdot$ | Propyl |
| $C_4H_{10}$ | Butane | $C_4H_9\cdot$ | Butyl |
| $C_5H_{12}$ | Pentane | $C_5H_{11}\cdot$ | Pentyl |
| $C_6H_{14}$ | Hexane | $C_6H_{13}\cdot$ | Hexyl |
| $C_7H_{16}$ | Heptane | $C_7H_{15}\cdot$ | Heptyl |
| $C_8H_{18}$ | Octane | $C_8H_{17}\cdot$ | Octyl |
| $C_9H_{20}$ | Nonane | $C_9H_{19}\cdot$ | Nonyl |
| $C_{10}H_{22}$ | Decane | $C_{10}H_{21}\cdot$ | Decyl |

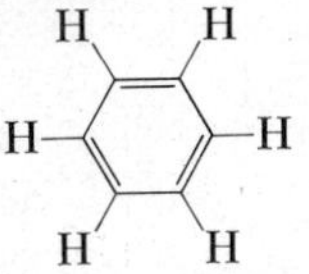

H—⬡—H + $(C_6H_6)\rightarrow$ Benzene   H—⬡·   Phenyl

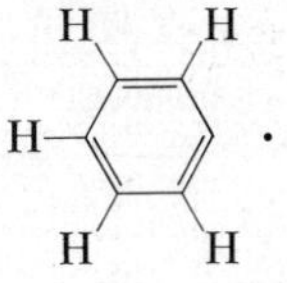

Carbon atoms in a molecule are counted in a straight line $C_1-C_2-C_3-C_4$.

Let us now consider

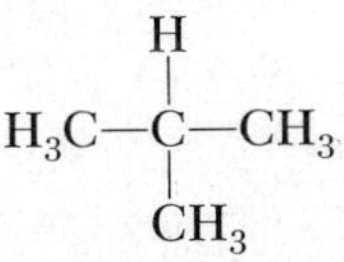

$$H_3C-\underset{\underset{CH_3}{|}}{\overset{\overset{H}{|}}{C}}-CH_3 \quad \text{to be} \quad H_3C-\underset{\underset{H}{|}}{\overset{\overset{H}{|}}{C}}-CH_3$$

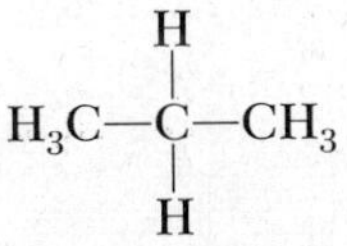

with a $CH_3$ group substituted for an H on $C_2$. Thus, it is a propane with a methyl group on $C_2$, or 2-methylpropane. One of the structures you can write for a $C_5H_{12}$ hydrocarbon is

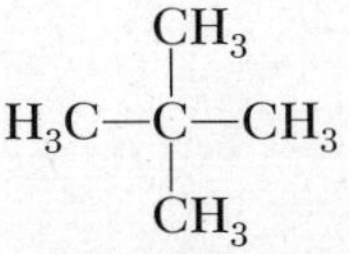

$$H_3C-\underset{\underset{CH_3}{|}}{\overset{\overset{CH_3}{|}}{C}}-CH_3$$

2,2-dimethylpropane

For alkenes the suffix *-ene* is used:

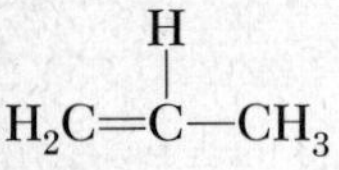

$$H_2C{=}\underset{|}{\overset{\overset{\displaystyle H}{|}}{C}}{-}CH_3 \qquad H_2C{=}\underset{1}{C}{-}\underset{2}{\overset{\overset{\displaystyle H}{|}}{\underset{|}{\overset{|}{C}}}}{-}\underset{4}{CH_3} \qquad H_3C{-}\underset{1}{\overset{\overset{\displaystyle H}{|}}{C}}{-}\underset{2}{\underset{3}{C}}{=}\underset{}{\overset{\overset{\displaystyle H}{|}}{C}}{-}\underset{4}{CH_3}$$

Propene 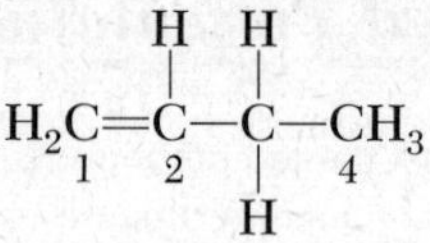            1-Butene 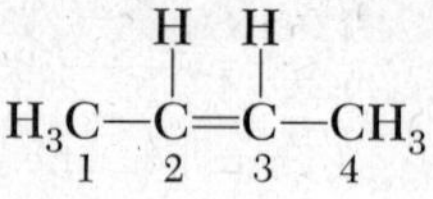                    2-Butene

Denotes position
of double bond

$$H_2C{=}CH_2$$

ethene
(commonly called ethylene)

Compounds that contain a carbon atom triply bonded to another carbon atom are called *alkynes*. To name these compounds, use the suffix *-yne*:

$$H{-}C{\equiv}C{-}H \qquad H_3C{-}C{\equiv}C{-}CH_3$$

Ethyne (acetylene)          2-Butyne

Joining two ends of a molecule together produces a cyclic compound. It can be saturated or unsaturated.

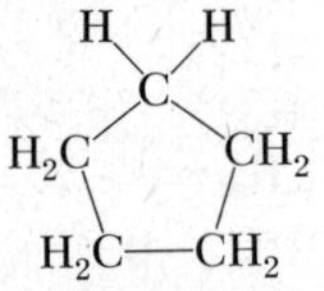

*Cyclopentane*

cyclic          saturated

5 C atoms in ring

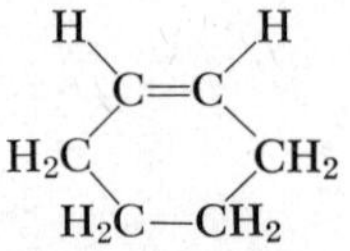

*Cyclohexene*

6C atoms    unsaturated

The nomenclature of aromatic compounds is varied but compounds are often named as substituted benzenes, for example,

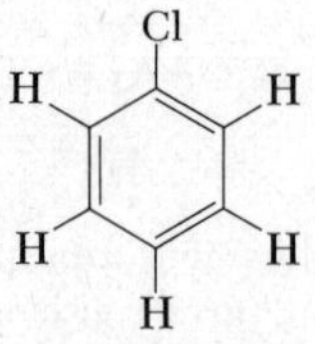

Chlorobenzene

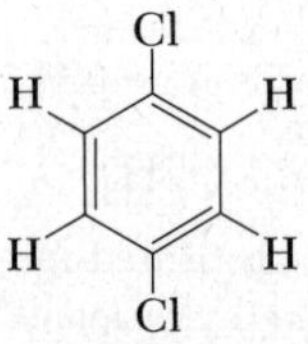

1,4-Dichlorobenzene
(para-dichlorobenzene)

A compound like DDT may be named as an ethane, with several groups substituting for the hydrogen atoms.

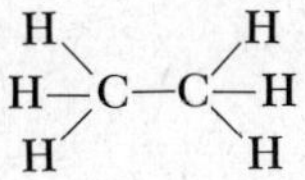

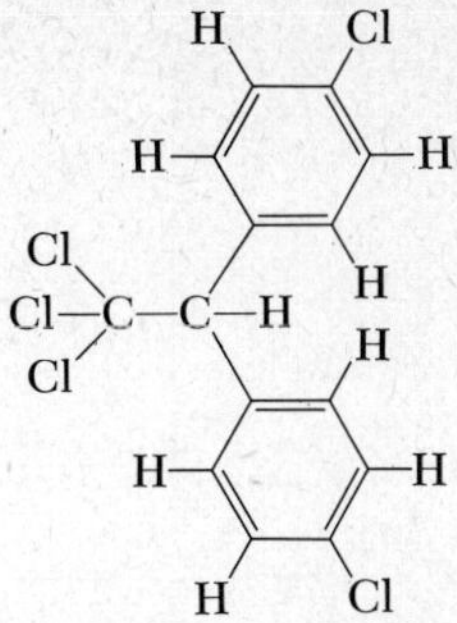

Para-chlorophenyl
group on $C_2$

Ethane

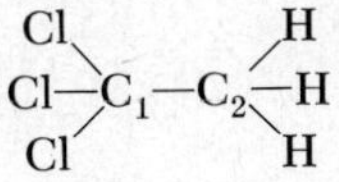

1,1,1-Trichloroethane

1,1,1-Trichloro-2,2-(para-
chlorophenyl)ethane, or DDT

Alcohols have the suffix *-ol*. Thus there are

$$H_3C—CH_2—CH_2—CH_2—OH \quad \text{and} \quad H_3C—CH_2—CH—CH_3$$

4    3    2    1

1-Butanol
Signifies that the
OH group is on $C_1$

2-Butanol

These two alcohols are isomeric.

For a more extensive consideration of the topic of nomenclature, a book specifically concerned with organic chemistry should be consulted.

# GLOSSARY

**Acid**  Any species that can provide a proton.

**Acidity**  Ability of a substance to release hydrogen ions into a solution.

**Active site**  That portion of an enzyme at which chemical reactions, such as hydrolysis, occur.

**Aerobiosis**  A condition of life in the presence of oxygen; often referring to the bacterial decomposition of organic matter in the presence of oxygen.

**Alkanes**  Hydrocarbons whose carbon atoms are always bound to four other atoms; general formula is $C_nH_{2n+2}$.

**Alkenes**  Hydrocarbons containing a carbon-carbon double bond; general formula is $C_nH_{2n}$.

**Alkynes**  Hydrocarbons containing a carbon-carbon triple bond; general formula is $C_nH_{2n-2}$.

**Amino Acids**  A group of molecules containing an amino group ($-NH_2$) and a carboxyl group $\left( \overset{\overset{\displaystyle O}{\|}}{-C}-OH \right)$. The constituents of proteins.

**Amphoteric species**  One capable of behaving as an acid or a base.

**Anaerobiosis**  A condition of life in the absence of oxygen; often refering to the bacterial decomposition of organic matter in the absence of oxygen.

**Analgesic**  A pain reliever.

**Animate**  Possessing life.

**Anion**  Negatively charged ion.

**Antagonistic relationship**  Condition that exists when the effect of two factors operating simultaneously is less than the combined effects of the factors operating separately.

**Antioxidants**  Substances that prevent reactions with oxygen in the air.

**Aromatic**  Pertaining to organic compounds that contain a benzene ring.

**Asymmetric**  Without symmetry.

**Atomic Number**  The number of protons in the nucleus of an atom.

**Base**  Any species that can accept a proton.

**Biosphere**  Region on earth in which life exists, usually considered as troposphere plus the oceans.

**Carbohydrate**  A class of organic compounds composed mainly of C, H, and O, which form the supporting tissues of plants.

**Carcinogen**  Cancer-producing substance.

**Chromosomes**  Bodies formed in the cell nucleus from chromatin network during mitosis as the means of passing genetic information to daughter cells.

**Colloid**  A suspension of finely divided suspended particles in a continuous medium that do not settle out of the medium rapidly and are not readily filtered.

**Configurations**  The different geometrical arrangements of atoms.

**Conformations**  The various shapes a molecule may have by rotation about its single bonds without breaking any bonds.

**Cracking**  Breaking down large organic molecules into smaller ones by heating them in the absence of air.

**Demographer**  Student of vital statistics of the population such as births, deaths, etc.

**Dynamic equilibrium**  Condition in which two changes exactly oppose each other, with no net change resulting.

**Elastomer**  A material that if stretched will return to its original shape when released.

**Electronegativity**  Measure of the ability of an atom involved in a chemical bond to draw the bonding electrons to it.

**Element**  Substance containing only one kind of atom.

**Fission**  Process in which atomic nuclei split apart.
**Functional group**  A group of atoms that is characteristic of a class of organic compounds and that determines the properties of the class.
**Fungicide**  Agent that destroys fungi.
**Fusion**  Process of joining, or fusing, two nuclei to form a third nucleus.

**Geometric Isomers**  Isomers having the same kind of atoms and bonds and the same two-dimensional sequence of attachment but a different spatial geometry.

**Half-life**  Time required for one-half of a given amount of material to decay.
**Hardness in water**  The presence of calcium and magnesium ions in water.
**Herbicide**  Agent that destroys or prevents the growth of plants (generally certain plants).
**Hydrocarbons**  Compounds composed solely of carbon and hydrogen.
**Hydrogen bonding**  Electrostatic attraction that a proton attached to an electronegative atom has for any center of high negative charge.
**Hydrolysis**  Splitting of a bond by the addition of water.

**Insecticide**  Agent that kills insects.
**Isomers**  Different compounds with the same molecular formula.
**Isotope**  One of several atoms with the same atomic number but different mass number, that is, the same number of protons but a different number of neutrons.

**Lipid**  Group of organic compounds that make up the fats, oils, and waxes.

**Lone pair**  A pair of valence electrons in a molecule not involved in bonding.

**Mass number**  Sum of the number of protons and neutrons in a nucleus.
**Metamorphosis**  Change of form or structure.
**Micelle**  Submicroscopic aggregation of molecules, such as a droplet in a colloidal system.
**Mitosis**  Process in which a cell divides, usually characterized by several distinct steps.
**Molecule**  Group of atoms held in place by forces resulting from the attraction of electrons for nearby nuclei.
**Mutagenic**  Causing genetic damage.

**Nematocide**  Substance that kills worms.
**Nucleotide**  Simplest repeating unit in a nucleoprotein, consisting of a base, ribose or deoxyribose, and a phosphate.

**Octane rating**  Measure of the performance of a fuel in an engine relative to the performance of isooctane.
**Ovulating**  Process of shedding eggs from an ovary or ovarian follicle.

**Peptide bond**  Linkage between the nitrogen portion of one amino acid and the carboxyl $-\overset{\displaystyle O}{\overset{\displaystyle \|}{C}}-\overset{\displaystyle H}{\overset{\displaystyle |}{N}}-$ portion of another.
**Pharmacologic**  Pertaining to the effects of chemical substances on living systems.
**Phase diagram**  Graph describing relationship between solid, liquid, and vapor states of a given substance.
**Polymerization**  The repetitive combination of small molecules to form very large molecules.
**Precipitate**  To fall out from solution.
**Primary pollutant**  Pollutant emitted by an identifiable source.

**Primary structure** The sequence of building blocks, such as the sequence of amino acids in proteins, or nucleotides in RNA or DNA.

**Protein** A combination of amino acids joined by peptide linkages.

**Proteolytic** Protein-splitting.

**Radioactivity** Spontaneous fission of an atomic nucleus.

**Radioisotopes** Those isotopes of an element that exhibit radioactivity.

**Rodenticide** Substance that kills rodents.

**Secondary pollutant** Pollutant not emitted directly into the environment, but formed by the reaction of two primary pollutants.

**Secondary structure** The twisting of a linear macromolecule, such as the helical nature of some protein.

**Semipermeable membrane** Membrane that allows some substances to pass through but not others.

**Sequence** Order of succession.

**Solute** Usually a minor component of a solution.

**Solution** A mixture of two or more substances dispersed as molecules (not restricted to the liquid phase).

**Solvated molecule** (or ion) A molecule (or ion) that is surrounded and bound by solvent molecules.

**Solvent** Major component of a solution.

**Structural isomers** Isomers that have the same kind of atoms, but differ in their order of attachment.

**Substrate** Substance being acted on.

**Synergistic relationship** Condition that exists when the effect of two factors operating simultaneously is greater than the combined effects of the factors operating separately.

**Template** Pattern or mold.

**Teratogenic** Causing birth defects.

**Tertiary structure** Amount of coiling and interweaving of an amino acid chain.

**Triple point** The particular condition of temperature and pressure at which the three states of a given substance are in equilibrium.

**Valence** Measure of the number of bonds an atom can form.

# DIRECTIONS FOR CONSTRUCTING MODELS

**Tetrahedron** Punch out the equilateral triangles and the holes at the centers. Fold on the dashed lines and tape at the red marks. The tetrahedra are then joined by pushing a pencil through the hole of one, out the opposite vertex, through the open vertex of the other and out its opposite hole. Hence, two tetrahedra are joined tip to tip. Rotation of the two tetrahedra demonstrates rotation in butane. See text, page 59.

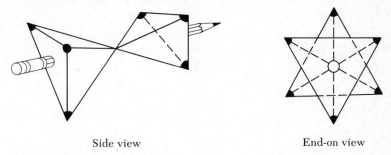

Side view                                                       End-on view

To make a model of ethene, place two tetrahedra edge to edge. The positions of the atoms in ethene (with the carbons at the center of the tetrahedra) are given in the following diagram.

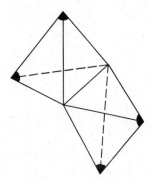

A rough model of ethyne can be made by placing the two tetrahedra face to face as shown. Again remember that the carbon atoms in ethyne are located at the center of each tetrahedron.

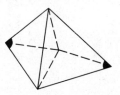

**"Wings"** Punch out the wings and slit the vertex of each as indicated by the red line. Join two wings together slit to slit to form a tetrahedron shaped molecule with one, two, three, or four substituents, depending on the selection of wings.

# Index

Abortion
  illegal, 253
  therapeutic, 253*t*
ABS, 181
Acetaldehyde, 62, 65
Acetic acid, 65, 69, 71, 72, 74, 78, 164
Acetone, 65
Acetylene, 62, 102, 103
Acetylsalicylaldehyde, 72
Acetylsalicylic acid, 76, 78, 164
Acids, 163–165
Acrolein, 71
Acrylonitrile, 93
Activated charcoal, 173
Adipic acid, 96
Addition polymers, 90, 91, 92, 93
Addition reactions, 61
Additives
  antioxidants, 268
  consumption of, 265
  early food, 263
  food, 263, 266*t*
  food flavoring, 268
  food sweeteners, 270*t*
  testing of food, 264
  unintentional, 269
AEC, 20
Aerobiosis, 173, 175
Air pollutants, 71
  primary, 120
  secondary, 121
Alanine, 72, 95
Alanylalanylalanine, 95
Alanylvaline, 95
Alcohol (class), 64, 65, 67, 68, 72
Aldehyde, 65, 67, 70, 71, 72
Alkanes, 57
Alkenes, 60
Alkynes, 60
Allantoin, 73, 74
Amides, 68, 71, 72, 94, 96
Amine, 65, 68, 71, 72, 96
Amino acids, 72, 94, 95, 204*t*. See

*also* Proteins
  biochemical codes for, 239, 240, 241
  essential, 204*t*
  importance of essential, 209
  mirror image isomerism in, 210
  sequence of in pancreatic ribonuclease A, 220
  structure of, 72, 202, 204
Amino caproic acid, 96
Ammonia, 68, 102
Amphetamine, 72, 73, 81, 82
Amphoteric specie, 164
Amylose, 217
Anaerobiosis, 173, 174
Anion exchangers, 169, 170
Antagonistic relation, 131
Antihistamines, 82
Antioxidant, 101
Aromatic compound, 63, 72
Aspirin, 76, 78, 164
Atmospheric composition, 113
Atom, 11
Atomic number, 11

Bactericides, 172
Barbital, 73, 74, 83
Barbiturates, 82, 83
Barbituric acid, 73, 74
Bases, 163–165
Benzaldehyde, 70, 71
Benzene, 62, 63
Benzocaine, 72, 73
Benzoic Acid, 71
Bicarbonate ion, 164, 165, 167–169
Biochemical oxygen demand (BOD), 174, 175, 177
Birth control, 251
  intrauterine loop, 251
  oral contraceptives, 223, 251
  vasectomy, 252
Boiling point, 151, 152, 155, 156
  elevation of, 162

Boiling water reactors, 180
Bromomethane, 65
Buffer, 164, 165
Butane, 57, 60
 normal, 58, 60
 iso-, 58
Butanols, 66
1-Butene, 61

Cadmium, 187–188, 189
Caffeine, 73, 74
Calcium carbide, 103
Carbohydrates, examples of, 215
 amylose, structure of, 217
 cellulose, structure of, 217
 chemical composition of, 200–201
Carbon
 asymmetric, 87–90, 209
 bonding, 55–56
 electronic structure, 55
 tetrahedral, 57, 60, 61
Carbonate ion, 164, 165, 167–169
Carbon cycle, 133
Carbon-14 dating, 13
Carbonic acid, 165, 168
Carbon oxides, 121, 126, 131
Carbon tetrachloride, 70
Carbon tetrafluoride, 65
Carboxylate salts, 67, 69, 71, 72
Carboxylic acid, 65, 67, 68, 69, 70,
  71, 72, 96
Carcinogens, 70, 101
Carothers, W. H., 96
Cation exchangers, 169, 170
Cell, components of, 200, 227
Cellulose, 90, 102, 217
Chain reactions, 14, 122
Changes of state, 154
Chelating agent, 170, 172, 182
Chemical bond, 36
Chemical calculations, 282
Chemical equations, balancing of,
  281
Chemical symbols, 10
Chloral hydrate, 82, 83
Chlordiazepoxide. See Librium
Chlorinated hydrocarbons, 70, 256
Chlorination, 172, 173
Chloroform, 65, 70
Cholesterol, structure of, 64, 213
Cinnamaldehyde, 70, 71

Citric acid, 70, 172
Class of compounds, 63, 65, 67
Cleopatra, 260
Cocaine, 81, 82
Codeine, 74, 75, 84
Coke, 102–103
Composition, percent, 282
Condensation polymer, 96–99
Conformations, 58, 59
Covalent bonding, 38, 55
Cracking (petroleum), 103
Cross-links, 98–100
Cuyahoga River, 149
Cyclazocine, 75, 76
Cyclohexene, 62
Cyclopentane, 62

Dacron, 98
DDT, 255
 and disease, 255
 structure of, 256, 287
Demerol. See Pethidine
Demographers, 252
Deoxyribose, structure of, 216
Desalination, 158, 162, 163
Detergent builder, 180
Detergents, 180–183
Detroit River, 188
Dew, 153
Dextromethorphan, 75, 76
Diamond, 55, 56
Diazepam. See Valium
Dibenzacridine, 70, 71
Dichloroethene, 60
Dickens, Charles, 247
Diethylether, 65
Diethylether peroxide, 65
Diethylmalonic acid, 73
Dimenhydrinate. See Dramamine
1,4-Dimethylbenzene, 63
Dimethyl ether, 64, 66
Dimethylsulfide, 65
Dimethyl sulfoxide, 79
Distillation, 152, 158
DMT, 86
DNA
 bases in, 229
 biosythesis of, 235
 helical nature of, 232
 hydrogen bonding in, 233
 replication of, 229, 234, 235

DNA (*continued*)
structure of, 229, 231
Dot diagram, 28
Double bond, 43
Dramamine, 83
Drugs, 72–87
terminology, 80–81
Dynamic equilibrium, 151

E. coli
RNA-amino acid codes, 239, 240, 241
molecular components in, 200*t*
Einstein, Albert, 102
Elastomer, 100–101
Electromagnetic radiation, 9
Electron, 11
Electronegativity, 42, 63, 64, 66
Electronic configuration, 34
Element, 28
Energy
kinetic, 8, 150–156
potential, 8, 154–156
Enzymes
active site of, 209
chemical composition of, 216
importance of, 216
inhibition of, 218
mechanism of action, 218, 219
pepsin, 207
EPA, 142, 174, 175
Ephedrine, 82
Epinephrine, 72, 73
Equanil, 83
Equilibrium, 153–158
Erie, Lake, 178, 186, 188, 192
Esters, 68, 69, 72
Ethanal. *See* Acetaldehyde
Ethane, 57
Ethanoic acid. *See* Acetic acid
Ethanol, 61, 64, 66, 78, 82, 83, 103
Ethene (ethylene), 60, 61, 90, 91, 103
Ether (class), 64, 65, 66, 67
Ethical drug, 78
Ethylamine, 65
Ethyl anthranilate, 72, 73
Ethylene glycol, 98, 162, 170
Eutrophication, 174, 179, 181
Evaporation of liquids, 150, 158
Excited level, 25

Exponential notation, 276
how to add and subtract in, 277
how to divide in, 277
how to multiply in, 277

Fats, 69, 166
polyunsaturated, 69
FDA, 77, 78, 186
Fermentation, 174
Fertilizer, 104
Filters, 171, 180
Fission, 14
Flocs, 171
Force
electrostatic, 7
gravitational, 7
nuclear, 7
Formaldehyde, 70, 71, 102
Formic acid, 69, 70
Formulas, empirical, 283
Fractional distillation, 152
Freeze-drying, 157
Freezing, 158, 159
Freezing point, 156
Freezing point depression, 162
Freon, 65
Frequency, 9
Fructose, structure of, 215
Functional groups, 63, 64, 65*t*, 72
Fusion, 20

Gamma-ray, 9
Gases, equation of state, 115
Gasoline, 62, 103, 152
nonleaded, 130
Geometrical progression, 250
Geometric isomers, 60, 99–100
Glucose, structure of, 215
Glycerol, 69, 98–99, 166
Graphite, 56
Greenhouse effect, 112
Ground level, 25
Group, 28

Half-life, 13
Hallucinogens, 79
"Hard" water, 166, 168. 181, 183
Hashish, 85
Heat of fusion, 154

Heat of vaporization, 155
Heroin, 74, 75, 84
Heterogeneous mixture, 46
Homogeneous, 46
Hormones, 222
  androgenic substances, examples of, 266
  estrogenic substances, examples of, 226
  estrogens, synthetic, 227
  oxytocin, structure of, 224
  progesterones, synthetic, 227
  structures of some, 223
  vasopressin, structure of, 225
Housefly, 261
Houston Ship Channel, 179
Humidity, 152
Hydration, 160
Hydrocarbons, 57
  saturated, 57, 62
  unsaturated, 60, 61, 62, 63, 101
Hydrogen atom spectrum, 24
Hydrogen bonding, 44–45, 66, 96–97, 160
  in proteins, 207
Hydrolysis
  of amides, 68
  of esters, 68
  of peptides, 95

Inert gas configurations, 37
Inner shell, 28, 34
Inorganic ions in cells, 200
Insect
  defensive secretions, 262
  differential susceptibility of, 258
  diseases, 262
  juvenile hormone, 262
  pathology, 262
Insect control. Also see Insect sex attractants
  bait crops, 258
  boll weevil, 260
  sterilization of male mosquitos, 260
Insecticides, 255
  biodegradability of, 255
  carbaryl, structure of, 258
  insect resistance to, 257
  isolan, structure of, 258
  natural pyrethrin, structure of, 259
  pyrolan, structure of, 258
  rotenone, structure of, 259
  synthetic pyrethrin, structure of, 259
Insect sex attractants, 260
  structure of gypsy moth, 261
  structure of housefly, 261
Insulin, amino acid sequence in bovine, 203
Internal combustion engine, 126
Iodine, 70, 172
Iodoethane, 65
Ionic bonding, 38, 55
Isoamylacetate, 68
Isomers, 57, 91
  See also Structural isomers; Geometric isomers; Mirror image isomers
Isoprene, 100
Isotope, 11

Kelvin, Lord, 275
Kerosene, 103, 152
Ketone, 65
Kilocalorie, 9
Kwoks disease, 265

LAS, 181
L-Dopa, 79, 80, 90
Length, units of, 275
Lewis diagram, 28
Librium, 83
Lipids
  bile acids, 214
  chemical composition, 200–201
  composition of, 210
  examples of waxes, 212
  preparation of soaps from, 212
  structure of vitamin $D_3$, 214
  triglycerides in, 211
Liquefaction point, 156
Liquids, properties of, 149–150
Lone pair, 40
LSD, 79, 80, 81, 85, 86, 90

MAC, 130
Marijuana, 79, 80, 84

Mass
  of the elements, 278
  of subatomic particles, 278
  units of, 275
Mass number, 11
Matter, 7
MDA, 86–87
Melting point, 154, 156
Meprobamate. *See* Equanil
Mercury, 183–187
  biological effects, 186–187
  industrial use, 184
Mescaline, 79, 86
Mesosphere, 113
Meteorology, 136
Methadone, 75, 76, 84
Methamphetamine, 82
Methane, 57
Methanol, 65, 102
Methylamine, 71
Methyl anthranilate, 72, 73
Methyl benzene. *See* Toluene
Methylmercury, 185–186
Methyl methacrylate, 93
Methyl nylon, 96–98
Methylphenidate, 82
Methyl salicylate, 78
Metric system, 275
Micelle, 166, 167
Microwaves, 9
Miller, John, 250
Miltown. *See* Equanil
Mirror image isomerism, 54, 87–90
Mole, 280
Molecule, 35
Monomer, 90, 92, 93
Morphine, 74, 75, 76, 84
m-RNA biosynthesis of, 238
Muscalure. *See* Housefly

Nader Task Force, 178, 180
Naphthalene, 63
α-Naphthylamine, 101
β-Naphthylamine, 101
Narcotics, 84
Neoprene, 101
Neutron, 11
Neutron activation analysis, 186
Nicotinamide, 68
Nicotine, 82

Nitrogen cycle, 134–135
Nitrogen oxides, 121, 122, 126, 131
Nitroglycerine, 69
NTA, 182–183
Nuclear reactor, 15
Nucleatides in DNA, 228
  structure of, 230
Nucleic acid–amino acid codes, 239
  chemical composition, 200–201
  differences between DNA and
    RNA, 237
Nucleus, 11
Nylon, 96–98

Oak Ridge National Laboratory,
    184
Octane rating, 129
Ohio River, 179
Oil pollution, 188, 190, 191
Opium, 74
Organobromide, 65
Organochloride, 65, 70
Organofluoride, 65
Organohalide, 65, 70
Organoiodide, 65, 70
Organomercury    compounds,
    184–187
Orlon, 93
Osmotic pressure, 162
Outer shell, 28, 34
Oxidants, 121
Oxidation reaction, 62
Ozone, 62, 112, 131

PAN, 122
Panogen, 185
Paracelsus, 263
Paraffins, 57
Particulates, 120
Penicillin G, structure of, 218
Pentane, 92, 93
2-Pentene, 62
Pentobarbital, 83
Peptide bond, 94–97
Peptides, 94–97
  *See also* Protein, structure of
Period, 28
Periodic table, 11, 28, 32
"Permanent hardness," 168

Peroxide (class), 65
Pesticides, 255
  biological, 262
Pethidine, 84
pH, 164, 165
Phase diagram, 156, 157
Phenazocine, 84
Phenobarbital, 73, 74, 83
Phenylmercury acetate (PMA), 185
Phytoplankton, 179, 186
Pipradol, 82
Plexiglas, 93
Pogo, 248
Polar bonds, 39, 63, 66
Polar molecule, 160
Polyamides, 96
Polyester, 98–99
Polyethylene, 90, 91, 94, 98
Polyisoprene. See Rubber
Polymers, 90–103
Polypeptides, 95
Polystyrene, 91, 92
Polyvinylchloride (PVC), 91, 92, 103
Population
  explosion, 250
  growth in America, 247
  history of, 248
  Massachusetts Institute of Technology study of, 253
  steady-state concentration, 251
  world, as a function of time, 249
Positive crankcase ventilation, 126
Principle of Le Chatelier, 153
Propane, 57, 61
Propene, 61
Protein
  amino acids in, 204t
  amino acid sequence of pancreatic ribonuclease A, 220
  composition of common, 208, 220
  digestion of, 207
  hydrolysis of, 207
  importance of amino acid sequence in, 202
  mirror image isomerism in, 209
  primary structure of, 220, 221
  secondary structure of, 220, 221
  structure of, 95, 202
  tertiary structure of, 220, 222
Proton, 11
Psilocybin, 86
Psychedelic drugs, 85–87

Putrefaction, 174
Pyrene, 70, 71

Radiation dosage, 18–19
Radiation hazards, 18
Radioactivity, 12
Radioisotopes, 13
Rayon, 102
Reactive site, 63
Relative humidity, 152
René Dubos, 192
Resources, effect of exhaustion of, 254t
Ribonuclease A, amino acid sequence in bovine pancreatic, 220
Ribose, structure of, 216
RNA, structure of, 234
  See also m-RNA
Rotenone. See Insecticide, rotenone
Rubber, 99–101

Salicylaldehyde, 72
Salicylic acid, 72, 78
San Diego Bay, 179
Santa Barbara Channel, 188, 190
Saran, 93
Sedatives and depressants, 82–83
Semipermeable membrane, 162
Sewage, 173–175
Sex pheromones, 260
Shell atomic structure, 26
Silk, 95–96
Soap, 69, 166, 180, 183
Sodium benzoate, 67
Sodium hypochlorite, 172
Sodium oleate, 166
Sodium tripolyphosphate, 170, 181
"Soft" water, 166, 169, 170, 181–183
Solar energy, 158
Solute, 159
Solutions, 159–162
  properties of, 161
Solvated molecule, 160
Solvent, 159
SST, 112
Starch, 90
Stationary combustion, 124
Stimulants (antidepressants), 81–82
STP, 86–87

Stratosphere, 112
Strip mining, 21
Structural isomers, 57, 58, 66
Styrene, 91, 92
Sucrose, structure of, 215
Sulfhydryl group, importance of, 206
Sulfide (class), 65
Sulfur oxides, 120, 122, 126, 130
Sulfur-sulfur bonds
  importance of in biological activity, 203, 206
  in bovine insulin, 203
Supercooling, 156
Superheating, 155
Superior, Lake, 178
Surfactant, 180
Synergistic relation, 132
Szent-Györgyi, Albert, 199

Tallow, 183
Teflon (polytetrafluoroethylene), 91, 92
Temperature, 115, 276
  Celsius, 115
  Fahrenheit, 115
  Kelvin, 115
"Temporary hardness," 168
Tetraethyl lead, 130
Tetrafluoroethylene, 91, 92
Tetrahydrocannabinols, 85
Thalidomide, 79, 80
Thebaine, 74, 75
Thermal inversion, 137
Thermosetting, 98
Thermosphere, 112
Thuricide. See Pesticides
Tobacco mosaic virus, diagram of, 238
Toluene, 63
Topography, 138

Torrey Canyon, 188, 190
Tranquilizers, 83
1,1,1-Trichloroethane, 70
Triple bond, 43
Triple point, 157, 158
Troposphere, 111

Ultraviolet radiation, 9, 111
Urea, 73
  synthesis of, 199
Uric acid, 73, 74
U.S. Bureau of Mines, 104
U.S. Public Health Service, 187

Valence, 41
Valine, 95
Valium, 83
Vapor pressure of liquids, 151, 152, 162
Vinylchloride, 91, 92, 103
Vinylidene chloride, 93
Viruses, action of, 234
Visible spectrum, 9
Volume, units, 276

Wastes, animal, 103–104
Water, cost of usable, 159
Water pollution, 176–192
Watson and Crick, 229
Wavelength, 8
Wöhler, Friedrich, 199
World Health Organization, 185

X ray, 9

Zeolite, 169, 170

1 2 3 4 5 6 7 8 9 10

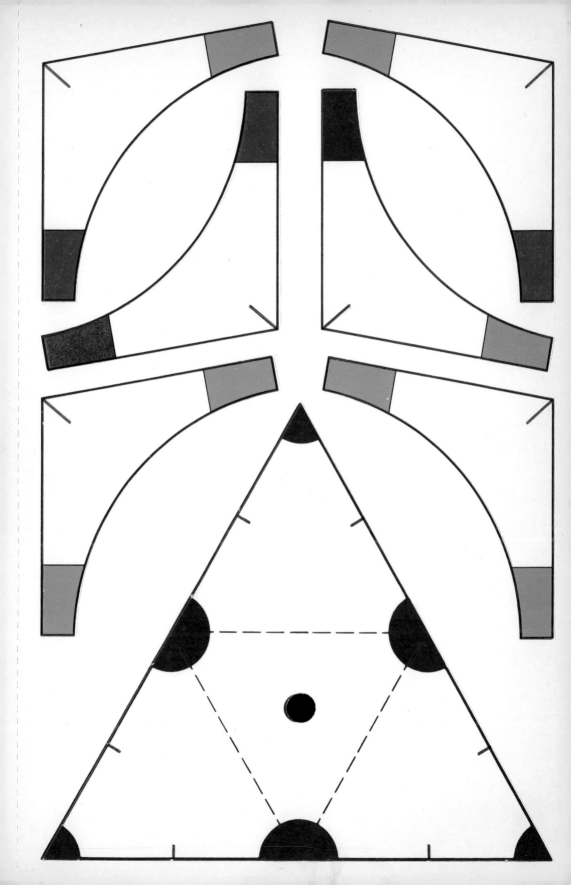

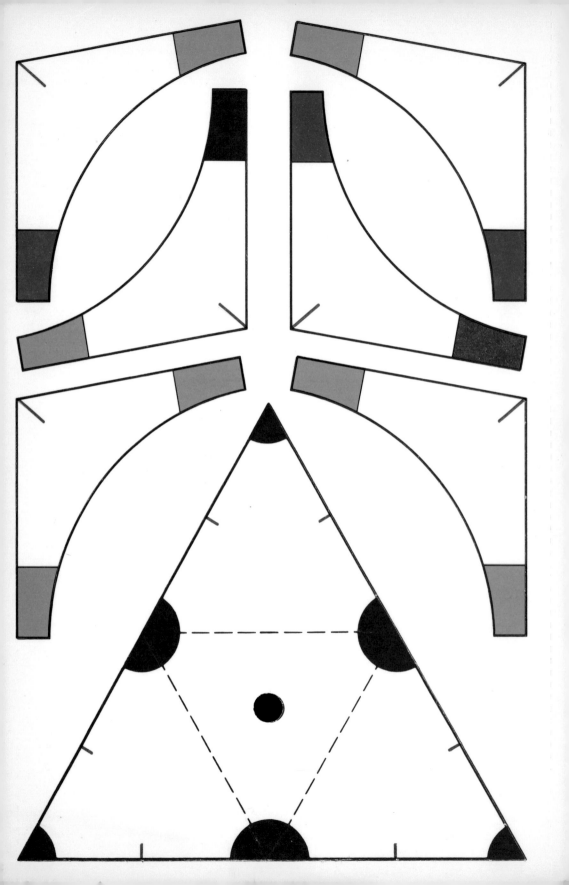